Wrestling with Nature

WRESTLING WITH NATURE

From Omens to Science

Edited by Peter Harrison,
Ronald L. Numbers,
and Michael H. Shank

The University of Chicago Press
Chicago and London

Peter Harrison is the Andreas Idreos Professor of Science and Religion at the University of Oxford, director of the Ian Ramsey Centre, and a Fellow of Harris Manchester College.

Ronald L. Numbers is the Hilldale Professor of the History of Science and Medicine and a member of the Department of Medical History and Bioethics at the University of Wisconsin–Madison.

Michael H. Shank is professor of the History of Science at the University of Wisconsin–Madison.

The University of Chicago Press, Chicago 60637
The University of Chicago Press, Ltd., London

Printed in the United States of America

20 19 18 17 16 15 14 13 12 11   1 2 3 4 5

ISBN-13: 978-0-226-31781-6 (cloth)
ISBN-13: 978-0-226-31783-0 (paper)
ISBN-10: 0-226-31781-1 (cloth)
ISBN-10: 0-226-31783-8 (paper)

Library of Congress Cataloging-in-Publication Data

Wrestling with nature : from omens to science / edited by Peter Harrison, Ronald L. Numbers, and Michael H. Shank.
p. cm.
Includes index.
ISBN-13: 978-0-226-31781-6 (cloth : alk. paper)
ISBN-10: 0-226-31781-1 (cloth : alk. paper)
ISBN-13: 978-0-226-31783-0 (pbk. : alk. paper)
ISBN-10: 0-226-31783-8 (pbk. : alk. paper) 1. Science—History. 2. Natural history—History. 3. Philosophy of nature—History. I. Harrison, Peter, 1955– II. Numbers, Ronald L. III. Shank, Michael H.
Q125.W86 2011
508.09—dc22
2010036527

∞ This paper meets the requirements of ANSI/NISO Z39.48-1992 (Permanence of Paper).

For David C. Lindberg

CONTENTS

ACKNOWLEDGMENTS

This book has benefited immensely from the generous advice of Thomas H. Broman, University of Wisconsin–Madison; Joan Cadden, University of California, Davis; Thomas Gieryn, Indiana University; Bernard Goldstein, University of Pittsburgh; Frederick Gregory, University of Florida; Florence Hsia, University of Wisconsin–Madison; Stuart W. Leslie, Johns Hopkins University; Geoffrey E. R. Lloyd, University of Cambridge; Lynn K. Nyhart, University of Wisconsin–Madison; John V. Pickstone, University of Manchester; Jamil Ragep, McGill University; Lisbet Rausing, Imperial College, London; A. I. Sabra, Harvard University; Eric Schatzberg, University of Wisconsin–Madison; Rennie B. Schoepflin, California State University, Los Angeles; and Robert S. Westman, University of California, San Diego. James Lancaster assisted with the preparation of the manuscript for publication, for which we are grateful. We would like to thank Karen Merikangas Darling at the University of Chicago Press for her encouragement and enthusiasm for this project.

We would also like to express our gratitude to the following institutions for their financial support: At the University of Wisconsin–Madison we received generous support from the Institute for Research in the Humanities through the Burdick-Vary Fund, the Anonymous Fund, the Department of the History of Medicine, and the Department of the History of Science. We also received assistance from the Evjue Foundation and the Brittingham Fund, Inc.

This book is dedicated to David C. Lindberg, who spent much of his distinguished career as the Hilldale Professor of the History of Science at Madison. His scholarship is well known to historians of science and students of the discipline around the world. To a privileged group of us, he has also been an enthusiastic teacher, a wise mentor, a congenial collaborator, a thoughtful and supportive colleague, and a good friend.

INTRODUCTION

One of the biggest gaps in the history of science is the paucity of studies of the history of the meanings of "science" and other labels used by investigators of nature to describe their own activities. Recent developments within the discipline suggest that such a study is long overdue, for it is now commonly claimed—rightly or wrongly—that "science" is an inadequate and unhelpful way of describing the systematic study of nature in the past. As a university discipline, the history of science is a relatively young field that traces its origins to the late nineteenth century. Early historians of science often tended to assume that while the science of the past differed in many ways from the science of the present, it was nonetheless essentially the same kind of enterprise. A key feature of these early histories of science was an emphasis on the progressive and cumulative nature of science. Past scientific achievements were understood in terms of the contribution they had made to an increasingly more sophisticated and rational understanding of the natural world. George Sarton (1884–1956), often regarded as the father of the history of science, thus insisted that all material and intellectual progress "can be traced back in each case to the discovery of some new secret of nature or to a deeper understanding of an old one."[1] For Sarton and a number of his fellow pioneers in the field, the importance of the history of science lay in the fact that it was the one discipline which more clearly than any other demonstrated the progress of human civilization. In his inaugural lecture at Harvard, Sarton ventured the bold claim that the history of science is "the only history which can illustrate the progress of mankind."[2]

In keeping with this vision, histories of science from the first half of the twentieth century traced humankind's gradual intellectual and material progress from modest beginnings in Egypt and Mesopotamia to the present. The key periods of history, according to this account, were the golden age of Greek science and the scientific revolution of the seventeenth century. It was taken for granted in these large-scale narratives that throughout history, "science," in spite of its changing context, was more or less the same kind of thing, namely, a quintessentially rational and systematic approach to the natural world. Given this conception of the nature of science, the purpose of setting out its history, as Sarton succinctly expressed it, "was to illustrate impartially the working of reason against unreason."[3] Science, in every age, was accordingly distinguished from myth, popular prejudice, superstition, religion, and magic. Today, these classical narratives of the history of science still hold considerable appeal for practicing scientists and continue to provide the basic plot lines for popular histories of science.

Over the past thirty years, however, historians of science have become increasingly dissatisfied with this story. Some have gone so far as to suggest that "science," as we now understand it, is a relatively recent phenomenon—specifically, a product of the nineteenth century. Certainly there is merit in the suggestion that in previous periods of history what we would presently regard as science was distributed across a number of distinct but related activities, such as "natural philosophy" and "natural history." These expressions, it might be argued, were not simply different labels for what later became known as science. In fact, as the use of these two particular terms—natural philosophy and natural history—might suggest, they retained important links with philosophy and history (broadly construed), and they included moral and religious elements that are now almost completely absent from science.

It is now often claimed that one tangible indication of the belated birth of modern science was the appearance of a new vocabulary. As one historian put it, "Our present use of the word 'science' was first coined in the nineteenth century and, strictly speaking, there was no such thing as 'science' in our sense in the early modern period."[4] This claim draws support from the *Oxford English Dictionary*, according to which the dominant sense of the term in modern use—branches of study that relate to the phenomena of the material universe and their laws, and which exclude reference to the theological and metaphysical—dates from April 1867. It is also significant that the now-familiar term "scientist" was used for the first time in the nineteenth century and was only fully accepted in the early twentieth.[5] Pushing the lexicographical evidence even further, specialists of different historical periods have often make similar claims about the meanings of "science." Sir Geoffrey Lloyd writes in relation to the study of nature in ancient Greece that "science is a modern category, not an ancient one: there is no one term that is exactly equivalent to our 'science' in Greek." Historian of medieval science David Lindberg observes that the modern term "science" has connotations that differentiate it somewhat from the earlier study of nature. If we come to the Middle Ages with our modern conceptions of science in mind, he cautions, the result is likely to be a distorted impression of the past. Expressing the point even more strongly for the sixteenth century, Nicholas Jardine has observed that "no Renaissance category even remotely corresponds to 'the sciences' or 'the natural sciences' in our senses of the terms."[6]

Yet the linguistic evidence alone is not conclusive, and the discontinuities between our present understandings of "science" and those of the past can be overstated. The ancient Greeks observed a clear distinction between "science" and mere "opinion," and from classical antiquity those intellectual practices labeled "science" enjoyed a special status.[7] Medieval discussions about whether, and in what sense, theology was a science reflect this, as do comparable discussions about the scientific status of medicine and certain of the mathematical disciplines. There also seem to be strong affinities between some of the usages of "science" in the seventeenth century and those of our own era, as the title of Galileo's *Discourses Concerning Two New Sciences* (1638) indicates.[8] The significance of the relatively late appearance of the English term "scientist" is not altogether clear, either, for it might be argued that there were much earlier expressions in other European languages that were equivalent to this label in certain respects. The French *savant* (or *sçavan*) dates from the seventeenth century. The Italian *scienziato,* which Italians still use for their scientists, goes back even further to the sixteenth century. The English used the expression "virtuoso," which had similar connotations. In no case are these expressions synonymous with "scientist," but they capture a similar sense of someone who is engaged in the serious and systematic study of nature, amongst other things.

The history of science, in any case, is more than the history of the meanings of particular terms, and there comes a point at which historians must lay aside their dictionaries and attend to the relevant activities themselves. In essence, that is the task we have set for the contributors to this volume. As noted at the outset, one of the biggest gaps in the literature is the lack of histories of the meanings of "science," "natural philosophy," "medicine," "mixed mathematics," and other activities concerned with the study of nature. One of our main aims is to bring together scholars who have grappled with the study of nature in various periods and cultures to collaborate in remedying this deficiency. There is now considerable interest in such questions, focused broadly on the identity of "science" and prompted by a growing awareness of the general problem of anachronism in the history of science and medicine. To date, however, discussions of these important questions have tended to be highly specialized and somewhat piecemeal. Treatments have typically focused upon discrete historical periods and have been conducted within the context of tightly defined disciplinary boundaries. This book seeks to range more widely, with contributions from experts representing different historical periods and different disciplinary specializations within the broad fields of the history of science and the history of medicine.

The general approach taken by the authors has been to examine how students of nature themselves have understood and represented their work. The aim of each chapter is to explain the content, goals, methods, practices, and institutions associated with the investigation of nature and to articulate the strengths, limitations, and boundaries of these efforts from the perspective of the actors themselves. Particular attention is paid to those features of the investigation of nature that might be distorted or misrepresented by the application of present-day connotations and categories, and every effort has been made to avoid anachronism and essentialism. Accordingly, the kinds of questions addressed in the various chapters are these: What did the investigators into nature take their work to be in the period under discussion? What did this activity encompass? What did it exclude? What was it called? What did its practitioners do to carry it out? What did it mean and how was it justified, perhaps personally but especially socially and culturally? How did this activity fit (or not) into the culture in which it took place? Particular attention is also devoted to the terminology used by the historical actors, and the chapters address questions such as these: What terms and categories did the actors use to describe their activity? What were their goals? What place did their explorations occupy in the intellectual and social world in which they found themselves?

The chapters have been arranged in rough chronological order and address four overlapping sets of concerns. First is the very general question of the ways in which, in particular cultures and historical periods, those who have grappled with nature have conceptualized their own activities. While in a sense the whole volume deals with that question, the first four chapters seek to tell this story by focusing respectively on the ancient Near

East, the classical world of ancient Greece and Rome, the Islamic Middle Ages, and the Latin Middle Ages. These chapters seek to bring to the traditional story about the rise of Western science the renewed historiographical sensitivities of the most recent scholarship.

The next three chapters (5–7) deal with basic historical divisions of the knowledge of nature and, in particular, those that have been of enduring importance in the West: natural history, the mixed mathematical sciences, and natural philosophy. These disciplines will have already made their appearance in the earlier chapters; accordingly, chapters 5, 6, and 7 will focus mainly on the meanings of these categories in the modern period—that is, from the sixteenth century onwards. Natural philosophy, which provided the basic rubric for the study of nature right up to the end of the nineteenth century, proved too large a subject for a single chapter and for this reason, while it is the particular focus of chapter 7, the preceding chapters on natural history and mixed mathematics will also discuss natural philosophy and its relation to the historical and mathematical approaches. These three chapters together deal with what used to be called the scientific revolution and its immediate aftermath.

A third perspective is to consider historical understandings of what is excluded by particular approaches to the study of nature. Some exclusionary principles are more or less neutral and reflect relatively innocuous divisions between theoretical and applied knowledge and, in the modern period, the boundaries between increasingly differentiated professions. The issue of the division between theoretical knowledge of nature and its various practical applications has been discussed since antiquity.[9] A central feature of such discussions has been the relative status of the purely theoretical in relation to the applied. Typically, the former has been accorded the higher status, but the case has also been made that the study of nature has value only insofar as it can be useful. The specific cases under consideration in chapters 8 and 9 are medicine and technology and their relations to "pure science."

Other exclusionary principles are more ideologically loaded. They reflect not so much professional differentiation or knowledge and its various applications but an attempt to give the formal study of nature identity by specifying what it is not. Throughout history various distinctions reflect this process: chemistry and alchemy, astronomy and astrology, medicine and quackery, technology and magic, science and superstition. Each of these sets of oppositions would be worth exploring in its own right, especially since the pejorative associations of some of these labels are quite recent. In this volume, we have chosen to restrict our attention to two general oppositions that still have a contemporary resonance: science and religion, and science and pseudoscience. Our intention with chapters 10 and 11 is to give particular consideration to the ways in which the contrasts between science and religion and science and pseudoscience have played a significant role in forging the identity of modern science.

The last three chapters (12–14) deal in various ways with the unique and privileged status of modern science and with the growth of its cultural authority. Chapter 12 charts the emergence of the idea of the scientific method, describing the manner in which it has been used to confer upon scientific knowledge a unique authority. Chapter 13 deals with science and the public, illustrating how in the nineteenth century the professionalization of science and the way in which it presented itself to the public were instrumental in contributing to the creation of science as we know it. Finally, while to a certain extent this volume upholds the current orthodoxy that what counts as science differs from one historical period to another, chapter 14 makes the convincing claim that geography is as important as chronology. Thus, when we speak of "science," we must take cognizance not only of historical context, but also of geographical place. Taken together, the chapters that deal with the modern period show how the question of the identity of science is almost as problematic for the present as for the distant past.[10]

Each of the chapters in this collection can stand alone, and there is nothing to prevent readers with interests in a particular theme or historical period from consulting chapters out of sequence. Given the limited space of the present format, there are inevitable omissions, but it is our conviction that none of these is egregious. As should be apparent by now, we have deliberately chosen to restrict ourselves to Western science but would hope that a similar approach might be adopted for histories of greater geographical range. For now, however, we hope to have made some contribution to an understanding of what it is that various individuals and groups, in different times and places, have imagined themselves to be doing as they have wrestled with nature.

P.H.

NOTES

1. George Sarton, "The New Humanism," *Isis* 6 (1924): 10.

2. George Sarton, *The Study of the History of Science* (Cambridge, MA: Harvard University Press, 1957), 5.

3. George Sarton, *A History of Science: Ancient Science through the Golden Age of Greece*

(Cambridge MA: Harvard University Press, 1952), xiv. Some of Sarton's contemporaries demurred from this vision. See, e.g., Otto Neugebauer, "On the Study of Wretched Subjects," *Isis* 42 (1951): 111.

4. John Henry, *The Scientific Revolution and the Origins of Modern Science* (London: Palgrave Macmillan, 2002), 4. For representative discussions of these issues see also Peter Dear, "What Is the History of Science the History *Of*? Early Modern Roots of the Ideology of Modern Science," *Isis* 96 (2005): 390–406; Andrew Cunningham, "Getting the Game Right: Some Plain Words on the Identity and Invention of Science," *Studies in the History and Philosophy of Science* 19 (1988): 365–89; and Simon Schaffer, "Scientific discoveries and the End of Natural Philosophy," *Social Studies of Science* 16 (1986): 387–420.

5. *Oxford English Dictionary*, s.v. "science," "scientist"; Sydney Ross, "'Scientist': The Story of a Word," *Annals of Science* 18 (1962): 65–86.

6. G. E. R. Lloyd, *Early Greek Science* (New York: Norton, 1970), xv, cf. 125; David C. Lindberg, *The Beginnings of Western Science* (Chicago: University of Chicago Press, 1992), 4; and Nicholas Jardine, "Epistemology of the Sciences," in *Cambridge History of Renaissance Philosophy*, ed. C. B. Schmitt, Quentin Skinner, Eckhard Kessler, and Jill Krave (Cambridge: Cambridge University Press, 1988), 685. For a strong claim about the constructed nature of the category "science," see Roy Harris, *The Semantics of Science* (London: Continuum, 2005).

7. On the contrast between science (Gk. *episteme*, Lat. *scientia*) and opinion (*doxa, opinio*), see Plato *Republic*, Book 5, 477a–478a; Plato *Timaeus* 37b5–c3; Aristotle *Posterior Analytics* 88b30–35; and Thomas Aquinas, *Summa theologiae* 2a2ae.1, 5. For a Renaissance restatement of this dichotomy, see Matteo Palmieri, "Civil Life, Book II" in *Renaissance Philosophical Texts*, vol II: Political Philosophy, ed. Jill Kraye (Cambridge: Cambridge University Press, 1997), 149–72 (quote, 153).

8. Galileo Galilei, *Discorsi e Dimostrazioni Matematiche, intorno a due nuove scienze* (1638). For seventeenth-century uses of the term, see the list of the sciences in D. Abercrombie, *Academia Scientiarum: Or the Academy of the Sciences, being a Short and Easy Introduction to the Knowledge of Liberal Arts and Sciences* (London: 1687); and the definition in Thomas Holyoake, *A Large Dictionary in Three Parts*, s.v. "scientia" (London: 1676). Illuminating discussions of some of these issues may be found in Lindberg, *Beginnings of Western Science*, 1–4; and Deborah E. Harkness, *The Jewel House: Elizabethan London and the Scientific Revolution* (New Haven, CT: Yale University Press, 2007), xv–xviii.

9. Pamela O. Long offers a comprehensive account of the relations among practical, technical, and discursive knowledge from antiquity to the Renaissance in *Openness, Secrecy, Authorship: Technical Arts and the Culture of Knowledge from Antiquity to the Renaissance* (Baltimore, MD: The Johns Hopkins University Press, 2001).

10. For the idea of modern science as overlapping sets of activities, see John Pickstone, *Ways of Knowing: A New History of Science, Technology and Medicine* (Manchester: Manchester University Press, 2000); and Peter Dear, *The Intelligibility of Nature; How Science Makes Sense of the Word* (Chicago: University of Chicago Press, 2006).

CHAPTER 1

Natural Knowledge in Ancient Mesopotamia

Francesca Rochberg

In the ancient Near East, long before the Greeks' earliest forays into the causes of natural phenomena, what we now call science provided methods for the systematic study of the world of perceptions, experience, and imagination. Such methods and their resulting bodies of knowledge, however, are inevitably historically and culturally determined, even where there is transmission and continuity along some well-defined lines, as, for example, in the history of Western astronomy. Changes in the practice and meaning of science in history and across cultures have been accompanied by changes in the perception and understanding of nature as well. In the context of the ancient Near East, and in particular Babylonia—which represents the most important locus of scientific activity before the classical period of ancient Greece—we can say at the outset that neither the words nor the concepts "science" and "nature" were part of the conceptual landscape. The same is true for astronomy, as we understand it today.

But it does not follow from this that celestial inquiry in ancient Mesopotamia has no place in the history of astronomy. In arguing against a unified image not only of science but also nature, Thomas Kuhn once stated that natural phenomena were not "the same for all cultures," and that "the heavens of the Greeks were irreducibly different from ours."[1] Yet he also insisted that this "does not mean that one cannot, with sufficient patience and effort, discover the categories of another culture or of an earlier stage of one's own."[2] This chapter will address the question of what the study of celestial phenomena meant for the ancient Babylonians, sug-

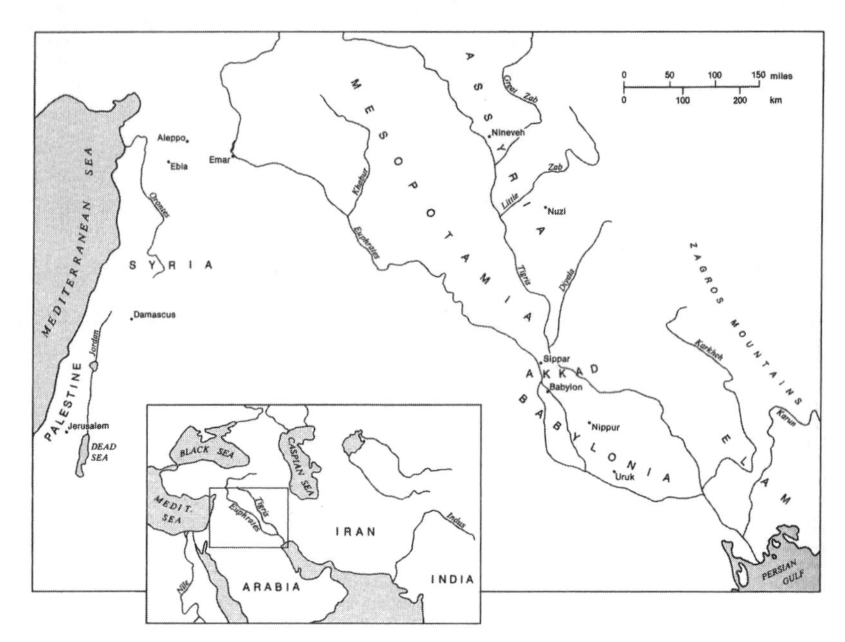

Figure 1.1. Ancient Mesopotamia

gesting that while their heavens, too, were "irreducibly different from ours," their investigations were nonetheless related to what we would recognize as scientific astronomy. Thus, although the ancients' interest in the stars and planets began with and was sustained within a context of celestial divination, it is still possible to discern a fully theorized astronomy that was produced in conjunction with this work.

Babylonian astronomical texts have special significance for the history of science because the foundation of Western astronomy is traceable in those sources. Just as Babylonian theoretical astronomy has a privileged place in the history of science, so does the observation of celestial (and other) physical phenomena for the purpose of determining portents belong to the long history of scientific observation itself. Accordingly, cuneiform evidence for the observation of physical phenomena in heaven and on Earth is a legitimate site for the study of the earliest selective attention to physical phenomena, meaningful within a certain body of knowledge and practice, and constitutive of an attitude about knowing the world through observation that falls within the bounds of what we understand to be science.

Mesopotamia, which corresponds geographically to modern Iraq, takes its name from the fact that it lies between two rivers, the Tigris and Euphrates. The extant written record from ancient Mesopotamia is made up of cuneiform texts in the languages of Sumerian, Akkadian, Babylonian, and Assyrian. It consists of hundreds of thousands of clay tablets and fragments and spans three thousand years, from the so-called Archaic Tablets of the late-fourth- and early-third-millennium BCE temple archives of the city of Uruk and other Sumerian city-states to the latest dated tablets of the first-century CE astronomical archive of the temple at Babylon (located some eighty-five kilometers south of the site of modern Baghdad). Bureaucratic administrative archives constitute the overwhelming bulk of the documentation. The next largest category of sources, many thousands of tablets, is made up of texts concerned with the systematic inquiry into elements of physical existence, or, to employ a native classificatory term, "whatever pertains to complementary elements of celestial and terrestrial parts of the universe (and) those of the cosmic subterranean waters." These sources span the Old Babylonian period (ca. 1800 BCE) to the Parthian period (second century BCE–first century CE). The latest dated cuneiform tablet is an astronomical text of 75 CE, and the Babylonian astronomical tradition would continue to influence quantitative astronomical science into the late Greco-Roman period and beyond.

It is worth emphasizing in the present context that virtually all our information for the Mesopotamian interest and understanding of natural phenomena derives from the cuneiform sources directly. Descriptions of Babylonian astronomy and astrology in works of Greek and Roman writers, such as Berossus, Strabo, Pliny, Diodorus Siculus, and Vettius Valens, prove to be unreliable, although these Greco-Roman authors preserved this legacy of ancient Mesopotamia in the West. From the rough numbers of tablets cited above it would appear as though there is an overabundance of material, more than enough to produce a full reconstruction and a clear insight into the mind of the ancient Mesopotamian "natural inquirer." Such a reconstruction, however, is fraught with problems both practical and interpretative. On the practical side is the fact that not all the works of central importance to the analysis of the ancient inquiry into natural phenomena have been reconstructed or edited, particularly those dealing with astrology. On the interpretative side, the relationships between the various parts of the ancient system of knowledge about the physical world is far from clear as the very classification of physical and metaphysical, material and divine, and the cosmological framework within which it has meaning appears to be quite different from that of the dominantly Greco-Roman Western tradition that flowed from it.

Since the subject of ancient Near Eastern inquiry into the phenomena of nature is vast, the present discussion will cut a narrow path along the following series of topics. A description of sources and their contents in the first section will highlight three areas: the divination corpus and its major classifications, celestial divination and horoscopy, and astronomy in relation to celestial divination and astrology. In the past, the late Babylonian astronomy, both tabular (ephemerides) and nontabular (procedures, diaries, almanacs, goal-year texts), occupied almost exclusively the attention of historians of science because of its significant relation to Greek astronomy.[3] Here I will consider it in the context of celestial divination and horoscopes. The second section will venture into analysis of "native" terminology, by means of which we may gain some access into the ancient Mesopotamians' views of the study of the phenomena. To establish a frame of reference, some aspects of Mesopotamian cosmology will be reviewed, followed by the nature of knowledge, and finally the notion of a "knower" considered as far as it is possible from a Mesopotamian point of view.

THE SOURCES

The cuneiform tablets that catalogue, systematize, describe, and predict observable natural phenomena provide us with sources for reconstructing the history of natural knowledge in the Mesopotamian Near East. We may divide these sources conveniently into two chronologically separate groups; the earlier represents the product of second and early-first millennia (ca. 1800–600 BCE) scribal scholarship, while the later stems from Achaemenid (ca. 500–300 BCE) and Hellenistic Babylonian (300 BCE–75 CE) archives, primarily those of the cities of Babylon and Uruk. Texts of the "early" tradition, however, continued to be copied in the later period. The persistence of tradition, even in the face of the progress of knowledge seen especially in the late astronomy, is a feature of the culture of Mesopotamian science. Our chronological scheme, therefore, does not imply an evolutionary process in which the contents of the more sophisticated late sources replace those of the simpler earlier phase.

The principal extant works of the earlier group come from the seventh-century BCE library housed within the palace of Assurbanipal at Nineveh. Part of this library was excavated at the mound of Kuyunjik on the Tigris by Austen Henry Layard from 1850 to 1851, another part during 1853 by Hormuzd Rassam. All the while, efforts to decipher cuneiform were bearing fruit, and by 1857, the successful test sponsored by the Royal Asiatic Society was carried out by Henry Creswicke Rawlinson, Edward Hincks, and W. H. Fox Talbot. Much of what we know today of the contents and character of the Mesopotamian tradition of scribal scholarship derives from the Nineveh library holdings, which, it appears, are an excellent representative of that tradition, as confirmed by scholarly tablets found outside Nineveh, even outside Mesopotamia, in foreign cities such as Susa (Elam), Hattušaš (Anatolia), and Alalakh (Syria).

The later sources, from Achaemenid and Hellenistic Babylonia, for the most part, have not been acquired by scientific excavation, but rather from the antiquities market going back to the late nineteenth and early twentieth centuries. Principally, the provenance of the scholarly tablets of interest here may be identified with the cities of Babylon, Uruk, and Sippar, whose major temples held archives for the scribe-scholars working within their ranks.

From the earlier corpus of texts, largely comprised of omens of a variety of kinds, the following are illustrative:

> If moths are seen in a person's house: The owner of that house will become important.
> If [in a dream] he eats the meat of a wild bull: his days will be long.
> If there is a mole to the right of his eyebrow: what his mind is set on, he will not attain.
> If the sick man turns his neck constantly to the right, his hands and feet are rigid and his eyes close and roll back, saliva flows from his mouth and he makes a croaking sound: epilepsy.
> If a malformed newborn [lamb] has two heads, and the second one is on its back, and its eyes look in different directions: the king's reign will end in exile.
> If on the fifteenth day the moon and sun are seen together: a strong enemy will raise his weapons against the land; the enemy will tear down the city gates.
> If the sun stays in the position of the moon: the king of the land will sit firmly on the throne.
> If Venus comes close to Scorpius: winds which are not good will blow towards the land; Adad will give his rains, and Ea his springs, to the Gutian land.
> If Regulus carries radiance: the king of Akkad will exercise complete dominion.
> If Jupiter passes to the west: the land will dwell in quiet.
> If the sky shouts and the earth quakes: Enlil will bring about the defeat of the land.

These "omens" seem to have little in common beyond their formulation as conditional statements: if *P*, then *Q*. Closer consideration will reveal the operation of two readily recognizable binary classifications: personal and public, earthly and heavenly. Beyond these exceedingly simple structures, however, are many more elements whose meanings are not self-evident, yet a certain descriptive quality seems characteristic of all the protases ("if"-clauses). The statements quoted above are, in fact, randomly selected examples of each of the major divination reference works in a particular classification of omens. This class is united by the feature that the ominous events, the phenomena, simply occur in the world. All physical phenomena occurring on Earth and in the sky, including the behavior of heavenly phenomena, animals, birds, and human beings, were potential signs indicating future mundane events as decided by the gods. In this form of divination, the diviners observed the world without specifically requesting signs. This technique contrasted with the classification of omens obtained by means of some material manipulation by the diviner. These so-called provoked forms of divination required that the diviner perform an action which served to invite a response from Šamaš and Adad, the gods of the sun and the storm who were associated with divination. The inspection of the entrails (extispicy) of a sacrificed lamb was perhaps the most frequently performed divination by provocation, and the most frequently examined of the exta was the liver. Clay liver models, probably used as teaching devices, some dating from the Old Babylonian period (ca. 1800 BCE), survive.[4] Other provoked omens were obtained by dropping oil on water (lecanomancy) or emitting smoke from a censer (libanomancy), as well as the casting of lots and scattering of flour. The notion of the god (often Šamaš) "writing" the signs on the exta of sheep is well known from prayers and incantations spoken by the diviner prior to the divination itself: "you (Šamaš) write upon the flesh inside the sheep (i.e., the entrails), you establish (there) an oracular decision."[5] Evidence of this practice is also found in the inscriptions of kings, who legitimized themselves or their activities by appeal to the affirmative answers of the gods, as here in an inscription of Esarhaddon of Assyria (680–669 BCE):

> Thus did he [my father] ask of Šamaš and Adad by divination: "Is this the heir to my throne?" and they replied to him with a strong affirmative: "He is your successor."[6]

The distinction between provoked and unprovoked omens reflects Mesopotamian thought about divination, as it appears from their use of different terms for types of diviners. The haruspex (the diviner who inspects the liver) as well as the oil and smoke diviner was the *bārû* and his special text corpus was designated *bārûtu*, an abstract noun also used as a term for the craft of this kind of diviner. The scholar trained in the interpretation of unprovoked omens was designated simply "scribe" or "literatus." By Cicero's classification of divination in *De Divinatione* I vi 12 (also xviii 34 and xxxiii 72), both Mesopotamian forms of divination would qualify as "artificial," as opposed to "natural." "Natural" divination referred to the technique of communication with the gods such as practiced by a prophet or seer, who obtained the divine word "directly" through ecstatic trance and auditory hallucination. Indeed, such prophets and ecstatics are attested in cuneiform literature as a wholly separate group from the scholars. The distinguishing feature of divination by provoked and unprovoked omens, then, was that it was an exercise in reason, in which the diviner deduced the interpretation of signs from the scholarly reference works in which he specialized.

The unprovoked omens were separated by type, arranged in series, and standardized as to tablet number and contents. The series are entitled by their first words as: *Šumma ālu*, "If a city" (daily life omens); "*Zaqīqu, Zaqīqu*, ᵈMA.MÚ, god of dreams" (dream omens); *Alandimmû*, "If the form" (physiognomic omens); "When the diagnostician is on his way to the patient's house" (medical diagnostic omens); *Šumma izbu*, "If the abnormality" (malformed fetus omens); and *Enūma Anu Enlil*, "When Anu Enlil" (celestial omens), which contained four sections for lunar, solar, planets and stars, and weather omens. These various omen series were written, copied, and commented upon over the course of centuries, roughly from the fourteenth century BCE to the last centuries BCE in Hellenistic Babylonian scribal centers. Some of them have even earlier forerunners to the later formulated and standardized versions. Though the remainder of this discussion will focus on the celestial omens, it should be clear that celestial divination differed from the other series just enumerated only in the location of the field of its phenomena, that is, the sky. The fact that celestial phenomena turned out to be more amenable to prediction than the moles on a person's face or the appearance of lizards on the wall of one's house was of no consequence as far as the Mesopotamian scribes' view of divination was concerned.

In the complete and standardized form known from Nineveh library recensions, *Enūma Anu Enlil* comprises seventy tablets devoted to "celestial" signs, that is, visible or anticipated phenomena occurring in the sky during the day or night. Thus, weather phenomena, especially cloud formations and other features of the daytime sky, also counted as celestial phenomena. Phenomena of the moon (such as duration of lunar visibility, halos, eclipses, and conjunctions with planets and fixed stars) are collected in the protases of Tablets 1–22. Solar phenomena (such as coronas, parhelia, and eclipses) are found in Tablets 23–36. Tablets 37–49/50 include such meteorological occurrences as lightning, thunder, rainbows, cloud formations, and winds; and finally, planetary appearances were organized into the last twenty tablets of *Enūma Anu Enlil*. Tablets 50–53 include omens for the constellations; 59–63, the planet Venus; 64–65, Jupiter. Perhaps one or two tablets were devoted to Mars, although their numbering in the sequence is unknown. Mercury and Saturn have omens as well, but are less well represented. Omens for fixed stars and constellations, contained in Tablets 50–51, sometimes include the appearances of the five planets, for example:

> If at Venus's rising the Red Star enters into it: the king's son will seize the throne.
> If at Venus's rising the same star enters into it and does not come forth: the king's son will enter his father's house and seize the throne.[7]

The apodoses ("then"-clauses) in *Enūma Anu Enlil* concern the health and life of the body politic rather than that of the individual. Predictions commonly found in *Enūma Anu Enlil* refer to concerns of an agricultural and political nature:

> The harvest of the irrigated land will prosper, the land will be happy.
> There will be scarcity of barley and straw in the land.
> The arable land will prosper.
> There will be rains and floods, the harvest of the land will prosper.
> Downfall of a large army.
> Adad will bring his rains, Ea his floods, king will send messages of reconciliation to king.

At an early date, the compendium was given the title *Enūma Anu Enlil*, after the three opening words of its bilingual (Sumerian and Akkadian) introduction:

> When Anu, Enlil, and Ea, the great gods, established by their true decision, the designs of heaven and earth, the increase of the day, the renewal of the month (= new moon), and the appearances (of celestial bodies), (then) humankind saw the sun going out from his gate and (the celestial bodies) regularly appearing in the midst of heaven and earth.[8]

While groups of sources for celestial divination are extant for the period from the late second millennium to the first century BCE, there are wide chronological as well as cultural gaps between these groups, as well as a great unevenness in the material preserved (for example, only a few letters in the Old Babylonian Mari archive relevant to the practice of celestial divination versus hundreds from the Sargonid archive of correspondence between king and scholars, or a handful of copies of various Old Babylonian celestial omen tablets versus several thousand extant copies and fragments from exemplars of the corpus from Nineveh). Only from the seventh century BCE is there enough evidence, beyond the copies of *Enūma Anu Enlil*, to gain some insight into the practical application of celestial divination. This evidence comes in the form of letters and reports from diviners, magicians, lamentation priests, and other scholars to two kings of the Sargonid dynasty, namely, Esarhaddon (680–669 BCE) and his son Assurbanipal (668–627 BCE).

The correspondence attests to the expertise of the diviners not only in the celestial and other omen literature, but also in incantations, rituals, and sacrifices necessitated by ominous signs. Since the affairs of state, not the least of which was the fate of the king himself, were the concerns of central importance in the omen apodoses, the institution of a regular watch of the heavens followed by interpretation of observed phenomena by reference to the apodoses of *Enūma Anu Enlil* was established in a number of Mesopotamian cities of the Assyrian Empire, such as Assur, Babylon, Nippur, Uruk, Cutha, Dilbat, and Borsippa. Anxiety seems to have hung over the king regarding eclipses in particular. To allay the king's fears, one scholar even quotes another, saying, "A certain Akkulānu has written: The sun made an eclipse of two fingers at the sunrise. There is no apotropaic ritual against it, it is not like a lunar eclipse. If you say, I'll write down the relevant interpretation and send it to you."[9]

The reason for anxiety over the king's safety stemmed from the association of lunar eclipses with the deaths of kings. Since the possibility of a lunar eclipse was predictable by the seventh century BCE, the danger this phenomenon indicated for the king could be addressed before the fact, as is implied in the letter quoted above. A predicted eclipse put in motion the ritual of the substitute king, whose purpose was to take upon himself the evil portents in place of the king and, when the danger period was over, to be put to death in order that the evil be carried with him to the Land-of-No-Return. One letter concerning the substitute king says:

> The substitute king, who on the 14th sat on the throne in [Ninev]eh and spent the night of the 15th in the palace o[f the kin]g, and on account of whom the eclipse took place, entered the city of Akkad safely on the night of the 20th and sat upon the throne. I made him recite the omen litanies before

> Šamaš; he took all the celestial and terrestrial portents on himself, and ruled all the countries. The king, my lord, should kn[ow] (this).

After some discussion of the pertinent eclipse omen from *Enūma Anu Enlil*, the letter concludes with the following advice: "Nevertheless, the king, my lord, should be on his guard and under strong protection. Apotropaic rituals, penitential psalms and rites against malaria and pestilence should be performed f[or the li]fe of the king, my lord, and of my lords the princes."[10]

A nightly watch of the sky, albeit of an entirely different nature from that instituted at Nineveh and throughout the Assyrian Empire for the express purpose of celestial divination, was undertaken at Babylon from the period of the reign of King Nabonassar (747–734 BCE). Although no eighth-century BCE examples are preserved, this archive of observational texts was established at Babylon from the middle of that century, as is indicated in later compilations of lunar eclipse reports. These so-called astronomical diaries collected lunar, planetary, meteorological, economic, and occasionally political events night after night, usually (at least in the later diaries) for six or seven months of a Babylonian year. The Babylonian astronomers classified these texts with the rubric "regular (celestial) observation which (extends) from the *x*th month of the *y*th year to the *z*th month of the *y*th year." The term "diary" is apt because the texts record daily positions of the moon and planets visible above the local horizon, as in the following lines from a diary dated in the year 331 BCE:

> Night of the 20th, last part of the night, the moon was [nn cubi]ts below β Geminorum, the moon being 2/3 cubit back to the west. The 21st, equinox; I did not watch. Ni[ght of the 22nd, last part of the night,] [the moon was] 6 cubits [below] ε Leonis, the moon having passed 1/2 cubit behind α Leonis. Night of the 24th, clouds were in the sky.[11]

In this diary of 331 BCE, in the section reserved for noteworthy political events, there is the report of the defeat of Darius III by Alexander the Great at Gaugamela. The diary relates that "(in Month VI on the 11th day) panic occurred in the camp before the king," and "(on the 24th) the troops deserted him . . . and fled." The report for the following month contains the statement, also in broken context, that "Alexander, king of the world, [came (?) in]to Babylon."[12] The astronomical diaries, therefore, compile a wealth of detail, of political and economic historical value, as well as being a prodigious source of contemporary dated astronomical observations, no doubt the source of the Babylonian observations utilized by Ptolemy in the *Almagest* (for example, those of Mercury in *Almagest* 9.7).

The regular and systematic observation of the heavens, exemplified by the astronomical diaries archive, and the focus upon lunar eclipses, represented by the eclipse reports and their compilations reaching back to the Nabonassar era (beginning 747 BCE and extending to the first century BCE), are testimony to the results, in purely astronomical terms, of the interest in astronomical prediction generated by celestial divination. Divination, however, is nowhere mentioned in the diaries. Following the Neo-Assyrian period, the locus of astronomical activity gradually shifted from the palace to the temple. Within the cuneiform scholarly repertoire, new elements emerged after ca. 500 BCE. The increase in importance of the individual with respect to the stars, the gods, and the cosmos is one such significant change. With the appearance of celestial omens for individuals, based on phenomena occurring on the date of birth, came a disappearance of evidence for the kind of celestial divination practiced under the Sargonid kings. Evidence is lacking for the Persian or Seleucid kings consulting *Enūma Anu Enlil* through advisors, and, so far, no nativities or horoscopes predate the fifth century BCE. Despite the lack of evidence for the practice of celestial divination in this period, *Enūma Anu Enlil* itself continued to be preserved, as exemplars dated to the Seleucid period clearly indicate.

It is possible, however, to view Babylonian horoscopy as an outgrowth of two traditional forms of Mesopotamian scholarly divination, one being celestial divination as represented by the omen series *Enūma Anu Enlil*, which always retained its concern with public matters (king and state), the other a tradition of birth omens, first attested in a series entitled *Iqqur īpuš*: "He tears down, he rebuilds."[13] *Iqqur īpuš* contains omens for phenomena occurring or activities undertaken on different dates, including the dates of births, and, in the last third of the series, occurrences in nature. In *Iqqur īpuš,* for example, we have the omen "If a child is born in the month of Ajaru (Month II), he will die suddenly," or "If a child is born in the month of Abu (Month V), that child will be despondent."[14] Later, during the Achaemenid period, omens from the occurrence of a variety of celestial phenomena on the dates of births emerged. From this type of "nativity omen" to the horoscope, which compiles the positions of all the planets on the date of birth, seems a rather short step. Only a few horoscopes contain personal predictions, and those that do contain very much the same repertoire of apodoses known from nativity omens. The subjects are generally concerned with family and fortune, such as, "he will be lacking in wealth," "his days will be long," and "he will have sons" or "he will have sons and daughters." The inclusion of such apodoses makes it appear as though the Babylonians viewed the situation of the heavens on the date of birth as a complex celestial omen. As such, horoscopes do not represent an entirely new form of astrology but a development closely tied to celestial divination. Babylonian horoscopes were perhaps the extension and elaboration of the late nativity omens, but, as the following example attests, depended on a far more sophisticated astronomical apparatus than did the earlier tradition of celestial omens or nativity omens:

> Year 77 (Seleucid Era), the 4th of Simanu, in the morning of the 5th(?), Aristocrates was born. That day, the moon was in Leo, the sun was in 12;30° Gemini. The moon goes with increasing positive latitude. "If (the moon) sets its face from the middle toward positive latitude: prosperity and greatness." Jupiter was in 18° Sagittarius . . . The place of Jupiter: (the native's life will be) prosperous, at peace(?), his wealth will be long-lasting, long days (i.e., life). Venus was in 4° Taurus. The place of Venus: he will find favor wherever he goes, he will have sons and daughters. Mercury was in Gemini with the sun. The place of Mercury: the brave one will be first in rank, he will be more important than his brothers; he will take over his father's house. Saturn in 6° Cancer. Mars in 24° Cancer . . . (remainder broken).[15]

The purpose of the Babylonian horoscope was to assemble the positions of the seven planets (moon, sun, Jupiter, Venus, Mars, Mercury, and Saturn) in the zodiac (the division of the ecliptic into twelve 30-degree "signs," developed by the late fifth century BCE) on the date of a birth. The question of the derivation of the astronomical data in the horoscopes is relevant to the problem of the relationship between "astrology" and "astronomy" in Babylonia. The terminology of the horoscope documents points to the nonmathematical astronomical texts, such as the astronomical diaries, as well as other genres of astronomical texts as the primary sources for this data. However, as in the example just given, some horoscopes computed planetary positions to degrees of zodiacal signs, indicating the use of some arithmetical scheme as a technique of generating the planetary positions. Despite changes in textual formalities and even specific content and methods, the continuity between Babylonian celestial divination and horoscopy may be found in the persistent belief that the sky could be "read" as a symbolic system and interpreted as relevant for the human realm.

The interest in celestial phenomena, from the earliest period of the omen texts to the latest period of the horoscopes, was a reflection of a particular relationship between human and divine in ancient Mesopotamia.

That is, the gods were thought to communicate with humankind through the behavior of physical phenomena, which in turn became intensely significant objects of observation and analysis. The results of such inquiry in the realm of celestial phenomena were the development of empiricism, mathematical theoretization of astronomical problems, and methods of predicting the phenomena.

An important astronomical work preceding the development of the late mathematical astronomy by many centuries is the two-tablet compendium MUL.APIN, which compiles the astronomical material required to understand *Enūma Anu Enlil*.[16] Stars and constellations were classified in accordance with paths in the sky in which they are seen to rise and set. Three such paths were demarcated and named for the gods Anu, Enlil, and Ea. In our terms, the Path of Anu is reckoned as the arc over the horizon where stars roughly 15 degrees on either side of the celestial equator rise and set, the Path of Enlil is to the north of this arc, and the Path of Ea is to the south. MUL.APIN also lists the simultaneous rising and setting of constellations and fixed stars in the three paths, season by season, and intervals of visibility and invisibility of planets.[17]

Perhaps most clearly representative of a desire to render celestial phenomena amenable to mathematical description in the period before 500 BCE, is *Enūma Anu Enlil* 14, which contains no omens but instead provides an arithmetical scheme concerning the duration of visibility of the moon each night.[18] Underlying the lunar scheme of *Enūma Anu Enlil* 14 was a simple scheme for the duration of daylight in relation to the schematic calendar month of thirty days. The length of day was not understood to be a function of the motion of the sun in the ecliptic, but the result of a division of the schematic year into four symmetrical "seasons" organized around the two days when day and night are of equal length (equinoxes) and the longest and shortest days (summer and winter solstices) which differ from the length of daylight at the equinoxes (3 units) by ±1 unit (that is, the length of day at equinox is 3, summer solstice is 4, and winter solstice 2). The ratio of longest to shortest day was therefore 2:1. In relation to this daylight length scheme, the duration of visibility of the moon at night, time between sunset and moon set for the first half of the month and between sunset and moon rise for the second half, was arithmetically schematized.[19] Suffice it to say here that *Enūma Anu Enlil* 14 is significant for the history of astronomy in that it represents the development of a quantitative methodology applied to solving several astronomical problems—here, the variation in the length of day (and night) and the variable duration of visibility of the moon each night due to its elongation from the sun as well as the variable length of night throughout the year. This arithmetic scheme is linear because it is characterized by a difference sequence of first order, that is, the entries in the table increase and decrease by a constant amount and the values are bounded by a maximum and a minimum value. The values in the table describe a so-called linear zigzag function, which is periodic. Much later and in a more sophisticated form, the zigzag function would still be used in the astronomy of the Hellenistic period, both in cuneiform mathematical astronomy and all those forms of Greek astronomy that derived from the Babylonian tradition.[20]

Elsewhere in *Enūma Anu Enlil*, omen protases reflect an underlying and systematic observation of many phenomena, as well as the recognition of some simple periodicities. The Old Babylonian (ca. 1800 BCE) lunar eclipse omen protases, for example, identify lunar eclipse days with the middle of the lunar month, or days 14–16, when lunar eclipses can occur. Additionally, these protases are constructed with elements such as the reddish color of the moon during total eclipses, the various times of night when the eclipse occurs (as well as its duration), and all appropriate phenomena for lunar eclipses. By the seventh century BCE, lunar eclipses in the omens became more descriptive and detailed, such as in the following example from *Enūma Anu Enlil* Tablet 20:

> If an eclipse occurs on the 14th day of Ṭebētu, and the god (= the moon), in his eclipse, becomes dark on the east upper part of the disk and clears on the west lower part; the west wind (rises and the eclipse) begins in the last watch and does not end (with the watch); his cusps are the same (size), neither one nor the other is wider or narrower.[21]

The prediction and exclusion of lunar eclipses was already an established practice in the letters and reports from court astrologers, mentioned above: "(I wrote to the king, my lord) '[The moon] will make an eclipse.' [Now] it will not pass by, it will occur."[22]

Only during the fifth century BCE was the eighteen-year eclipse period, or "Saros," used for predicting eclipse possibilities, but during the period of the Sargonids, letters from scholars attest to the fact that they kept a regular watch for predicted lunar eclipses[23] and even for anticipated solar eclipses.[24] The knowledge of astronomical periods during the seventh century BCE, however, was characterized by crude intervals between characteristic planetary visibility phenomena (for example, one year between a first visibility of Jupiter and a last visibility). The interest in prediction was circumscribed by celestial divination, in which only first visibilities and first invisibilities were ominous, as were the planets in retrograde, but not their stationary points.

Probably before the formulation of mathematical predictive schemes, goal-year texts exemplify a method of prediction by the use of synodic periods. These tablets give dates of characteristic planetary (synodic) phenomena and of planets' passing by bright ecliptical stars for a certain year (the goal year) by tallying those positions, culled from astronomical diaries, an appropriate number of years before the goal year for the planet (for example, Venus eight years before, or Jupiter seventy-one years). In seventh-century BCE Assyrian sources as well, the use of planetary periods as a way of anticipating certain celestial phenomena deemed ominous is well documented in the scholars' correspondence with the Sargonids. This "nonmathematical" method of astronomical prediction was different from that of the mathematical astronomy developed later, in which the behavior of celestial phenomena was subject to more rigorous mathematical description. Aside from a few exceptional planetary daily motion ephemerides, the Babylonian mathematical astronomical texts did not tabulate positions of celestial bodies between synodic "events."[25] Consequently they do not appear to have been designed with horoscopy in mind, since a horoscope requires the positions of the seven planets for an arbitrary time, namely, the date of birth.

Because the zodiac provided the basic and primary reference point for astronomical data recorded in horoscopes, the usefulness of astronomical sources to those who wrote the horoscopes can be assessed in terms of their use of this reference system. Because most of the planetary positions recorded in the diaries represent observations, these are reckoned some number of cubits (1 cubit = 2 1/2°) with respect to (in front of, behind, above, and below) a group of about thirty ecliptical stars (e.g. "night of the 7th beginning of the night, the moon was 2/3 cubit in front of ε Leonis")[26] A systematic relationship between such observational statements and positions with respect to degrees of the zodiac, which are not strictly speaking observable, has not been determined, and so we are not certain if the ancients had a method of converting cubits with respect to the "normal stars" to degrees of the zodiac.[27] The diaries, however, include an enumeration of the zodiacal signs in which the planets are found during a given month, with dates of entries into signs, and so could conceivably have been practical for the construction of horoscopes.

Some horoscopes give planetary positions in degrees of the zodiac. The only astronomical text genre in which data are given in this form are the ephemerides, although, as mentioned above, the methods employed in these tables are designed for computation of positions for synodic appearances, not the appearances on arbitrary dates. Interpolation schemes would be necessary to adapt the methods of the ephemerides to use for horoscopes.

THE ANCIENT SCHOLARS' PERCEPTION OF THE NATURAL WORLD AND THEIR CONCEPTION OF KNOWLEDGE OF IT

Having introduced the relevant sources and their particular concerns, we may now address the question of the nature of the scribes' knowledge of the phenomenal world. While the category "nature" has no direct counterpart in the Mesopotamian sources, natural phenomena were clearly a central focus for observation and systematic thought. We shall therefore attempt to define, in terms appropriate to the cuneiform material, a picture of the phenomenal world, which was the object of inquiry in ancient Mesopotamian scribal scholarship. In the following, an appeal to mythological texts will aid in the definition of the ideological background for scholarly inquiry, not because a religious mode of thought about the world is characteristic of the ancient scribes, but only because a continuity of thought is evidenced between divination and cosmological mythology. We do not see such a continuity of thought as grounds for a claim about a nonrational or animistic or mythopoeic mode of cognition.

According to our major source for Babylonian cosmology, the poem *Enūma Eliš*, as well as other mythological sources in the background of *Enūma Eliš*, form and order in the world came to be from a prior state of formlessness. Cosmogony did not focus on a natural order, but a divine and political one that included the natural. The divine order provided a structure of authority, with AN ruling remote heaven, Ea (Sumerian, EN.KI) ruling the waters around and below the earth, and EN.LÍL the space between the great above and the great below with its atmosphere and wind. In the revisionist cosmogony of *Enūma Eliš*, Marduk is made the preeminent authority; the seven gods of the "destinies" and the fifty Anunnaki and three hundred Igigi come after.[28] This last divine cosmic arrangement was part of a fixed order that came into existence with creation and became a permanent characteristic of the cosmos. Included with this fixed order—indeed, made possible by it—were omens and portents: "the norms had been fixed (and) all [their] portents."[29] That physical phenomena were signs meant to portend the future for humankind was laid down as part of creation. The channel of communication between divine and human, evidenced by the belief in the efficacy of divination, was therefore seen as part of the original structure of the world. As stated in the concluding paragraph of a source for *Enūma Anu Enlil* Tablet 22, celestial signs came into being with the creation and organization of the heavenly bodies themselves:

> When Anu, Enlil, and Ea, the great gods, created heaven and earth and made manifest the celestial signs, [they fixed the stations and estab]lished the positions [of the gods of the night . . . they divided the co]urses [of the stars and drew the constellations as their likenesses.][30]

The physical connection between the cosmic regions above and below, established by celestial signs above portending matters below, is symbolized by the motif of the "bonds of heaven and earth." The cosmic bonds were imagined as ropes or cables that tied down and controlled the flow of waters from the heavens. Underpinning this notion is the mythological story in *Enūma Eliš* of the heavens' creation from the slain sea monster Tiāmat, whose body Marduk splits in two "like a bivalve," as the text puts it. The conception of the cosmic bonds thereby relates to the image of the gates locking in the waters of Tiāmat after half of her body had been installed as the roof of the uppermost heaven. In the cosmic sense, the rope- or cable-like bond became a linking device which could be held by a deity, as in the description of Ištar as the goddess "who holds the connecting link of all heaven and earth." This bipartite cosmos persisted in later Babylonian thought, where religious and scholarly texts of the first millennium BCE refer to cosmic designs, literally "plans of the above and below." These cosmic designs, perhaps the image of universal order, are frequently associated with what are called cosmic "destinies," for example, as in the divine epithet "lord of cosmic destinies and designs." The two are integrally connected by virtue of the semantic force of the Akkadian term "destinies," which signified the order or nature of things, and so did not exist before the creation of the universe itself. *Enūma Eliš* 1.8 says of the time before creation, "no destinies [order of things] had been determined."

Obviously, in the context of mythology, the underlying causes or reasons for regularity in the cosmos would be expressed in terms of divine agency, not natural law. But outside the boundaries of mythological texts, that is, in omen literature, the patterns observed in nature were still taken as physical manifestations, or manipulations, of the divine. Here, if a law-like behavior was to be attributed to physical phenomena, it would be in the sense of being subject to the judgments and rulings of the gods. In accordance with such a conception of divine influence in the world, the order perceived in phenomena could just as easily be disrupted as maintained by divine will, as in *Enūma Eliš* 4.23–4, which attests to this very power of the god Marduk to create as to destroy order in the heavens: "By your utterance let the star be destroyed, command again and let the star be restored." The deliberate disruption of order by an angry god is also found in the *Erra Epic*: "When I (Marduk) left my dwelling, the regulation of heaven and earth disintegrated; the shaking of heaven meant the positions of the heavenly bodies changed, nor did I restore them."[31] Again, the prayer literature provides us with a parallel, this time with specific reference to an omen: "You (Nabû) are able to turn an untoward physiognomic omen into (one that is) propitious."[32]

The instrumental role of the gods manifested in phenomena reflects a nonmechanistic element in the Mesopotamian cosmos. Even in the context of the mythological "designs of heaven and earth," which suggest something of a fixed structure in the world, the designs were drawn by gods who were never conceived of as simply setting things in motion only to step away and leave the machinery running, but as active participants in the world. Within such a cosmology, signs in nature produced by gods could not be viewed as occurring out of deterministic necessity. The most compelling evidence against determinism in Babylonian divination and cosmology, however, was the viability of magic and ritual for dispelling bad omens (apotropaic magic). This further dimension of Mesopotamian divination—the human response to an omen's meaning—is entirely what one would expect of a system understood as fundamentally one of communication between divine and human.

The Mesopotamian "natural world" was therefore conceived of as the arena within which human beings could observe the workings of the gods. The orderliness of the universe was viewed as a product of divine creation, but religious texts do not hesitate to express the fear that, were a god of a mind to change any of that order, he could certainly do so. The basic order and regularity of phenomena was given expression in the omen compendia, where phenomena in all their diversity were arranged in systematic fashion. Here, the contingent universe, changeable by divine will, was amenable to divination because the ill consequences—divine decisions associated with particular phenomena—were potentially alterable by magic.

Given that the scribal repertoire of texts consisted in anonymous compendia of words, omens, medical symptoms, and magical ritual prescriptions, the only place in such works where one can obtain a glimpse of scribal identity and how the contents of the tablet are classified or conceptualized is in the colophon, the final paragraph devoted to facts relating to the production of the text, such as the name of the copyist, date and place where the inscription was copied, and any comments the scribe might have about the contents of that text. Unlike the contents of the texts, which do not vary from one copy to another, the colophons are unique and within certain parameters contain vastly different information, as the following examples illustrate. The first is from the terrestrial omen series, the second from the celestial:

> 52nd Tablet (of the series) "If a city is situated on a height." According to its [original (it was) written an]d collated. Hand of Sillâ, son of Šu[. . . the app] rentice scribe, the junior exorcist. Tablet of Nabû-mudammiq, the [. . .][33]

Enūma Anu Enlil series:

> Tablet of Anu-aha-iddin, son of Nidinti-Ani, the son of Anu-bēl-šunu, the descendant of Ekur-zākir, the Exorcist of Anu and Antu, the high priest of the Rēš Temple [. . .]
> Hand of Anu-aha-ušabši, son of Ina-qibīt-Ani, son of Anu-uballiṭ, the descendant of Ekur-zākir, the Exorcist of Anu and Antu, the high priest of the Rēš Temple, Scribe of *Enūma Anu Enlil.*
> He wrote [the tablet] for his study, long days, good health, that he not become sick, and for the reverence of his lordship, and placed it in Uruk. Whoever worships Anu and Antu, may he not remove it [the tablet]! Uruk, Month XI, day 26, year 117 of Antio[chus, k]ing of kings.[34]

It has long been recognized that in colophons some texts are designated as "secret" or "exclusive" knowledge (Akkadian *pirištu*, from the root *prs*, meaning "cut off" or "separate"), as in this colophon from a scholarly commentary:

> Exclusive knowledge: The one who knows may show it to another one who knows. [The tablet was] completed and collated; old original.
> Hand of Kidin-Sin, junior scribe, son of Sutu, scribe of the king.[35]

The colophon of an incantation text of the exorcist of the temple of Assur is similar:

> Exclusive knowledge of the great gods. The one who knows may show [it] to another one who knows. The one who does not know may not see it. It [belongs] to the forbidden things of the great gods.
> Written according to its original and collated.[36]

Summarizing the genres of texts so designated, Joan Goodnick Westenholz found lists of gods, stars, cult symbols, incantations, rituals, omens, medical texts, and astronomical texts, all of which genres belong to *ṭupšarrūtu* "scholarship."[37] While the notion of secret knowledge has generally been assumed to be late, (that is, first millennium in origin), Westenholz's particular thesis was "that the earliest lexical compilations may have been more than a utilitarian convenience for the scribes who wrote them; that they contained a systematization of the world order; and that at least one was considered as containing 'secret lore.'"[38] In contrast to the view that lexical lists were the practical result of the need to store data and to preserve the scribal school curriculum, that is, the lists were simply inventories and teaching tools, Westenholz sees already in the early third millennium BCE the notion of privileged, exclusive knowledge within the community of scholar-scribes.[39] Whereas *pirištu* denoted exclusivity, another term used in reference to the texts of the scribal repertoire even more clearly connoted secrecy. This was the term *niṣirtu* "guarded" (from the root *nṣr*, meaning "to guard" or "to protect"), which often referred to a "(hidden, guarded) treasure," hence "secret knowledge." Another word for "knowledge" or "wisdom" is *nēmequ*. Often, but not exclusively, associated with *nēmequ* was the god Ea, who, as creator and ally of humankind, was willing to not only divulge his secrets in the form of magic and *arcana mundi* to human beings but to make messages available as omens for the benefit of the human race. Indeed, the scribes' view that the authority behind the series *Enūma Anu Enlil*, as well as other omen, incantation, and ritual texts, was the god Ea, appears in a catalog of texts and "authors":

> [The Exorcists'] Series (*āšipûtu*), The Lamentation Priests' Series (*kalûtu*), The Celestial Omen Series (*Enūma Anu Enlil*), [(If) a] Form (*alamdimmû*), Not Completing the Months, Diseased Sinews; [(If)] the Utterance [of the Mouth], The King, The Storm(?), Whose Aura is Heroic, Fashioned like An: These (works) are from the mouth of Ea.[40]

That the contents of secret texts stem ultimately from the "mouth" of a god seems to imply some kind of divine revelation in the process of transmission of these traditions to the ranks of scholars. An explicit story of such revelation is told about the antediluvian sage Enmeduranki and the gods of divination:

> Šamaš and Adad honored him, Šamaš and Adad [set him] on a large throne of gold, they showed him how to observe oil on water, a mystery of Anu, [Enlil and Ea], they gave him the tablet of the gods, the liver, a secret of heaven and [underworld].[41]

The sage Enmeduranki in turn shared with "the men of Nippur, Sippar, and Babylon" the knowledge he had obtained from the gods, and the text continues with the promise that the scholar, the one who knows, who guards the secrets of the great gods, will bind his son whom he loves with an oath before Šamaš and Adad by tablet and stylus and will instruct him.[42]

This brief look at the subject of "knowledge"—and more specifically and problematically, "secret knowledge"—barely scratches the surface of available sources with which one might more thoroughly investigate the cuneiform scribal phenomenon. Perhaps we can observe here, however, that the object of knowledge that seems to concern the scribes, while ultimately referring to the phenomena of heaven and Earth, is found in the contents of texts. Viewed in this way, what one knows is not nature, but texts. The expressions "secret knowledge" and "belonging to the forbidden things of the gods" referred to the sanctity of texts, not of nature. William Eamon has described the "secrets of nature" as "one of the most prominent and most powerful metaphors in the history of science."[43] In his analysis, "secret" knowledge becomes a way of expressing not only a deeper knowledge of the phenomena beyond our ordinary sense perceptions but also comments on the privileged status of the one who has come to know the deeper meaning of nature. From a Mesopotamian point of view, the deeper meaning would refer, I think, both to the knowledge of the significance of the phenomena as omens (that is, the material contained in the texts) and the understanding of what and when to observe phenomena deemed ominous. As shown above, *Enūma Anu Enlil* presented a systematic organization of celestial phenomena, reflecting at least a rudimentary understanding of the behavior of the moon, sun, planets, and fixed stars. Such privileged knowledge was clearly a cumulative product necessitating development over time, even though it was claimed to have originated "in the mouth of Ea."

As seen in several examples cited above, the colophons make it clear that access to the tablets containing secret, exclusive knowledge was limited to persons termed *mūdû*, "the one who knows." Contrariwise, the one who had not entered the privileged group of the knowledgeable was *la mūdû*, "the one who does not know." An interdiction against the *la mūdû* is further specified in the colophons with the statement that for the uninformed to see the tablet belonged to the forbidden things of one or more of the gods. This taboo seems to go back to Middle Babylonian times in a text giving a list of gods and their divine symbols.[44] The latest sources in which this interdiction is expressed are the astronomical texts of the Hellenistic period. Consider the following colophon from a lunar ephemeris written about 190 BCE:

> On eclipses of the moon.
> Tablet of Anu-bēl-šunu, lamentation priest of Anu, son of Nidintu-Anu, descendant of Sin-lēqi-unninni of Uruk. Hand of Anu-[aba-utêr, his son, scri]be of *Enūma Anu Enlil* of Uruk. Uruk, month I, year 12[1?] Antiochus [. . .]
> Whoever reveres Anu and Antu [. . .]
> Computational table. The wisdom of Anu-ship, exclusive knowledge of the god [. . .]
> Secret knowledge of the masters. The one who knows may show (it) to an [other one who knows]. One who does not know may not [see it. It belongs to the forbidden things] of Anu, Enlil [and Ea, the great gods].[45]

In view of the interdiction against persons outside the circle of "knowers," as well as the implication that the knowledge contained in the tablets of divine secrets and wisdom was transmitted from a divine source, the nature of knowledge, in terms of the scribes' own references to the contents of specific texts, was considered arcane.

On the other hand, our investigation of the practice of celestial divination and of observational and mathematical astronomy would have us conclude that knowledge about celestial phenomena in the Mesopotamian practice was derived by textual hermeneutics, empiricism, theoretization, and prediction. When a celestial omen specialist interpreted the meaning of a phenomenon by reference to the omen compendium, the authority of the interpretation was grounded in the text, not on a claim to divine inspiration. Mesopotamian divination was *not* prophecy; rather, it operated on the basis of interpreting the divinely decided consequences of phenomena. In some cases, signs were taken to counterbalance or annul other signs, and such exegetical techniques allowed for much latitude in interpretation.[46] In addition, before we conjure up too extreme a picture of an exclusive learned society, it is worth noting that, as M. Stolper showed in the case of Bēl-ittannu of the fourth century BCE, one and the same scribe has been discovered to participate in both learned scholarship and clerical record-keeping. Stolper observed that "the same men sometimes wrote texts of both sorts and sometimes stored their archives and their libraries together. Excavated and reconstructed groups of late Achaemenid tablets from Ur, Uruk, and Nippur that included legal and administrative as well as scholarly and practical texts foster the same view."[47]

To return specifically to the astronomically informed, one of the professional titles encountered in the colophons of astronomical texts is *ṭupšar Enūma Anu Enlil*, literally, "scribe of *Enūma Anu Enlil*." The meaning of this title may seem to speak for itself, but it has been rendered into English variously as "astrologer" or "astronomer," or even, in an attempt not to differentiate between the two, "expert in celestial matters."[48] Surprisingly, the colophons of copies of *Enūma Anu Enlil* never indicate that the copyist is a member of this class of scribes. In fact, there are only four Neo-Assyrian documents in which this title appears, one of which is a letter mentioning the "reports of the scribes of *Enūma Anu Enlil*"[49] and another of which is an administrative document listing employees of the court.[50] Another mentions two scribes of *Enūma Anu Enlil* who "look day and night at the sky."[51] Scribes not holding this title, but who were exorcists or priests, could quote celestial omens, as was the case with the priest Akkullānu, an "Enterer of the Temple of Assur," who carried out celestial observation and research in the *Enūma Anu Enlil* series, counseled the king on this basis, and personally supervised the apotropaic rites necessitated by celestial omens which he recommended be performed.

The literary activities of the "scribe of *Enūma Anu Enlil*" changed in the Hellenistic period, when the term comes to be associated with scribes who produced mathematical astronomical texts. As in the Neo-Assyrian correspondence, the evidence from colophons of Seleucid astronomical texts shows that the scribes who either copied ("hand of so-and-so") or owned the tablets ("tablet of so-and-so") were not always designated "scribe of *Enūma Anu Enlil*," but were oftentimes identified by the professions lamentation priest (*kalû*)[52] or exorcist (*āšipu*). As was once fashionable, it would be easy to characterize ancient Mesopotamian knowledge not as "scientific," but as "religious" or "priestly" by virtue of the fact that the scribes held "priestly" titles. This merely substitutes one anachronism for another, since the problem we create in importing the word "science" into the ancient Mesopotamian milieu applies to the word "priest" as well.

As to the employment of the scholars who dealt with celestial sciences, from Achaemenid times onward, we may suppose that they were no longer employed by the king—at least there is no evidence for such a claim. On the other hand, whether they were all in the service of the major temples is also difficult to pin down, although the available evidence points in this direction. During the Neo-Babylonian period (sixth century BCE), for example, it appears that the scholars' workplace was a separate institution called the "temple academy."[53] The scholars producing ephemerides and procedure texts for which colophons remain appeared to be working within the temple institution during the Seleucid and Arsacid (Parthian) periods.[54] In Babylon, scribal scholarship seems to have been attached to the Marduk (Bēl) temple Esagila, and in Uruk to the Anu temple, the so-called Rēš sanctuary. Accordingly, on the upper edge of the astronomical tablets from Babylon, an invocation "according to the command of Bēl and Bēltīja, may it go well" is frequently preserved. The Uruk tablets have the same, only the gods of the Uruk temple, Anu and Antu, are invoked.

Why the exorcists or lamentation text scribes, who were also scribes of *Enūma Anu Enlil*, became functionaries of the temple may be tied to their authority in matters of ritual. While earlier, in the Neo-Assyrian period, exorcists and lamentation text scribes served the king, the association of these functionaries with the temple in this period is also attested. Some Neo-Assyrian lamentation experts, and possibly also exorcists, were consecrated members of the temple. These Assyrian officials, however, did not bear the title "priest" (LÚŠID = *šangû*).[55] According to a list of names from Uruk in the Hellenistic period, exorcists were classified as "enterers of the temple (Eanna)."[56] Among the exorcists listed in this text, Ekur-zākir and Hunzū both appear in the colophons of astronomical and astrological texts as ancestors of scribes. But Ekur-zākir is also found with the title scribe of *Enūma Anu Enlil* in a mathematical text. The relationship to the cult of such exorcists who also engaged in astronomical activity is not at all clear, as the class "enterers of the temple" was rather broad, encompassing any member of the temple personnel who had access to areas of the temple that were closed to others. By itself, the term "enterer of the temple" carried no special sacred status; the English word "priest," as Brinkman has pointed out, implies much more than does the designation "enterer of the temple."[57]

Perhaps it is futile to attempt from this prosopographical evidence to answer an epistemological question. Our interest is in the Mesopotamian scribes' understanding of the nature of "natural knowledge." We have seen, at least, that the evidence sheds no special light on celestial phenomena as a separate category, except that an astronomical corpus was eventually produced of a quality that ensured its preservation and influence well beyond antiquity into the European astronomical tradition. It seems clear that the idea of (natural) knowledge was inseparable from ideas about the role of the gods in the physical environment, as well as in the encyclopedic reference works devoted to the phenomena. There seems, however, to be a wide gap between the practice of scholarly divination (and astronomy) and that of religion itself. If the temple became the preserve of cuneiform scholarship, it does not imply that the work of transcribing and preserving of texts had cultic applications; it may simply mean that it belonged to the *traditum* as a whole. We do not know how or if the celestial omen compendium *Enūma Anu Enlil* was still used, only that it was still copied and collated by the scribes. Regardless of the way celestial divination, or astronomy, functioned within the late Babylonian temple institution, association with the temple was without doubt the key to the survival of Babylonian astronomy and celestial divination for so many centuries after it had become defunct in the political sphere. As a further consequence, the maintenance of Babylonian astronomy and celestial divination by the temple scholars made possible its transmission to Greeks interested, as it is put in one Greek horoscope, in the science of "ancient wise men, that is the Chaldeans."[58]

NOTES

I wish to thank Professor Bernard R. Goldstein for his always valuable comments on an early draft of this chapter. My gratitude also goes to Michael Shank, Ronald Numbers, and Peter Harrison for their constructive and insightful questions and recommendations.

1. T. S. Kuhn, "The Natural and the Human Sciences," in *The Interpretive Turn: Philosophy, Science, Culture*, ed. David R. Hiley, James F. Bohman, and Richard Shusterman (Ithaca: Cornell University Press, 1991), 19, 21.

2. Ibid., 21.

3. For internal analysis of the mathematical structures of these texts, see Otto Neugebauer, *A History of Ancient Mathematical Astronomy*, 3 vols. (Berlin: Springer Verlag, 1975); see also A. Aaboe, *Episodes from the Early History of Astronomy* (Berlin: Springer Verlag, 2001), and N. M. Swerdlow, *The Babylonian Theory of the Planets* (Princeton, NJ: Princeton University Press, 1998).

4. For the Old Babylonian liver models, see M. Rutten, "Trente-deux modèles de foies en argile provenant de Tell-Hariri (Mari)," *Revue d'Assyriologie et d'Archéologie Orientale* 35 (1938): 36–52.

5. See W. Mayer, *Untersuchungen zur Formensprache der babylonischen "Gebetsbeschwörungen"* (Rome: Biblical Institute Press, 1976), 505: 111.

6. Riekele Borger, *Die Inschriften Asarhaddons König von Assyrien* (Graz: Archiv für Orientforchung, Beiheft 9, 1965), 40: 12–14.

7. E. Reiner and D. Pingree, *Babylonian Planetary Omens Part Two: Enūma Anu Enlil, Tablets 50–51* (Malibu, CA: Undena Publications, 1981), 49.

8. L. W. King, *The Seven Tablets of Creation*, vol. 1 (London: Luzac, 1902) 124; L. W. King, *The Seven Tablets of Creation*, vol. 2 (London: Luzac, 1902) pl. 49: 9–14 (Akkadian version).

9. S. Parpola, *Letters from Assyrian Scholars to the Kings Esarhaddon and Assurbanipal Part II: Commentary and Appendices* (Neukirchen-Vluyn: Verlag Butzon & Bercker Kevelaer, 1983), 113.

10. Ibid., 287.

11. A. J. Sachs and Hermann Hunger, *Astronomical Diaries and Related texts from Babylonia*, vol. 1: *Diaries from 652 B.C. to 262 B.C.* (Vienna: Verlag der Österreichischen Akademie der Wissenschaften, 1988), 177, No. 330.

12. Ibid.

13. R. Labat, *Un calendrier babylonien des travaux des signes et des mois* (Séries Iqqur Îpuš) (Paris: Librairie Honoré Champion, 1965).

14. Ibid., 132–32, §64: 2 and 5.

15. F. Rochberg, *Babylonian Horoscopes* (Philadelphia: American Philosophical Society, 1998), 82–85.

16. The dating of MUL.APIN is summarized in Hunger and Pingree, *Astral Sciences in Mesopotamia* (Leiden: Brill, 1999), 57.

17. Hermann Hunger and David Pingree, *MUL.APIN: An Astronomical Compendium in Cuneiform* (Horn, Austria: Verlag Ferdinand Berger & Söhne, 1989), 148–49.

18. F. N. H. Al-Rawi and A.R. George, "*Enūma Anu Enlil* XIV and Other Early Astronomical Tables," *Archiv für Orientforschung* 38/39 (1991/1992): 52–73.

19. For an exposition of the text, see Hunger and Pingree, *Astral Sciences*, 44–50.

20. Al-Rawi and George, "*Enūma Anu Enlil* XIV and Other Early Astronomical Tables."

21. Adapted from F. Rochberg-Halton, *Aspects of Babylonian Celestial Divination: The Lunar Eclipse Tablets of Enūma Anu Enlil* (Horn, Austria: Verlag Ferdinand Berger & Söhne, 1988), 209.

22. Hermann Hunger, *Astrological Reports to Assyrian Kings* (Helsinki: Helsinki University Press, 1992), 269.

23. Hunger and Pingree, *Astral Sciences*, 120–21.

24. Ibid., 122–23.

25. O. Neugebauer, *Astronomical Cuneiform Texts*, vol. 2: *The Planets* (London: Lund Humphries, 1955), Nos. 310, 654, 655, 810.

26. Sachs and Hunger, *Astronomical Diaries and Related texts from Babylonia*, vol. 1, 220–21, obv. 4.

27. See G.Grasshof, "Normal Star Observations in Late Babylonian Astronomical Diaries," in *Ancient Astronomy and Celestial Divination*, ed. N. M. Swerdlow (Cambridge, MA: MIT Press, 1999), 97–148.

28. *Enūma Eliš* VI, 80–81.

29. Ibid., VI, 78.

30. Rochberg-Halton, *Aspects of Babylonian Celestial Divination*, 270–71.

31. *Erra Epic* I, 133–34.

32. O. R. Gurney and J. J. Finkelstein, *The Sultantepe Tablets*, 2 vols. (London: The British Institute of Archaeology at Ankara, 1957), No. 71:20.

33. Hermann Hunger, *Babylonische und assyrische Kolophone* (Neukirchen-Vluyn: Verlag Butzon & Bercker Kevelaer, 1968), No. 77.

34. Ibid., No. 93.

35. Ibid., No. 50.

36. Ibid., No. 206.

37. Joan Goodnick Westenholz, "Thoughts on Esoteric Knowledge and Secret Lore," in *Intellectual Life of the Ancient Near East: Papers Presented at the 43rd Rencontre*

Assyriologique Internationale Prague, July 1–5, 1996, ed. Jiří Prosecký (Prague: Academy of Sciences of the Czech Republic Oriental Institute, 1998), 456n28.

38. Ibid., 451.

39. A. L. Oppenheim, *Ancient Mesopotamia: Portrait of a Dead Civilization*, rev. ed., completed by E. Reiner (Chicago: University of Chicago Press, 1976), 246–48.

40. W. G. Lambert, "A Catalogue of Texts and Authors," *Journal of Cuneiform Studies* 16 (1962): 64 I (K.2248), on lines 1–4.

41. W. G. Lambert, "Enmeduranki and Related Matters," *Journal of Cuneiform Studies* 21 (1967): 126–38 (132–33).

42. Ibid., 126–38.

43. William Eamon, *Science and the Secrets of Nature: Books of Secrets in Medieval and Early Modern Culture* (Princeton, NJ: Princeton University Press, 1994), 351.

44. Goodnick Westenholz, "Thoughts on Esoteric Knowledge and Secret Lore," 455n26.

45. Neugebauer, *Astronomical Cuneiform Texts*, No. 135 (reading of the date is uncertain, see p. 19 for discussion); see also Hunger, *Babylonische und assyrische Kolophone*, No. 98.

46. See the case study of the omens taken with reference to an observed late appearance of full moon, and the conjunction of Saturn and Mars with the moon in Virgo of March 15, 669 BCE, in Ulla Koch-Westenholz, *Mesopotamian Astrology: An Introduction to Babylonian and Assyrian Celestial Divination* (Copenhagen: Museum Tusculanum Press, 1995), 140–47.

47. M. Stolper, "Lurindu the Maiden, Bēl-ittannu the Dreamer, and Artaritassu the King," in *Munuscula Mesopotamica: Festschrift für Johannes Renger*, ed. B. Böck, E. Cancik-Kirschbaum, and T. Richter (Münster: Ugarit-Verlag, 1999), 595; on the Nippur scribes of the Achaemenid period, see F. Joannès, "Les archives de Ninurta-ahhê-bullit," in *Nippur at the Centennial: Papers Read at the 35e Rencontre Assyriologique Internationale, Philadelphia, 1988*, ed. Maria deJong Ellis (Philadelphia: Occasional Publications of the Samuel Noah Kramer Fund, 1992), 87–100.

48. E. Reiner, *Astral Magic in Babylonia* (Philadelphia: American Philosophical Society, 1995), 63.

49. Parpola, *Letters*, No. 76; cf. Hunger, *Astrological Reports to Assyrian Kings*, No. 499.

50. C. H. W. Johns, *Assyrian Deeds and Documents* (London: George Bell and Sons, 1898), No. 851 obv. i 8.

51. S. Parpola, "A Letter from Šamaš-šumu-ukīn to Esarhaddon," *Iraq* 34 (1972): 22.

52. P. A. Beaulieu, "The Descendants of Sîn-lēqi-unninni," in *Assyriologica et Semitica: Festschrift für Joachim Oelsner*, ed. J. Marzahn and H. Neumann (Münster: Ugarit-Verlag, 2000), 1–16.

53. P. A. Beaulieu, *The Reign of Nabonidus King of Babylon 556–539 B.C.* (New Haven, CT: Yale University Press, 1989), 7–8.

54. F. Rochberg, "Scribes and Scholars: the ṭupšar Enūma Anu Enlil," in Marzahn and Neumann, *Assyriologica et Semitica: Festschrift für Joachim Oelsner*, 359–76.

55. S. Parpola has argued that in Neo-Assyrian, the writing ^{LÚ}SANGA (ŠID) = *šangû* is reserved for "priest," while "scribe" is consistently written ^{LÚ}DUB.SAR or ^{LÚ}A.BA; see

Parpola, *Letters*, 319. It should be noted that in Seleucid texts, the distinction between scribe and priest, both written ^{LÚ}ŠID, read either SANGA (*šangû* "priest") or UMBISAG (*ṭupšarru* "scribe") is often made in translation by context and can be misleading.

56. W. G. Lambert, "Ancestors, Authors, and Canonicity," *Journal of Cuneiform Studies* 11 (1957), Appendix 2, p. 10.

57. J. P. Brinkman, review of Priest and Temple in Hellenistic Babylonia, by G. J. P. McEwan, Journal of Cuneiform Studies 35 (1983): 232.

58. O. Neugebauer and H. B. van Hoesen, *Greek Horoscopes* (Philadelphia: American Philosophical Society, 1959), 42, No. 137C, col. i 3.

CHAPTER 2

Natural Knowledge in the Classical World

Daryn Lehoux

What did the ancients know about objects in nature? What even was an object in nature for them? What was nature? Who were the experts? Then, as now, the answers to these questions depend on time, place, interests, and individuals. Nor is the knowledge base ever settled or static; it contains many uncertainties, disagreements, areas for further exploration, dead ends, and—dare I say it—misperceptions. Nevertheless, what individual investigators know about nature, how they approach it, the questions they ask, the authorities they look to, the debates they engage, all of this forms an intertwined multifaceted whole. Only by keeping of all of this in mind can we ourselves be said to know much about ancient science.

What we find is that there is profound change over time. From the first investigations into something explicitly called "nature" in the sixth and fifth centuries BCE, through the massively synthetic systematization of Aristotle in the fourth century BCE, we see methods, questions, priorities, and assumptions all changing. As Athens gave birth at just about this time to the main philosophical schools (Platonism, Stoicism, Epicureanism, and others) that would color and shape the questions asked of nature and the answers given, we see further change and refinement. At the same time, the tradition to which each school was responding was communal, and so the variability was confined and constrained both by what went before, and by the different schools' debates with themselves and with each other. The kinds of answers each school was looking for are shaped not just by the methods of investigation, but also by the larger reasons people saw themselves as having for investigating nature in the first place. Theological and ethical concerns played central roles for almost everyone, though the specifics differed from person to person, as we shall see. Finally, as we approach the end of classical antiquity, we will see how it was under the Roman Empire, in the work of two of antiquity's most prolific investigators—Claudius Ptolemy and his contemporary Galen—that this intertwining of concerns produced some of the most sophisticated and technical inquiry of the whole ancient period.

THE WORLD ALL AROUND

The great first-century CE Roman encyclopedist and polymath, Pliny the Elder, composed a massive collection of as many facts as he could find, all in one book, called the *Natural History*,[1] often touted as the world's oldest extant encyclopedia. Immensely popular right through to the Renaissance and beyond, it has in more recent times fallen into an unfortunate obscurity among nonspecialist readers, an obscurity that belies its incredible importance to the history of the sciences. Pliny begins this formidable work with a definition, and at the same time a paean:

> The world, THIS (according to whatever other name you want to call the heavens by which everything is embraced round), is rightly believed to be a god, eternal, immeasurable, never born nor ever perishing. What is outside its bounds is of no concern to men—nor is it even within the reach of the human mind. It is sacred, eternal, immeasurable, everything in everything. Indeed it is itself the everything, finite but as though infinite, certain in all things but as though uncertain, the whole within and without encompassed in itself, both the product of the nature of things, and the nature of things itself.[2]

If we parse out the separate claims in this passage, we see that Pliny's universe is, among other things, (1) eternal (both unbeginning and unending); (2) a finite but very large sphere (the outside of which is both unknowable and irrelevant); (3) a complete whole (with strong undertones of it being a rational, or at least a rationally organized, whole); and (4) a god.

None of these propositions is unique to Pliny, but then the combination of the four is not exactly universal either. Pliny, like many of his Roman contemporaries, was a widely educated but also very individualistic thinker. He had thought through these propositions and made deliberate decisions on each of them. Then again the choices he made and indeed the very framework of his questions were deeply entrenched in sets of ancient debates and traditions. These debates and traditions were themselves rooted in common stocks of experience and observation as well as common understandings and common methods of experience and observation, and these factors made Pliny's particular universe characteristically Roman. No practicing scientist today would or could use the evidence available to them to prove just these four propositions, let alone these four in combination with the rest of what Pliny believed about the cosmos. So Pliny was innovative, but he was innovative within a particular tradition. This combination of tradition and innovation is, I suggest, as inherent to science as to any other intellectual pursuit—it is the result of individuals living in and participating in cultures: scientific, intellectual, professional, social.[3]

Now, Pliny was not himself a practicing scientist in the modern sense, nor even did he perform the kinds of research and direct observation we find in, say, his near-contemporaries the astronomer Ptolemy or the anatomist and physician Galen (both second century CE). Indeed, there is no real ancient analog of the modern scientist, whose research is funded by government or industry, and who works in a laboratory producing research publications on a full-time basis. There are ancients who can be said to "do science" (forgiving the anachronism of the nineteenth-century word "science" here), but the institutional contexts are so very different from the modern ones as to make it difficult to describe this doing as what we would think of as a job. Instead, what we have is a situation where well-educated (and therefore typically upper-class, almost universally male) individuals performed various investigations into nature whenever they could afford the time. This "affording of the time" had very different meanings over the centuries and cultures covered by the rather broad phrase "classical antiquity." For example, one of the most important philosophers of antiquity, Aristotle (fourth century BCE), began under royal patronage as tutor to the young heir designate to the Macedonian throne (Alexander the soon-to-be Great).[4] Later he founded a school (called the Lyceum) where he and his students investigated and debated a wide range of problems. Everything from fundamental questions of logic to theories of poetry, rhetoric, politics, ethics, and natural philosophy was on the agenda. The investigations undertaken into "natural philosophy" also covered a much wider range than would be found in any one university department or research facility today: astronomy, cosmology, physics, biology (including systematics, anatomy, physiology, reproduction, and inheritance theory), meteorology, psychology, mathematics, acoustics, and more. Just a glance at an index to Aristotle's complete works will suffice to show the breadth of the project. In one column alone, chosen entirely at random, we find the following entries: planets, plants, pleasure, plot, poetry, point (mathematical), political science, policy, population, pores, porpoise, possibility, potentiality, poverty, practical thinking, prawn, predication, pregnancy, priests.[5] The importance of natural philosophy in the school's curriculum meant that fundamental scientific investigation was part and parcel of the work undertaken there, although the means of this investigation was variable. Sometimes it meant reading everything written on a particular subject and analyzing and synthesizing it critically. Sometimes it meant performing deliberate and difficult observations (anatomy, for example). Sometimes, a combination. The institutional context of the school, however, provided the locus of the discussion and determined the agenda for investigation, as well as the intellectual framework within which results could be understood. That intellectual framework was never static, but then it was not randomly variable either. What emerged was a complex and very influential set of ideas and methods that would form a kind of backdrop against which much later science developed.

If we look toward the other end of our period, from Aristotle to Pliny and then forward by a century or two, we are dealing with completely different political and cultural contexts. The great learning of the day was now happening in the cultural context of the huge and incredibly diverse Roman Empire. Advanced learning was still largely restricted to fairly well-to-do individuals, but the Roman political dominance, the increased interconnectedness of the empire, and the political and cultural centrality of the empire's capital city, conditioned many of the relationships of scholars to each other and to their work. Education meant not only bilingualism (Latin and Greek) but also a familiarity with both Latin and Greek scholarly traditions, with a general emphasis on law and rhetoric.[6] Philosophy, which had been seen as a questionable or even subversive pursuit in the generation before Plato, had become respectable but also diverse: upper-class Romans had the choice of sending their children to learn under the best teachers of several different philosophical schools, and Romans had a tendency to combine what they thought of as the best ideas of each into their own individual amalgams rather than becoming strict followers of any one school.[7] It was in this context that Pliny found both the motivation and the market for his massive thirty-seven–book collection of all the world's known facts. So why then did Pliny make the four claims we saw earlier, that the cosmos is eternal, spherical, complete, and divine? Context is everything.

SPHERES

It was common in antiquity, if not quite universal, to accept that the cosmos was spherical. The roots of the idea go at least as far back as Parmenides (fifth century BCE), although it was the work of Aristotle that gave it its most compelling formulations. Nevertheless, many of Aristotle's individual arguments, observations, and considerations—and above all many of his questions—can be traced back to the greats among his predecessors, Plato, Empedocles, Parmenides, and others. The perfection of the sphere, its omnidirectional uniformity from the center, as well as its ability to turn in place without projecting into the space around it, all had for some time contributed to its plausibility as a model for the complete perfection of the cosmos itself. These arguments in turn meshed very well with the increasing amount of observational evidence accumulated over the course of the fifth and fourth centuries BCE about the behavior of the stars and planets. Tantalizing bits of knowledge about the heavens had been trickling in from Babylon during this period, stimulating and challenging homegrown Greek investigations (this Babylonian stream was to become a flood in the next couple of centuries).[8] To the Greek mind, all of it pointed with increasing force toward spherical rotation, such that several astronomers before Aristotle then used spheres to model the motions of the heavenly bodies, including arguing for a finite sphere of the heavens itself.

Enter Aristotle. His argument for the sphericity of the cosmos was complex and multifaceted, extending (or being reworked) across several books. Individual arguments in any one book are coherent and self-sufficient, but by the end of his oeuvre, a reader can look back and see the complex of arguments that reinforce and interact with each other as well. In short, there are a lot of good reasons for thinking of the cosmos as spherical, with much cross-fertilization between them, and important consequences extend out from the individual arguments and their combinations into the history of later cosmology. One thing in particular worth paying attention to is the complex interaction of observational and experiential claims with highly developed (if often extremely unfamiliar) theory. Such interaction and mutual reinforcement is a characteristic not just of ancient science, but of all science. The weirdness we see in Aristotle's theoretical understanding is not a measure of its intrinsic inadequacy—it was very adequate for a very long time—so much as a function of historical and cultural distance.

We know the universe to be finite, said Aristotle. But he offered different (if compatible) arguments to support this claim in different works. In *On the Heavens*, he began with the observation that we see the heavens revolving around us each day, and argued that if the heavens were infinite then they could not complete a revolution in any finite period of time.[9] In the *Physics*, he began with the fact that we know from observation that all bodies have proper places to which they tend: heavy bodies fall downward, and fire and air tend upward.[10] This claim builds on the Aristotelian arguments for a four-element theory, which sees all physical objects as composed of some combination of four basic elements, called earth, air, water, and fire, which have two pairs of opposing qualities through which they act (hot and cold, wet and dry).[11] A consequence of this theory is that there must be a center to the cosmos as a whole, which turns out to be the center of Earth, toward which heavy bodies can fall and away from which light bodies can rise. If the cosmos were infinite, the center could not be defined, and so bodies would have nowhere or everywhere to fall toward, which is contradicted by experience. That the finite cosmos must be spherical emerged as a result of Aristotle's analysis of motion and his argument that the universe must be both eternal and uncreated in time. He divided all motion into three types: circular, rectilinear, and some mixture of the two.[12] Any motion that is along a straight line cannot be continuous, since in a finite universe all straight lines must be finite, and so the only possible eternal motion would involve eventually getting to the end of the line and then moving back again (thus coming to a brief moment of standstill in between the two directions, and suffering periods of acceleration and deceleration toward these rest points). But if eternal rectilinear motion is thus discontinuous, then any motion composed even partly of rectilinear motion (that is, the third type of motion, mixed motion) must also be discontinuous, and so the only type of continuous and eternal motion possible is circular.[13] Circular motion also makes sense in this context, Aristotle was quick to point out, insofar as it is both the simplest eternal motion, and is "complete" in itself, with any one point on the circumference serving simultaneously as a direction-toward-which motion is occurring and a direction-away-from-which.[14] Furthermore, unlike an eternal rectilinear motion back and forth along a finite line, circular motion does not require an extra cause to account for the changes of direction. Once it's going, it just keeps going. That the cosmos as a whole rotates in place is derived from a number of considerations, and Aristotle returned to the idea repeatedly in his cosmological works. We have already seen the argument that eternal and continuous motion is necessarily circular, and Aristotle showed in *On the Heavens* that the cosmos is in motion eternally and continuously. There are also, as we shall see in a later section, powerful theological considerations that further compelled Aristotle to these conclusions.

By Pliny's day, Aristotle's eternal, spherical cosmos had been subjected to much scrutiny and revision. One school that was very prominent from the third century BCE and for half a millennium afterward, that of the Stoics, accepted the sphericity and divinity of the cosmos, as did many of their contemporary Platonists. The Stoics, however, offered a variation on its eternity, arguing that the cosmos was subject to periodic conflagration and rebirth following a strict chain of natural causes. The matter in the cosmos persisted through the destruction (how could it do otherwise—how could something become nothing?). Because there were no random events but only cause following cause in a strict chain, the Stoics argued that the reordering of the cosmos, its birth after any one conflagration, was necessarily identical to its reordering after any other conflagration. This meant that the unfolding of the universe, all its causes and all its events, happened in exactly the same way again and again, a finite period of cosmic history recurring identically to eternity—the eternal recurrence of the same, to use Nietzsche's wording for the idea he borrowed from them. What is interesting here is that the eternal recurrence—and with it a complex fatalism—emerges from a belief in strict causal necessity, that identical matter subjected to identical causes must always react in identical ways.[15] The periodicity further tied in neatly with contemporary astronomical contexts where cosmic periodicity, the eventual return of planets to identical configurations, often was seen to imply an astrological return to identical cosmic influences and effects.

But one school rejected all versions of the finite world in favor of an infinite cosmos with no inhering or interfering divinity whatsoever. Dating back to before Aristotle's time, several versions of atomism had been proposed, with Epicurean atomism eventually taking the predominant place from the third century BCE onward. Atomists argued that motion would be impossible in a universe that was completely full of matter, and so there must be void spaces between things if anything were to move at all. Furthermore, the particles of matter would have to be perfectly indivisible (in Greek *a-tomos*), in order to explain the existence of any perceivable objects. The argument was that destruction happens faster than construction, and unless there were a stopping point—indivisibility—for destruction, there could be no macroscopic cohesions of atoms for us to observe.

The infinity of the cosmos likewise emerges elegantly from a thought experiment: if the world were finite, what would happen to a spear thrown directly against its inside edge? If the spear bounced back, then there must be something outside of the cosmos resisting it. If the spear flew on through, then there must also be something outside. Therefore the cosmos cannot be finite, and if not finite, then not spherical or any other definite shape.[16] The apparent rotation of the heavens around Earth on a day-by-day basis created some problems for this idea (problems exploited by competing schools), and the Epicureans attempted to get around this by arguing that several explanations can be forwarded to account for these phenomena, one of which must be right. Moreover, since the goal of life is happiness (defined by Epicurus as the absence of fear), and since knowledge of minutiae about stellar motions can contribute nothing to easing fear, then we need not waste time on these problems.[17] Here we see how a shift of emphasis in one area (in this case ethics: the particular definition of happiness espoused by Epicurus) can have a profound effect on the framing of the problem set that a group of thinkers considers worth engaging. Secondly, we see how privileging different questions and different kinds of explanations can lead to radically different physics between the different schools.

CERTAINTY

A spherical cosmos was also accepted universally by ancient mathematical astronomers, scientists whose aim was not to understand elemental causes (physics) but to mathematically understand the very complex motions of the heavenly bodies. Classical antiquity offers no greater practitioner of the exact sciences (those with a mathematical substructure) than Claudius Ptolemy (second century CE). Writing definitive works on astronomy, astrology, optics, harmonics, geometry, geography, and more, Ptolemy combined exceedingly careful observation and experimentation with sophisticated mathematical analyses to produce one of antiquity's greatest scientific legacies. If we had the now-old-fashioned aim of looking back to antiquity to try and find practitioners whose work most closely approaches the methods and aims of the modern sciences, we could hardly do better. This is not to modernize Ptolemy—my aim is quite the contrary—but to clarify what sets his work slightly apart from the projects of the natural philosophers and philosophical schools we have been discussing so far. Ptolemy's work in astronomy, for example, involved the careful design of astronomical instruments, the taking of fine-grained observations, the collection and organization of massive amounts of data (his own and the work of previous researchers, going back as far as the historical record permitted and across several languages), and the use and further development of very sophisticated and highly accurate mathematical modeling based on that data. This is a different kind of project than the one Aristotle was engaged in. Unfortunately, we know next to nothing about Ptolemy's life, and so we are not in a position to speculate with any certainty on who, if anyone, patronized his work, what the institutional context was, and so on. But we are lucky enough to have his own account of the intellectual justification for his work, and he paints an interesting picture.

For Ptolemy, the glorious thing about mathematical sciences was their certainty. Of course, any ancient mathematician would agree: mathematics was the very paradigm of certainty in antiquity, and other sciences sometimes looked to it as a model. Galen, for example, consciously tried to model his medical and physiological arguments and demonstrations on geometrical proofs. But he did this in a formal sense, using the logical tools and structures of geometrical demonstration, rather than in the literal sense of applying rulers and compasses to the human body. But Ptolemy actually went this one step further to explore the point where mathematics meets the physical world. Here is what he had to say about the certainty this offers:

> Hence we thought it fitting . . . to devote most of our time to intellectual matters, to teach especially those theories—many and beautiful that they are—to which the name "mathematical" is particularly applied. For Aristotle divides theoretical philosophy very fittingly into three primary categories: physics, mathematics, and theology . . . Now the first cause of the first motion in the universe, if one considers it plainly, can be thought of as an invisible and motionless god. The division [of philosophy] concerned with investigating this is theology, and up in the highest reaches of the Cosmos, it can only be examined by thought . . . The division that investigates material and ever-changing nature . . . one may call "physics" and its realm is situated for the most part amongst the corruptible bodies below the lunar sphere.[18]

So theology's object of study is so remote as to be only accessible in the mind, and the object of the study of physics is inherently corruptible and always changing. Neither of them, then, can provide certainty: "theology because of its completely invisible and ungraspable nature, and physics because of the unstable and unclear nature of matter."[19] In comparison to mathematics, these two branches of philosophy pale to such an extent that Ptolemy called them "guesswork." He despaired: "there is no hope that philosophers will ever be agreed about them!" But mathematics!—now there is a subject to inspire the mind:

> Only mathematics can provide sure and unshakeable knowledge to its devotees, provided one approaches it rigorously . . . Hence we were drawn to the investigation of that part of theoretical philosophy, to all of it insofar as that is possible, but especially to the theory concerning divine and heavenly things, for that alone is devoted to the investigation of the eternally unchanging . . . It is also the best science to help theology along its way, for it is the only one that can speculate about that action that is unmoved and apart, because mathematics is familiar both with perceptible bodies . . . as well as with what is eternal and unchanging.[20]

If theology by itself has problems with certainty, Ptolemy could now offer a remedy: applied mathematics as a tool for a more certain understanding of the eternal divinity—a divinity that is best accessed by the science of astronomy which studies divine and unchangeable motions in the heavens themselves. What better tool than mathematics to access this eternal divinity through consideration of its tangible effects? For mathematics is situated exactly between the material world, where the effects of the divine cause ripples ever outwards, and the world of the eternal and unchanging.

Or so runs the argument. But Ptolemy did not himself pursue this theological project beyond the mere framing of his inquiry in his preface. Instead, it is the mathematical analysis of sensory data that he was interested in. In the *Almagest*, the mathematical work of systematizing and modeling the planetary motions, as well as the provision of more accurate calculating devices (tables) seems to be an end in itself, although he said in the *Tetrabiblos* that such accuracy benefits the much more useful science of astrology. In the *Optics* and the *Harmonics*, he used mathematics not only to systematize his discoveries in nature and to compare phenomena with each other, but also—and this is a theme that recurs again and again in his works in very sophisticated ways—as a means of using reason to correct for the vagaries of the senses and their ever-changing objects.[21] Through mathematics as one of its primary tools, reason allows us to understand the world better, to know it more accurately than the bare senses would or could permit. Mathematics provides certainty and accuracy in our understanding of the perceptible world.

But notice the problem sets: Ptolemy did not try to colonize any new territory, did not geometrize biology, did not mathematize medicine or alchemy. Instead he chose to work in fields where mathematization was already established. The lay of the scientific land, the established interconnectedness of certain kinds of problem sets and the tools for attacking them, all this meant that there was more than enough work to be done as it was. Moreover, his training and his reading, his very enculturation as a scientist, immersed him in these problem sets as sets, a package with known questions and promising avenues of approach already sketched in, just waiting for further exploration. These sets were rich and varied: a heady combination of mathematical tools, cosmologies, theologies, and a set of well-defined goals, all of which Ptolemy would build on and develop.

But not all sciences are mathematized or mathematizable, and there was work to be done in those fields as well. If Ptolemy could plead for the high status of his own subjects on the basis of mathematical certainty, that did not mean that work could or should not proceed apace in fields where mathematics was not applicable. Nor did it mean that the investigators in those fields could not bring sophisticated logical and analytical tools to bear on their subjects, tools that could, when used well, provide nonmathematical kinds of knowledge.

TREES, FOR EXAMPLE

So what about nonmathematized sciences? If we jump back in time again to Aristotle, we find that he did a vast amount of work in biology, work that was carried on and further developed by his successor in the Lyceum, Theophrastus of Eresus.[22] In addition to classificatory work, one of the areas of fundamental innovation in Aristotle was his analysis of growth and development. How, the old question goes, do acorns become oaks?

Aristotle did not spend a lot of time on plants in his oeuvre, but he did do a lot of work on animals (about a quarter of his extant work is zoological). Where he did discuss plant development specifically, it does not differ fundamentally from his discussion of animal development. He began by looking at the qualities of the mature organism and saw these as actually existing in the adult. Here, "actual" is a technical term meaning that the particular qualities both can be present and are present at the particular time of analysis. In the case of a deciduous tree at higher latitudes, for example, green leaves are actually present only part of the year, but they are potentially present in winter. This winter potential is not a way of saying they could grow in winter, but that in winter the green leaves are not actually present, even though under different circumstances (at a different time of year) they can be. The qualification of potential covers the fact that something is hiding in there in abeyance. But it does not cover an ignorance of what that something is. Potential turns out to be an explanatorily rich notion, playing a prominent role in Aristotelian accounts of change. In development, it explains how a small seed can grow to become a large (and very different) adult plant. The seed does not actually contain leaves and a trunk, but it does in potential, and the process of development is just the actualization of that potential. But this process of actualization depends on a kind of end-driven process, where the seed has some power within it that pushes it toward that actualization, that end product. One may be tempted to ask chicken-and-egg questions at this point, but Aristotle put a stop to that by arguing that the mature form of the plant or animal is prior (not just temporally, but hierarchically) to the seed it produces.[23] Moreover, the process of change from seed to adult, even though aiming at an end, was one driven internally, within the seed itself, by nature—where nature just is this process of self-change.[24] This means that concepts of nature and of (for example, human) artifice are both defined in terms of change: the one driven internally, within the organism or object itself (the growth and decay of plants, the heaviness of rocks), and the other driven externally, by the imposition of force from without (human craftsmanship for example, but also a dog picking up a stick against the stick's natural inclination to fall down).

This way of understanding development was not universally adopted in antiquity, but aspects of it were widespread. In particular, the use of teleological explanations (explanations that use the end-toward-which something happens as some kind of cause) in biology had—and continues to have—a very long history.[25] One prominent principle that emerged in tandem with teleology and that dominated much ancient thought, not just in biology but in the sciences in general, was the maxim that nature does nothing in vain.[26] This maxim works as a hermeneutic, guiding the investigator to pay attention to every smallest detail, since all are relevant to an understanding of nature, and all are purposive.

EXPERIMENT AND OBSERVATION

Both this purposiveness and this demand for extreme attention to detail come out very clearly in the work of Ptolemy's contemporary, the second-century CE anatomist and physician (perhaps the single greatest of each in antiquity), Galen.[27] Five centuries before Galen, Aristotle himself had performed anatomies on a wide range of animals and fishes, although his interests were biological rather than strictly medical. Within the medical traditions, we find some animal anatomies in Hippocratic texts (fourth to third centuries BCE), and a brief but very important period of research into human anatomy during the third century BCE in the medical school at Alexandria.[28] By Galen's day, however, it was no longer possible to perform human anatomies, so he worked on animals instead, and extrapolated from animals to humans. It is worth remembering in this context that the cutting open of human bodies requires the opening up of a very complex cultural space in which it might be seen as permissible and desirable to procure and systematically mutilate the bodies of the dead. Even today, when the idea of anatomy as a basic research and training tool is absolutely central to our medical system, not just anyone gets to perform or even attend the cutting open of human corpses, and the contexts in which it can be performed are very limited and highly regulated—do not try this at home. Indeed, we may now see human anatomy as a fundamental part of the education of the physician where Galen's contemporaries objected to it for moral reasons, but on the other hand Galen's contemporaries had other useful teaching tools available to them that we now object to for moral reasons—animal vivisection for example. Morals matter, and morals change.

Galen's careful attention to detail and his skill in anatomy emerge repeatedly in his writings. His ability to isolate, explore, and devise experiments to determine the functions of the smallest structures was unsurpassed for many centuries. (The word "experiment" is used deliberately here. Experiment is not as pervasive in ancient science as it is in later science, but the frequent reports of its absence are grossly exaggerated.[29]) In his anatomical work, Galen noted again and again how astoundingly complex the body is, and how very well suited each part is both for its particular function and for the functioning of the whole body. But this functional analysis is highly teleological. The hand, for example, is shaped the way it is because of the functions it is supposed to perform. And this is not just at the level of macroscopic description (five fingers, so many articulations, fleshiness here, rigidity there), but also at the level of almost microscopic internal analysis. To begin with the macroscopic:

> The fingernails were made for the sake of bettering the action of the hands because, although even without their aid the hands would certainly be able to grasp, they could not handle objects of all sizes or grasp as well as they can now. For I have pointed out that small, hard objects would readily escape them if some hard substance capable of supporting the flesh did not underlie

> the tips of the fingers . . . But I have not yet told why they were made just so hard and no harder or why they are rounded on all sides, but now is the proper time to do so.[30]

And so on for two full pages: they are just firm enough to offer resistance without being brittle; they cushion blows to the backs of the fingers; they are rounded so that parts do not break off; they grow in length but not in width because only the ends wear away. "Thus everything about the fingernails shows the utmost foresight on the part of Nature."[31] The modern reader may be forgiven for finding Galen's paean to the fingernail strange to the point of being almost amusing, but the reader's hasty amusement is quickly trumped by respect as Galen proceeds to expose and explain the internal structures of the hand in the most rigorous, skilled, and sophisticated manner. In all cases, though, and at all levels, the structure of the hand is goal-driven. Not only this, but the extreme complexity of the living organism, and this incredible match between structure and function in all its parts, leads Galen to conclude that not only is nature purpose-driven, but that Nature (and here it seems best to capitalize the "N") is also deliberately providential—a divinity that has rationally and consciously shaped the structures of the bodies to be best suited to their particular ends. He moves from teleology to theology and back again, and the argument for a providential divinity both guides his inquiry and frames his understanding of the whole.

By providing a window through which Galen saw his subject, the providence of Nature at one and the same time feeds into how he sees, as well as providing an incentive for undertaking increasingly difficult, increasingly careful, and increasingly invasive observation. I do not wish to suggest that it provides the only such reason (prestige and public performance, for example, also play large roles),[32] but it does stand in a potent feedback relation to Galen's observational project as a whole. The temptation from a modern perspective is to see the imposition of a personified deity as interfering in the pursuit of objectivity, but if anything the converse is true. Galen was not seeking his own justifications for the usefulness of the parts of the body, he was seeking Nature's. This forced him to take as broad and unencumbered a stance as possible in order to see what the roles of a part in the overall physiological system might be, and to try and catalog and understand not just the obvious function of the part but all the functions, especially the hidden ones. We may today want to replace Galen's deified Nature as an explanatory mechanism, with a depersonified natural selection, but the end products (the fingernails, the tendons, the nerves) are still the same, and on many readings the analysis from natural selection is no less teleological than Galen's physiology of the body had been.[33]

In his pursuit of a complete analysis of Nature's wisdom and beneficence, Galen moved farther and farther into the body. His skill in finding, exposing, and analyzing the interior of the body, down to the very limits of visibility (and beyond—he correctly argued for the existence of invisible capillaries, for example) was unsurpassed for centuries before and after (indeed, one testament to his stature is the fact that we have more of his writings preserved—considerably more—than of any other ancient Greek author).[34] His careful technique can be seen in his tracing of the nerves to the larynx:

> In the larynx there are three other pairs of muscles most necessary for the production of the voice . . . Let us see, then, how perfectly Nature has provided for these two considerations: namely, supplying what was necessary for [muscular] action, while avoiding the injustice of supplying inappropriate nerves to the muscles. She decided to bring nerves down from the brain, like the others I have spoken of earlier, by way of the sixth pair, from which nerves must also be given to the heart, stomach, and liver, but to make them run a sort of double course, carrying them first to parts below the larynx, and then bringing them back up again to its most important muscles. But they could not run back without making a turn, so that Nature was forced to seek a turning-post, so to speak, for the nerves around which she might bend them.[35]

Here we see the combination of very skilled observation (when Galen traced the nerves responsible for the motion of particular muscles through their complex paths) with his experimental physiological work (the determining of particular nerve functions in the first place) all overarched by his argument on the providence and foresight of Nature. In other investigations, the complexity is multiplied by the need to observe the functioning of very deep organs without killing the anatomical subject. So his description of the valves of the heart at the origin of the pulmonary artery (which he calls the "arterial vein"):

> They lie facing outward from within, surrounding the whole circumference of the orifice, and each has such an accurate shape and size that when they are all tensed and stand erect, they become a single large membrane sealing the whole orifice. When they are opened by an outward flow from within

> [the heart] they fall outward against the surface of the vein itself and permit a free flow by opening the orifice wide and to its greatest extent. If, however, there is a flow from outside back in again, this very flow pushes the valves together so that they overlap each other and make of themselves a gate, as it were, perfectly sealed . . . For Nature certainly did not wish to tire the heart with useless labour by causing it sometimes to send material out when it is supposed to be coming in from some part of the body, or to draw material in when it is supposed to be going out to some part.[36]

And so, part by part, structure by structure, he explores the entire body over the course of two very large volumes.

Why such thoroughness? In part it is complete for completeness' sake—a flurry of almost superhuman virtuosity—but it is also framed as an argument against a rival interpretation of the body, and indeed of the universe as a whole, that of the Epicurean atomists. Where I have been arguing in this chapter for a general teleological and theological frame to ancient science, I have not been telling the whole story. There was a good degree of diversity of opinion on many matters, some fine grained (the function of the veins, mechanism of the nerves, precise value for the equinoctial year, physical makeup of the moon) and some more general. Indeed, by Galen's day there were a wide range of philosophical schools to choose from, Stoicism, Platonism (whose proponents were called "Academics" after Plato's "Academy"), Aristotelianism ("Peripatetics"), Epicureanism, or even Scepticism and Cynicism. Many of these schools agreed on many points, teleology and theology crossing many boundaries in this respect. But some schools blocked even this move, Epicureans, by the Roman period, very prominent among them.

As we saw briefly above, the Epicurean universe was made up of atoms separated from each other by spaces of pure nothingness, pure void (an idea that would see an interesting revival in the early-modern period).[37] Atoms interacted only by crashing into each other, and all physical phenomena that we encounter in the macroscopic world are caused ultimately by thousands upon thousands of these tiny collisions. This chaos of crashing atoms, combined with a technical detail of Epicurean physics called the swerve (an occasional randomness that affects the motion of some atoms at some times), meant that the particular combinations of atoms that we see—dogs, trees, stars, rocks—could perfectly well have happened otherwise. There is, in short, no intervening deity (although the Epicureans did insist on the existence of gods, they qualified it by arguing that the gods did not interfere at all in the functioning of the universe—such concern would be beneath them). It is to this randomness argument, this could-have-been-otherwise argument, that Galen sees himself as responding in *On the Usefulness of the Parts of the Body:*

> The atoms or indivisible bodies which some posit as elements are not, according to these very men, naturally formed into whole [bodies] by an external force or by anything permeating them. They are left, then, to build the structure of perceptible bodies by a random interweaving with one another. But a random interweaving almost never produces a useful artifact, but usually something useless and silly. This is why men who claim that the primary elements are the kinds of things atomists say they are, are unwilling to think that Nature is skillful; for in spite of the fact that they clearly see as soon as they look at an animal's body that it has no part without a use, they try to find just one thing that will serve for contradiction, either at first glance or from dissection. Consequently, they have imposed on me the necessity of explaining all the parts and of extending the discussion even to things that are not necessary for the treatment, prognosis, or diagnosis of disease.[38]

Hence his completeness: if a single counterexample would satisfy his opponents, Galen could afford to leave no stone unturned. Notice here that he referred to a knowledge of anatomy as also being useful for the physician in his day-to-day practice, a claim that not all of Galen's contemporaries would take for granted, and a claim for which he argued explicitly in other works.[39]

GODS

The centrality of divinity in Galen's teleological argument was direct and immediate. The purposiveness of Nature, her wisdom, her foresight, her careful planning emerged as more than metaphors in his work. The obviousness of intentional design was demonstrated at every level of observation, and the force of this idea seemed to strengthen the farther and farther Galen moved into the body. For Aristotle before him, divinity had also played a central role in nature, even though it came in only after a number of other considerations. Aristotle argued that the stuff of the heavens, with its eternal and unchanging natural circular motion, was superior to the stuff of our changeable sublunary world, even claiming that the farther away from our changeable world, the better the matter would be.[40] But all motion requires a cause, either internal to the object itself (as when animals walk) or external (as when balls are kicked by children). When heavy bodies fall, that is a natural motion internal to the body itself. Nevertheless, because there is a terminus for falling—a bottom toward which heavy bodies move—we need an explanation for why the cosmos has not simply achieved a state of stagnancy: all the earth having fallen to the center, all the water at rest around that hard core, all the air above the water, and all the fire having risen up to the outside. We may be tempted to look to the large class of self-moving beings, living things, for keeping motion in play, but as Aristotle pointed out these things are only self-moving in virtue of a large number of prior motions among the elements themselves. Instead it is the movement of the heavens (and the sun primary among the heavenly bodies) that produces changes in the qualities (hot/cold/wet/dry)—and therefore the elements—down here. But now what keeps the heavens moving? Here Aristotle turned to an analysis of what we may think of as "chain reactions," where one thing moves another that moves another, and so on. If I push a billiard ball with a cue, the immediate cause of the ball's motion is the thing that touched it directly, a bit of chalk on a small piece of felt. But to say the chalk caused the ball to move is only to tell an insignificant part of the story. The chalk moved because the felt it was rubbed into moved, and the felt moved because the glue binding it to the cue moved, and the glue moved . . . ultimately because I pushed the body of the cue forward. So where does the chain of causation (and of explanation) end for the cosmos? That there must be one first, unmoved, mover emerges from a number of considerations.[41] Where the account gets most richly fleshed out, however, is in *Metaphysics*, book 12. The argument runs as follows: there must be some eternal motion (in our eternal and currently moving universe). This combines with an argument that there must be some unmoved mover (lest the chain of causation regress to infinity) to produce some necessary first (primary) cause in the chain which is itself unmoved. But what can move without being itself moved by that which it moves? Does the billiard ball not push back against me to some extent when I strike it with the cue? Here the answer gets very interesting. "Objects of desire," Aristotle says, can cause motion without pushing back. Three paragraphs later he is into a full-fledged theological account of the prime mover, which is now characterized as "that for the sake of which" the heavens rotate, as an unmoved divinity existing without magnitude at the circumference of the very cosmos and causing all heavenly spherical motion. Aristotle's physics necessitates a god as a first and eternal cause for all motion.

The Stoics took this one step further, identifying god not with the beginning of a causal chain, but as physically identical to the entirety of the cosmos itself. Moreover, since the cosmos is the highest order possible, and since it is itself divine, it must also be rational. Not just rationally understandable, mind you, but itself capable of reasoning—and therefore alive. There are hints of these ideas in some passages in Aristotle (and some of his predecessors), but only hints, never fully developed. It is only later, in Stoicism, that all the consequences of the equation of divinity and order get fully explored, and it is through physics that this rational order is found: theology is again a consequence of sophisticated physical investigation. The complex order demonstrated by the motions of the planets and stars, for example, betrays the hand of a creator, far superior in ability and wisdom to humans, and it does this so obviously that to doubt the divinity behind it would be like doubting the existence of the sun, as the great statesman and philosopher Cicero would put it not long before the assassination of Julius Caesar.[42]

Again, the Epicureans disagreed, limiting causation strictly to physical interaction between atoms (plus the infamous "swerve"). Gods existed, but they did not interfere in the universe except insofar as they provided models of virtue for humans to contemplate. This emphasis on physical causation and bracketing of divinity may look on the face of it more like the disciplines we now think of as "scientific," but there was a curious upshot: other than arguments about the most basic levels of causation, Epicureans did not tend to engage in what we would think of as scientific investigation—we have no long list of prominent Epicurean anatomists, geometers, biologists, or astronomers. Via ethics, Epicurean physics has made overly careful scientific observation essentially irrelevant. Epicurus's treatment of astronomy, for example, makes it clear that minute investigation in the sciences was beyond the scope of atomism, because atomism was ultimately an ethically driven philosophy.[43] The observation and explanation of the actual motions of the stars was irrelevant to the main aim of the school, which was after all to provide for human happiness. So long as the philosopher could be sure that some causal mechanism was at play, he need not trouble himself about precisely why or how the stars moved. Detailed empirical investigation could, in this sense, hang itself. This looks to be a curious result, that a billiard-ball physics which bracketed the gods out of the cosmos should turn away from an emphasis on too close an investigation into nature, while the highly theologized versions such as Aristotle's and Galen's should produce such a profound curiosity about the tiniest details of nature. We should further note that these theological arguments cannot be seen as attempts by ancient religion to provide justifications for itself, since ancient religion was really a very different animal to what we are seeing here. Ancient religion was not highly theologized, but instead stressed ritual devotion rather than intellectual awe. What we have here is a philosophically motivated and empirically undergirded theological movement that is largely independent of contemporary religious practice.[44]

KNOWING NATURE

Ancient science approached nature in a remarkably wide variety of ways, with a correspondingly wide variety of motivations. But all philosophers, all investigators came to their subjects not in a vacuum but as understood within developing . . . what?—one almost wants to say disciplines although that would not quite be the right word; perhaps fields, broadly conceived, is better. They came to their subjects as integrated parts of established fields; fields that had ways of approaching nature, ways of understanding, ways of analyzing, ways of synthesizing that were effective and impressive in their scope and power. Not hidebound by tradition, but not infinitely variable either. The development within and across subject areas could and often did cross-fertilize in fruitful and interesting ways.

The breadth of the projects is remarkable: ethics feeding into logic feeding into mathematics feeding into theology feeding into physics and back again. The diversity of these encounters with nature, likewise. We have seen physics reverberate out into divine reverence in some schools, but not in others. From the third century BCE, ethics plays very direct and important roles in the Hellenistic schools (Stoicism and Epicureanism) and their physics. Finally, plain old curiosity, that motivation almost too mundane to mention, finds an elevation and an outlet in the extraordinarily wide range of phenomena as witnessed and experienced in the world—in Nature—where the sophisticated methods and tools of classical science found that they could provide knowledge.

NOTES

1. Good overviews can be found in M. Beagon, *Roman Nature: The Thought of Pliny the Elder* (Oxford: Oxford University Press, 1992); and J. F. Healy, *Pliny the Elder on Science and Technology* (Oxford: Oxford University Press, 1999).

2. Pliny, *Natural History* 2.1 (translation mine).

3. One could point here to Kuhn's "essential tension"; see T. S. Kuhn, *The Essential Tension* (Chicago: Chicago University Press, 1977).

4. On Aristotle, see e.g., C. Shields, *Aristotle* (New York: Routledge, 2007); J. Barnes, ed., *The Cambridge Companion to Aristotle* (Cambridge: Cambridge University Press, 1995).

5. J. Barnes, ed., *The Complete Works of Aristotle* (Princeton, NJ: Princeton University Press, 1984), 2483.

6. Y. L. Too, *Education in Greek and Roman Antiquity* (Leiden: Brill, 2001).

7. M. Griffin and J. Barnes, eds., *Philosophia togata* (Oxford: Oxford University Press, 1989); K. Algra, et al., eds., *The Cambridge History of Hellenistic Philosophy* (Cambridge: Cambridge University Press, 1999); J. M. Dillon and A. A. Long, eds., *The Question of "Eclecticism"* (Berkeley: University of California Press, 1988).

8. See, e.g., A. Jones, "The Adaptation of Babylonian Methods in Greek Numerical Astronomy," *Isis* 82 (1991): 441–53.

9. Aristotle, *On the Heavens* 1.5.

10. Aristotle, *Physics* 3.5; cf. Aristotle *On the Heavens* 3.6.

11. For an overview, see G. Lloyd, *Early Greek Science: Thales to Aristotle* (London: Chatto and Windus, 1970).

12. Aristotle, *Physics* 8.8; Aristotle *On the Heavens* 1.2.

13. Aristotle, *Physics* 8.8.

14. See also Aristotle, *On the Heavens* 2.4.

15. See S. Bobzien, *Determinism and Freedom in Stoic Philosophy* (Oxford: Oxford University Press, 1998).

16. See Lucretius, *On the Nature of Things* 1.958–1051.

17. Epicurus, *Letter to Herodotus*, in Diogenes Laertius 10.79f; Lucretius *On the Nature of Things* 5.509–770.

18. Ptolemy, *Almagest* 1.1. Here and throughout, translation modified slightly from G. J. Toomer, *Ptolemy's Almagest* (London: Duckworth, 1984).

19. Ptolemy, *Almagest* 1.1.

20. Ibid.

21. A. Barker, *Scientific Method in Ptolemy's Harmonics* (Cambridge: Cambridge University Press, 2000); D. Lehoux, 'Observers, Objects, and the Embedded Eye; Or, Seeing and Knowing in Ptolemy and Galen,' *Isis* 98 (2007): 447–67.

22. On Aristotle's biology, see J. G. Lennox, *Aristotle's Philosophy of Biology: Studies in the Origins of Life Science* (Cambridge: Cambridge University Press, 2001).

23. Aristotle, *Metaphysics* 9.8. See also *Metaphysics* 12.6.

24. Aristotle, *Physics* 2.1.

25. See, e.g., A. Ariew, R. Cummins, and M. Perlman, eds., *Functions: New Essays in the Philosophy of Psychology and Biology* (Oxford: Oxford University Press, 2002).

26. See, e.g., Theophrastus, *On the Causes of Plants* 1.

27. See V. Nutton, *Ancient Medicine* (New York: Taylor and Francis, 2004).

28. H. von Staden, *Herophilus: The Art of Medicine in Early Alexandria* (Cambridge: Cambridge University Press, 1989).

29. See, e.g., A. Barker, *Scientific Method in Ptolemy's* Harmonics (Cambridge: Cambridge University Press, 2000).

30. Galen, *On the Usefulness of the Parts of the Body* 1.10–11. Here and throughout, translation from M. T. May, *Galen: On the Usefulness of the Parts of the Body* (Ithaca, NY: Cornell University Press, 1968), slightly modified.

31. Galen, *On the Usefulness of the Parts of the Body* 1.11.

32. H. von Staden, "Anatomy as Rhetoric: Galen on Dissection and Persuasion," *Journal of the History of Medicine and Allied Sciences* 50 (1995): 47–66.

33. See, e.g., Richard England, "Natural Selection, Teleology, and the Logos: From Darwin to the Oxford Neo-Darwinists, 1859–1909," *Osiris* 16 (2001): 270–87; C. Allen, M. Bekoff, and G. Lauder, eds., *Nature's Purposes* (Cambridge, MA: Harvard University Press, 1998).

34. Vivian Nutton (plausibly) estimates Galen's surviving work as constituting about 10 percent of *all extant ancient Greek literature*; see Nutton, *Ancient Medicine*, 390n22.

35. Galen *On the Usefulness of the Parts of the Body* 16.4.

36. Ibid., 6.11.

37. On atomism and its revivals, see H. Jones, *The Epicurean Tradition* (London: Routledge, 1992).

38. Galen, *On the Usefulness of the Parts of the Body* 17.1.

39. See, e.g., his comments on Methodism—a rival medical school that disparaged anatomy—in *On the Sects, for Beginners* 6–7, but a constantly repeated emphasis on the importance of anatomy recurs throughout his corpus.

40. Aristotole, *On the Heavens* 1.2.

41. The arguments for this are complex, and differ to some degree between the versions in *Metaphysics* 12 and *Physics* 8, not to mention some of the shorter versions in other texts (e.g., *On Generation and Corruption*; *On the Movement of Animals*).

42. See Cicero, *On the Nature of the Gods* 2.

43. Indeed, most Hellenistic schools saw themselves as centrally concerned with ethics; see, e.g., K. Algra et al., eds., *The Cambridge History of Hellenistic Philosophy* (Cambridge: Cambridge University Press, 1999); P. Hadot, *What Is Ancient Phisosophy?* (Cambridge, MA: Harvard University Press, 2002).

44. See, e.g., C. Ando, *Roman Religion* (Edinburgh: Edinburgh University Press, 2003).

CHAPTER 3

Natural Knowledge in the Arabic Middle Ages

Jon McGinnis

The medieval Arabic-speaking world had southern Spain, or Andalusia, as its far western border and then stretched across North Africa eastward to include all of modern-day Iran. Its two major intellectual centers were Cordova in the west and Baghdad in the east. As for its temporal extent, what might be termed "the classical period" of Arabic philosophy and science roughly began in the first half of the ninth century with the "first" Arabic philosopher, al-Kindī, and continued until the end of the twelfth, when Persian began to emerge as a rival to Arabic for writing and thinking about philosophy and the "Aristotelian" approach to science began gradually to be abandoned. Certainly one of the significant contributions of those working in the medieval Arabic-speaking world was the continuation of a scientific tradition going back to the earliest Greek natural philosophers, which attempted to explain the various natural phenomena and physical features that make up our world.

The study of nature in the medieval Arabic-speaking world was characterized by two currents that usually flowed in parallel, while occasionally crossing over and feeding one another: these were the intellectual traditions associated with *kalām* and *falsafa*. Although one is tempted to translate these terms respectively as "theology" and "philosophy," it is not clear how helpful such labels are for understanding the differences between the two, since both traditions were interested in roughly the same set of questions, and their answers often shared common intuitions. Perhaps a better way to distinguish between the two is to consider how the historical actors viewed themselves and what they thought the differences were. The proponents of *falsafa* saw themselves as adopting, adapting, and generally extending the Greek philosophical and scientific tradition, while the advocates of *kalām* envisioned themselves as promoting a way of thought intimately linked with the Arabic language and the Islamic religion. The emphasis of this characterization is on the two groups' own perceptions of themselves rather than whether the perceived differences were as real as they thought.

This chapter focuses primarily on the notion of nature as it appears in the *falsafa* tradition, namely, as a continuation of Aristotle's discussion of nature as well as that by the Greek Aristotelian commentary tradition. At the end of this survey, however, there is also a brief discussion of *kalām* accounts of nature and its response to the Greco-Arabic conception of nature. To this end, we shall begin with the Arabic vocabulary used for nature as well as various definitions of nature either taken over from or inspired by Aristotle. This section is followed by some brief notes on certain post-Aristotelian Greek developments that would affect the discussion of nature in the medieval Islamic world. In the next two sections it is argued that the desire of Arabic-speaking natural philosophers to address these later Greek developments led first to what might be considered a uniquely Arabic conception of the coming to be of the various natures at particular times, culminating in Avicenna's "Giver of Forms." This is followed by a section that considers the reaction among Andalusian Peripatetics to these new theories, where the focus is primarily on Averroës' response to Avicenna's thesis. The chapter concludes with a brief look at *kalām* conceptions of nature and the general critique of an Aristotelian understanding of nature considered as an internal cause of motion and rest.

THE VOCABULARY OF NATURE

The English term "nature" comes from the Latin *natura*, which itself is derived from the Latin verb *nascor*, "to be born, spring forth, originate." Latin-speaking philosophers themselves frequently understood the philosophical sense of *natura* by reference to Aristotle's definition of the Greek term *phusis*, which, like its Latin cousin, comes from a verb (*phuō*) meaning to bring forth, produce, or engender. What is common to both the Greek *phusis* and the Latin *natura* is that a nature has the sense of something arising from within a thing itself rather than coming from without. It was in this vein that Aristotle provided what would become the classical definition of nature as "a certain principle and cause of being moved and being at rest, belonging primarily to that in which it is essentially, not accidentally."[1] Thus a nature, according to Aristotle, is something wholly internal to a thing that accounts for the various activities (or motions) of that thing.

In the Arabic-speaking world, although the philosophers sometimes used *ḥaqīqa* ("truth or reality") to characterize a thing's nature—and indeed this term was the preferred term in *kalām*—by far the most common philosophical term for nature was *ṭabīʿa* (and sometimes the etymologically linked *ṭabʿ*). Indeed, when we turn to the Arabic translation of Aristotle's *Physics*, the rendering of the definition for nature is practically verbatim with its Greek counterpart: "Nature [*ṭabīʿa*] is a certain principle and cause on account of which the thing in which it is primarily is essentially, not accidentally, moved and at rest."[2]

Virtually every Arabic-speaking philosopher simply assumed this definition, either implicitly or explicitly. Thus consider the first Arabic philosiopher, al-Kindī (ca. 800–870 CE), who was associated with the ʿAbbāsid court in Baghdad during the caliphal reigns of al-Ma'mūn (r. 813–833 CE), al-Muʿtaṣim (r. 833–842 CE), and al-Wāthiq (r. 842–847 CE) and was intimately involved in the earliest interpretations and dissemination of the newly acquired Greek sciences within the Arabic world. In his *The Definition and Description of Things*, he defined nature as "a starting point [*ibtidā'*] of motion and resting from motion, where the most important [starting point] is the powers of the soul."[3] (Here, one should note that *ibtidā'*, "starting point," is etymologically linked to *mabda'*, the term for "principle" occurring in the Arabic translation of Aristotle's definition.)

Also relying heavily on Aristotle's definition of "nature" was al-Fārābī (ca. 870–950 CE), who was active within the circle of philosophers known as the Baghdad Peripatetics and was certainly one of the most important philosophical system builders in the medieval Islamic world. Unfortunately, despite his significant role in the history of philosophy done in Arabic, very little is known about the details of his life. As for a thing's nature, he identified it with a thing's essence (*māhīya*) and then immediately described the essence as "that on account of which that species does the activity generated from it as well as the cause of the rest of the essential accidents belonging to it, whether motion, quantity, quality, position, or the like."[4] In other words, like Aristotle before him, al-Fārābī understood a thing's nature as an internal cause of the activities associated with it. Also, pseudo-Fārābī would define nature as "the principle of motion and rest, when [that motion or rest] is neither from something external nor a result of volition."[5]

This strong reliance on Aristotle's definition of nature is also seen among the Baghdad Peripatetics, a group of philosophers whose activity extended roughly from 870 CE to 1023 CE and who focused primarily, though not exclusively, on aspects of Aristotelian logic. Thus both the Syrian Christian Yaḥyá ibn ʿAdī[6] (893–974 CE), who studied with al-Fārābī and subsequently became titular head of the Baghdad Peripatetics, and his student Ibn as-Samh[7] (d. 1027 CE), for whom we have little biographical information, offered Aristotle's definition verbatim in their discussions of nature.

Also drawing heavily on Aristotle's *Physics* was Avicenna (980–1037 CE). Known in both the East and West for his unique philosophical system as well as his work on medicine, *The Canon*, Avicenna was associated in varying capacities—sometimes as court physician, sometimes vizier—with a number of short-lived sultanates in Iran. Like many before him, Avicenna approvingly cited and commented upon Aristotle's definition of nature and further noted that nature in the strict sense (so as to be differentiated from the vegetative, animal, and celestial souls) is "a power that brings about motion and change and from which the action proceeds according to a single course without volition."[8]

Ibn Bājja (1085 or 1090 to 1139 CE) was the first of the great Andalusian philosophers as well as vizier of the governor of Granada for twenty years. In his commentary on the *Physics*, he gave this abridged definition of nature—"a principle of motion and rest in the thing"—neither mentioning nor commenting on the idea that the principle belongs to the thing essentially and not accidentally.[9] Unlike earlier thinkers within the *falsafa* tradition, Ibn Ṭufayl (ca. 1110–1185 CE), the next in the line of great Andalusian philosophers, did not use *ṭabīʿa* when he spoke of nature in his philosophical novel *Ḥayy ibn Yaqẓān* (a genre of doing philosophy, one might add, that apparently had no earlier precursor). Instead, he used *ḥaqīqa* ("true nature") to speak of a thing's nature, saying, "the true nature of any body's existence is due only to its form, which is its predisposition for the various sorts of motion, while the existence that it has due to its matter is a weak existence that is barely perceivable."[10] (Here it is worth noting that although Ibn Ṭufayl's account, with its introduction of "form" and "matter," might seem to go beyond Aristotle's definition, his addition in fact encapsulates Aristotle's later identification of nature with form and matter.) The final figure in the triumvirate of Andalusian Peripatetics is the great Aristotelian commentator, Averroës (1126–1198 CE), who in addition to expositing the works of Aristotle was chief *Qāḍī*, or judge, of Cordoba and court physician to the Spanish Caliph, Abū Yaʿqūb Yūsuf.

As one might expect, in both his *Epitome* and *Long Commentary* on the *Physics* he cited Aristotle's definition of nature verbatim and then commented upon it.[11]

Despite the obvious similarities between Aristotle's original definition and its Arabic variants, there is a difference between them, not so much with the definitions themselves, but with the implicit connotations of the Greek and Arabic terms being defined. Again, the Greek *phusis* is derived from a verb that carries with it the connotation of coming forth from within. In contrast, *ṭabīʿa* is derived from the Arabic verb *ṭabaʿa, yaṭbaʿu, ṭabʿ*, which means to be sealed, stamped, or impressed (from without) and so also conveys the sense of being made or created so as to act in a determined way. Consequently, while the notion that a nature is a principle and cause is explicit in both the Greek and Arabic philosophical definitions of nature, the Arabic account additionally carries with it an implicit sense that a nature is imposed from without, whether by God or some other agent, and that it is only once a thing is so impressed that its nature acts as a cause of the various natural activities that arise from it.[12] This shift in emphasis may in part be explained by the fact that Aristotle did not see his "god" as a creator of the very existence of the physical world, but only as the explanation of the motion of an independently existing world, whereas later thinkers, particular those working within one of the various monotheistic religious traditions, viewed God as the Creator in the sense of the efficient cause of the world's very existence, a point to which we shall return in the next section.

This implicit connotation of the Arabic *ṭabīʿa*, namely, that it is impressed upon a thing by an external agent, can be seen in the very earliest discussions of nature by Arabic-speaking philosophers. Thus, according to al-Kindī, "natural science is the science of moved things precisely because nature is the thing that *God has made* as a cause and a reason for the cause of all things subject to motion and rest."[13] Similarly the iconoclast and renowned physician Abū Bakr Muḥammad ar-Rāzī (born ca. 864 CE) complains of Aristotle and certain Greek commentators, asking "Why do you deny that God, great and mighty, in Himself is what necessitates [and so makes exist] the powers of all other actions and *the natures of things*?"[14] Here we see at least two of the earliest Arabic-speaking philosophers ascribing to God the explicit role of creating natures and the implicit role of impressing them into physical things. In the Islamic east, later philosophers, such as al-Fārābī and Avicenna, would relegate this task to an immaterial substance or angel below God, namely, the "Active Intellect" or "Giver of Forms." Before we can appreciate their theories, however, we must consider certain developments within the Greek scientific tradition that were to influence Arabic discussions concerning nature.

HISTORICAL BACKGROUND

The Arabic translators' choice of *ṭabīʿa* to render Aristotle's notion of *phusis* was not simply happenstance; rather, it seemed to be the product of developments in the Greek Aristotelian commentary tradition itself. Aristotle, having defined nature as a principle of change, further identified a thing's nature with its matter and form.[15] He additionally argued that although the individual instances of a form-matter composite—such as a particular person, a given tree, a quantity of water, and the like—inevitably come to be at some time and cease to be at some time, matter and form absolutely—that is, the underlying stuff and what it is to be human, tree, water, and the like—are eternal and exist necessarily.[16] Consequently, for Aristotle the matter of and forms in the universe need no efficient cause to explain their existence; rather, what needs explanation according to Aristotle is the cause of the changes in the universe, which Aristotle explained by appealing to an unmoved mover as an ultimate object of desire.[17] In this respect Aristotle's unmoved mover, or "God," is not an efficient cause of the universe's existence at all, but only a final cause of its motion.

Such a position came to be unacceptable to a number of later Neoplatonists. Neoplatonism had its origins in the *Enneads* of Plotinus (205–270/71 CE) with its appeal to "the One," which later thinkers would identify with God, and which in a real sense was considered to be beyond existence and being, but from whom all being or existence emanates. Thus Proclus (412–485 CE), whose own thought was much indebted to that of Plotinus, complained against Aristotle that it was not enough that God should be the final cause of the universe, as Aristotle had maintained; one must also show that God is the efficient cause, the very source, of the universe's existence. Unlike earlier Neoplatonists, such as Plotinus and Proclus, later Neoplatonists were quite keen to show the harmony of the thought between Plato and Aristotle. Accordingly, as part of their attempt to reconcile these two, later Neoplatonists wrote commentaries on a number of Aristotle's works, which were in turn either translated into Arabic or were known in paraphrastic versions. These works greatly shaped the reception of Aristotle in the Arabic world.[18] Thus in response to Proclus's complaint, his own student, Ammonius (ca. 440–520 CE), maintained that despite appearances to the contrary, and notwithstanding what Aristotle's earlier commentators might have thought, Aristotle himself had in fact held that God is both final and efficient cause of the very existence of the universe. Ammonius specifically argued as much in a treatise on Aristotle's "creator"—a treatise which is now lost, although we have hints of its contents from Greek and Arabic sources.[19] It was this Ammonian interpretation of Aristotle that the Arabic-speaking world inherited and which in part may explain the choice of *ṭabīʿa* as the translation for nature; for again Aristotle identified nature with matter and form, and yet if God is the efficient cause of the existence of the universe as a form-matter composite, as Ammonius had suggested, God would be such precisely by creating and then impressing the various specific forms into matter.

In addition to this issue of God's causal relation to the universe were developments concerning the question of how Aristotle's formal and material natures interacted. Latent in some of Aristotle's physical treatises (such as *On the Heavens*, *On Generation and Corruption*, and the *Meteorology*) is the idea that the specific natures of things supervene on their elemental or humoral mixtures.[20] This idea was articulated more fully by later thinkers, particularly Galen (ca. 129–210 CE) in his medical writings (such as *The Elements* and *Mixtures*). These Galenic treatises made their way into Islamic lands via the Persian city of Jundishapur—situated in the southwest region of modern Iran—when the city saw an influx of Greek scholars in the wake of the persecution of heterodox Christian sects and the closing of the Academy at Athens in 529 CE. These scholars brought with them the works of Galen and other medical authors, which provided the theoretical framework for medical practice in the Islamic world. Jundishapur was home to the first "teaching hospital," founded around 550 CE, and remained the center of medical learning in the region even after Muslims took control of the former Sassanid, or Persian, Empire. Eventually its position was usurped by Baghdad, after the ʿAbbāsid caliph al-Manṣūr (r. 754–775 CE) asked the then head of the Jundishapur medical school to treat him. The caliph's request precipitated a migration of physicians to Baghdad and the gradual rise of Baghdad as the preeminent center of medical learning.

Galen had taught that the different proportions of the elements (earth, water, air, and fire) and the more complex elemental mixtures such as the humors (blood, phlegm, yellow bile, and black bile) determined both what species form or nature a physical thing would have as well as the characteristic differences among individuals within a species, for example, why a particular person is sanguine, phlegmatic, bilious, or melancholic. While this simplified account looks broadly Aristotelian, Galen, drawing on a Stoic (materialist) natural philosophy and his own findings, also happily criticized Aristotle on points of natural and biological science. Indeed, at least one project among philosophers in Islamic lands was to reconcile, or at least adjudicate between, the best natural philosophy of the time as presented in Aristotle and the best medical theory of the time as seen in Galen. So, for example, from the point of view of natural philosophy, one challenge that medicine posed for philosophers in the Islamic world was to situate Galen's physiognomy within Aristotle's physics and show how a thing's underlying elemental mixture or temperament was related to its nature. This challenge, one might add, called for a reassessment of Galen's own philosophical assumptions. The one-time head of the teaching hospital in Baghdad, Abū Bakr Muḥammad ar-Rāzī, is credited with being the unsurpassed physician of Islam; he was among the first to rise to this challenge.[21] While certainly indebted to Galen, ar-Rāzī's own close observations, emendations, and advancements went well beyond Galen in virtually all areas of medical learning and practice, such as anatomy, diagnosis, and pharmacology, and it was in light of his own independent speculation that ar-Rāzī wrote his *Doubts Concerning Galen.*[22] Despite ar-Rāzī's rightly earned renown, it was Avicenna's *Canon* that would become the culmination of Arabic medicine; for in it not only did Avicenna present Galen's and ar-Rāzī's voluminous medical writings in a synoptic form, but he also attempted to provide for the science of medicine a theoretical basis that was grounded in Aristotelian natural philosophy.

Astronomy provided yet another discipline where advancements both in the later Greek and Arabic worlds went beyond Aristotle. Most notably, Aristotle had argued that the motions of the heavens accounted for changes in the elemental mixtures themselves and even suggested an astronomical model based on the system of Eudoxus (ca. 400–347 BCE), the best astronomer of his time. Unfortunately, Eudoxus's theory of rotating concentric spheres with Earth at the center was, within a generation of Aristotle, seen to be empirically inadequate. It was ultimately replaced by Ptolemy's (ca. 85–165 CE) astronomical system, with its appeal to eccentric and deferent-epicycle models.[23] Thus one issue facing philosophers in the Islamic world was how the physical principles and celestial motions assumed by Ptolemy's system (which were quite different from those assumed by Aristotle) could be incorporated into an Aristotelian natural philosophy with its explanation of changes in elemental mixtures, where those elemental mixtures in their turn determined a thing's specific nature and particular temperament—and all this while remaining sensitive to the conviction that God must be the ultimate cause of the natures that are impressed upon matter.

NATURE IN THE EARLY EASTERN ISLAMIC WORLD

The first Arabic philosopher to attempt this synthesis was al-Kindī, who, unlike most of his Aristotelian predecessors and successors, argued that God created the world along with all of its various motions from nothing and did so at some first moment in time in the finite past. Clearly, then, for al-Kindī, God, who created the existence of all things *ex nihilo*, is the cause of the absolute existence of natures. Al-Kindī added, however, that God uses the motion of the heavens in the generation and corruption of the individual instances of those natures thereafter. Al-Kindī's general strategy was something like this: the different motions of the elements—whether away or toward the center of the universe, that is the center of Earth itself, as well as the relative speeds away or toward the center—determine their natures. Heat is the cause of something's moving away from the center, while cold causes motion toward the center, whereas the dryer an element is the faster it moves and the wetter it is the slower it moves.[24] So, for example, the nature of the element fire is a combination of hot and dry, and as such fire naturally moves upward quickly. These motions, which again are linked to the qualities that determine the natures of the elements, are themselves affected by the size, speed, and proximity of the celestial bodies moving over them. In a popular survey of Ptolemaic astronomy and Galenic medicine, al-Kindī observed the following:

> We see that the body of every animal comes to have a humor commensurate with its elemental mixture. Thus humors follow upon the proximity and distance from us of the [celestial] individuals and how high or low, or fast or slow they are as well as whether they are in conjunction or opposition. Moreover, [our humor] is proportionate to the elemental mixtures of our bodies at the time that the semen is produced as well as when it settles in the wombs.[25]

Following Ptolemy, he then went on to describe the providential design of the heavens and their motions—including the sun's eccentric motion along the elliptic, as well as the various planetary motions produced by the combined effects of eccentrics, deferents, and epicycles. The myriad varying celestial motions, al-Kindī insisted, function together to give rise to numerous combined motions here on Earth, which themselves give rise to the various elemental and humoral mixtures of Aristotelian physics and Galenic medicine so as to account for the different specific natures that we find in the world as well as the particular temperaments of individuals. In summary, then, for al-Kindī, God is the proximate cause of the existence of the heavens and their motions, creating them from nothing at some first moment of time, while their motions, in turn, are the proximate causes of the generation and corruption of natures here on Earth.

Many subsequent Arabic-speaking philosophers would accept, at least in outline, this synthesis of Aristotle, Galen, and Ptolemy, but with one major alteration. Al-Kindī's account seemingly had the celestial motions educe natures out of an underlying elemental mixture by affecting certain basic qualities in the elements—a feature that is in fact in keeping with Aristotle's own account of elemental change. Consequently, this account makes it appear as if accidental qualitative changes in the matter causally explain the existence of the various species forms, and yet for most later thinkers the causal explanation was just the reverse: form explains the actualized existence of matter, and species forms are causally prior to accidental forms.[26] Hence al-Fārābī maintained:

> It would seem that the existence of forms is the primary aim, but since they subsist only in a given subject, matter was made a subject for bearing forms. For this reason, as long as forms do not exist, the existence of matter is in vain, but no natural being is in vain. Therefore, matter cannot exist devoid of a given form. Matter, then, is a principle and cause solely by way of being the subject for bearing the form; it is not an agent, nor an end, nor something that can exist independently of some form. Matter and form are both called "nature," although form is more aptly named such.[27]

If matter alone cannot explain the existence of natures, understood as forms, whereas the celestial motions merely produce accidental qualitative changes in matter, then the question becomes "From whence do the natures or species forms arise and what impresses them into matter?" The question does not concern the ultimate cause of the absolute existence of natures, which all took to be God, but instead is "What causes the particular existence of a given nature in some bit of matter at a precise time?"

Although al-Fārābī suggested that natures temporally come to be in matter as a result of the "Active Intellect,"[28] which is the immaterial substance associated with the mover of the moon, Avicenna explicitly said as much and integrated this element into his overall theory of generation and corruption.[29] Avicenna summarized his account thus:

> There is a single account about all of that, namely that through the mixture of the compound body it was prepared to receive a certain disposition or form or specific property [in other words, the natures] and that comes to be

in it as a result of an emanation from nothing other than the Giver of Forms and Powers. They emanate from it on account of its goodness and because it does not stint on providing [forms or natures] to whatever is deservingly prepared.[30]

To be more specific, according to Avicenna every natural substance has an elemental disposition suitable to the nature informing it, where this elemental disposition is determined by how hot, cold, wet, or dry the substance is. Moreover, as in al-Kindī's system, elemental dispositions are constantly undergoing alteration as a result of the motions of the heavenly bodies. When, in a given natural substance, the alteration of its elemental disposition is significant enough, the matter is no longer suitable to the nature informing it, and so the matter receives a new nature that better accords with its new elemental disposition. Again it is the motions of the heavenly bodies that are the causes for the changes in a material substance's elemental dispositions. However, as such, the heavenly bodies are only preparatory or auxiliary causes for the occurrence of the new nature. The cause that imparts the new nature, that is, the new form, is "The Giver of Forms," which Avicenna identified, following al-Fārābī, with the last of the separate substances or Intellects, namely, the so-called "Active Intellect." The Giver of Forms, then, causes the suitable elemental disposition to receive the new form by emanating the appropriate form or nature into the prepared matter.[31]

Avicenna's conception of the role of the Giver of Forms in the temporal coming to be of natures and their concomitant actions would basically become the standard theory for later Muslim philosophers working in the east.[32] So, for example, as-Suhrawardī (ca. 1154–1191 CE) embraced Avicenna's account, albeit recast in his preferred light imagery, and as such the theory became a mainstay of later Illuminationist philosophy in the Islamic east. (One should be careful, however, not to confuse the Illuminationist philosophy mentioned here with the tradition, frequently associated with the work of Ibn al-Haytham, that treats theoretical optics in the medieval Arabic-speaking world.)[33] As-Suhrawardī wrote, "Lights become the cause of motions and heat, where both motion and heat obviously belong to light, not that they are its cause, rather, they prepare the recipient so that it [a light] occurs in it from the dominating light that emanates through its substance onto the recipients properly prepared for it."[34] Here "light" is a trope for "form" or "nature," and "dominating light" is as-Suhrawardī's terminology for a separate, immaterial substance, such as al-Fārābī's "Active Intellect" or Avicenna's "Giver of Forms." Thus we see as-Suhrawardī in effect repeating the Avicennan position that certain accidental changes in motion and heat prepare matter such that it is impressed with a nature or form by a separate, immaterial substance.

NATURE IN THE LATER WESTERN ISLAMIC WORLD

There is evidence that the idea of a separate substance's impressing natures into matter reached philosophers working in the Islamic Empire in Spain at a fairly early date. In his *Inquiry into the Active Intellect*, the first major Iberian Peripatetic, Ibn Bājja, asserted the following:

> The bodies subject to generation and corruption are subordinate to bodies that move circularly insofar as these are neither generated nor corrupted, where the former is like the elements. The elements, taken in their entirety, are not subject to generation, while their particular instances, namely the species of things existing materially, are generable. When we consider their particular instances, namely, the things subject to generation, it follows necessarily that there is a form that is not in a matter at all [namely, the Active Intellect], but which is intimately related to material forms and is a cause of their existence.[35]

Here we see all the salient features of Avicenna's theory of the Giver of Forms—that the elements are subject to the motion of the celestial bodies, but that the cause of the existence of a particular species, or nature, in the matter, is due to a separate immaterial substance, identified, following al-Fārābī, with the Active Intellect.

Despite hints of this theory in the later western Islamic world, it never really seemed to capture the imagination of the Spanish Muslim philosophers, who on the whole preferred a theory of nature and the generation of natural things more closely aligned with the historical Aristotle. Thus even though Ibn Bājja, in his commentary on Aristotle's *Physics*, asserted that generation is the most significant part of the science of physics, there was no immediately apparent reference in that work to the role of the Active Intellect in generation as is suggested in his *Inquiry into the Active Intellect*.[36] Similarly, Ibn Ṭufayl, in his only extant philosophical work, *Ḥayy ibn Yaqẓān*, said nothing about a possible role of either the Active Intellect or the Giver of Forms in generation, even though he had much to say about generation and the role of the celestial motions in the formation of elemental mixtures and readily admitted that the philosophy of Avicenna had influenced his own philosophical thought. Finally, although Averroës would mention Avicenna and al-Fārābī by name, noting the role that they had assigned to the Giver of Forms (that is, the Active Intellect) in generation, he did so only to indicate what he considered to be an aberration of the moderns which, he claimed, belied a fundamental misunderstanding of Aristotle's position. (We shall return to Averroës' criticism of Avicenna and al-Fārābī shortly.)

In general the Andalusian Peripatetics seemed happier to explain the emergence of natures either, as in the case of the elements and nonliving things, in terms of qualitative changes brought about by celestial motions, or, as in the case of living things, through the activity of seeds and semen on a recipient matter. In Averroës' commentary on Galen's *Elements*, he wrote of the simple bodies:

> It has become clear in the science of physics that every body is a composite of matter and form. The matter of the simple bodies is their common component that exists only in potency, as will become clear, while their forms are the four simple qualities, which are at the extreme. (I mean the two of them that are active and passive, for example, the hot and dry that are in fire and the cold and wet that are in water.)[37]

Averroës identified the basic primary qualities, hot, cold, dry, and wet with the elemental forms or natures themselves, by which he probably meant that different natures are to be associated with different ratios between hot and cold and wet and dry. Consequently, as a result of the motions of the heavenly bodies, there would be changes in these primary qualities and their ratios, which in turn would explain the emergence of a new form or nature in a particular instance.[38] Thus, concluded Averroës, there is no reason to appeal to a separate substance that gives forms.

Similarly, according to these western Arabic-speaking philosophers, the species form or nature arises in living things when something possessing an active principle, namely, a specific type of semen or seed, brings about a change in the matter. Ibn Bājja gave a series of examples to make this point—"the embryo does not result from the [menstrual] blood until the semen unites with it . . . and the plant does not come from the mixture of water and earth until the seed unites with them."[39] On this point, Averroës wholly concurred. For these philosophers, one did not need to posit some separate, immaterial substance that impresses natures onto the prepared matter; rather, the seeds and semen that are part of our physical world can impart their own nature to a suitably disposed material when they come into direct contact with it.

Averroës further argued that the introduction of the Giver of Forms indicated a fundamental misunderstanding of the relation between matter and form; for if the matter's being prepared were different from the form impressed onto it, then one must assume that matter and form would be really distinct, when in fact they are merely conceptually different. For example, if one considers an actually existing bed, one might conceive of the shape of the bed as different from the stuff that has that shape, but the shape and stuff of the bed are not really distinct such that there could be both a subsisting shape and subsisting matter. Yet, objected Averroës, such an opinion seems to be exactly what is assumed when one maintains that the Giver of Forms has certain forms that it impresses into prepared matter.

In the end, Averroës complained that both al-Fārābī and Avicenna had been misled about the generation or temporal coming-to-be of natures "because it was an opinion very much like the account upon which the practitioners of *kalām* in our religion rely, namely that the agent of all [generated] things is one and that some of the [generated] things do not bring about an effect in others."[40] This criticism is interesting. One of its key complaints concerns the assumption that there must be some single efficient cause of all things—that is, that there is one agent who generated all things, a premise that Averroës would in fact deny. In denying the need for such an agent, Averroës in effect was rejecting the Ammonian interpretation of Aristotle, which made God both a final and efficient cause of everything in the universe. This, as we have seen, was the very issue that motivated earlier eastern accounts of nature. In fact Averroës considered it an open interpretative question within Islam whether God is the efficient, rather than just the final, cause of all things.[41] Indeed, Averroës himself sided with Aristotle on this point, maintaining that certain substances other than God, such as the heavenly bodies, are eternal and so do not need an efficient cause; rather, God is precisely the final cause of the world's existence and as such brings about celestial motions, which, as we have seen, were for Averroës the causes of elemental changes here in our world.[42]

Here, then, we see that the discussion concerning nature within the *falsafa* tradition, which had its origins in Aristotle, was affected by later developments within the Greek intellectual tradition, underwent significant modifications at the hands of Arabic-speaking philosophers in the east, finally to come full circle in the thought of Averroës, who reestablished Aristotle's account of nature and natural change.

NATURE IN *KALĀM*

Whereas the cast of players in the *falsafa* tradition might disagree about whether the nature arose from within or without a natural thing, they all agreed that once existing in such a thing, the nature is a cause and principle of that thing's actions and motions. In contrast, the dominant position in *kalām* came to be that while a thing's true nature (*ḥaqīqa*) came directly from God (or perhaps through the intermediacy of an angel), such a nature had no causal efficacy considered in itself, and indeed God was the true and only cause of all things—both the cause of existence itself as well as any particular actions and motions or changes. This position, which culminated in a type of occasionalism, was not a matter of blind religious faith but was the conclusion of a series of arguments.

Before turning to those arguments, however, we should briefly consider sources for *kalām* conceptions of nature. Whereas the *falsafa* tradition was clearly indebted to Aristotle and his later Neoplatonic commentators for its understanding of nature, the sources for early *kalām* conceptions of nature are more obscure. Certainly many of the "theological" issues treated by practitioners of *kalām* had been part of the philosophical and theological systems of the Greek world. Moreover, there is evidence that part of the impetus for the early Greco-Arabic translation movement of Greek philosophical and scientific works was to provide factual information, particularly concerning natural philosophy, for theological debates between Muslim and Christian theologians.[43] Consequently, it is not surprising that at least one significant early *mutakallim* (a practitioner of *kalām*, plural, *mutakallimūn*), al-Jubbā'ī, wrote a treatise discussing and refuting arguments from Aristotle's corpus on natural philosophy.[44] Thus it seems likely that those working in the *falsafa* and the *kalām* traditions were in part drawing upon the same body of literature, except that whereas the former more openly embraced Greek learning, the latter seem to have been more hostile toward it. Perhaps one source for this difference in orientation toward Greek science was the Arabic language itself, or more particularly its grammar. Many *mutakallimūn* were leery of the new Greek science precisely because of its heavy reliance on Aristotelian logic, which they took to be nothing more than thinly disguised Greek grammar.[45] It was common to question whether Greek grammatical categories could provide a better way of conceptualizing the world than the categories that Arabic grammarians used, especially, it was argued, since the philosophizing was taking place in the Arabic language itself.

It may have been these linguistic concerns that motivated those working in the *kalām* tradition to adopt "true nature" (*ḥaqīqa*) for their notion of nature; for *ḥaqīqa* can simply mean the proper or strict sense or use of a word, and so a *ḥaqīqa* can be merely that which fixes the referent of some term without having any deeper metaphysical implications beyond this linguistic role. Thus Abū Rashīd (who flourished during the first half of the eleventh century CE) wrote of a thing's true nature:

> The thing itself inevitably is specified by a certain description by which it is distinguished from other [things], where that description inevitably has a characteristic by which [the thing] is known[46] and that characteristic is, as it were, its true nature (*ḥaqīqa*) and a necessary condition of [the thing's] existence.[47]

Here the emphasis is on a certain characterization or description by which one can pick out or identify a thing. Thus *ḥaqīqa,* far from identifying the causal principle of something's proper actions as it does in the *falsafa* tradition, indicates in the *kalām* tradition the feature(s) by which we sensibly recognize something and fix a referent in the language.

This is not say that theories of causal interaction among physical things were absent within the *kalām* tradition. They were not. Some of the earliest *kalām* thinkers maintained a theory by which one thing might "be engendered" (*tawallud*) by another and so caused. For example, the movement of the hand engenders the movement of the ring on the hand. A response to the theory of engenderment came at the hands of no less than al-Ghazālī (1058–1111 CE) himself. Al-Ghazālī's significance in the Islamic intellectual tradition cannot be understated. He was born in Ṭūs in the province of Khurasan in northeastern Iran and taught in both Baghdad and Nīshapūr. Among his intellectual accomplishments are his legitimization of Aristotelian logic among the *mutakallimūn,* his trenchant critique of *falsafa,* and his integration of Sufism, *kalām,* and even elements of *falsafa* into a systematic whole. When responding to the theory of engenderment, he presented what would become the dominant opinion within *kalām,* pointing out that such a theory, while perhaps capturing the imagination, lacked philosophical precision.

> Now in our opinion what is known concerning the expression "to be engendered" is that some body emerges from inside of another body, as the fetus emerges from the mother's belly and plants from the belly of the Earth. This is absurd with respect to accidents, since the motion of the hand has neither an inside such that from it the motion of the ring emerges nor is it something containing things such that from it part of what is in it emerges. So if the motion of the ring is not concealed in the very motion of the hand, then what is the meaning of its being engendered by it?[48]

In addition, this early *kalām* causal theory of engenderment seemed liable to the same type of criticism that *kalām* opponents of Aristotelian natural causation would raise, to which we shall now turn.

Aristotle and most (although not all) of those working within the *falsafa* tradition took the existence of natures, understood as causes of species-specific actions and motions, as virtually self evident and not in need of proof.[49] Aristotle wrote, "Trying to prove that there is nature is ridiculous; for it is obvious that there are many such things, whereas proving obvious things through what is not obvious belongs to one who is incapable of distinguishing between what is known in itself and what is not."[50] We have seen al-Kindī appealing to the regular movements of the elements, whether away from or toward the center, as witness to the existence of natures, and other philosophers point to the regularity of fire burning, alcohol's intoxicating, and the like as evidence that these physical things have certain innate causal powers, which the philosophers identified with those things' natures.

The first in a chain of *kalām* arguments directed against philosophers' conception of natures was intended to undermine the claim that the existence of natures, understood as internal causes, is self-evident. One *mutakallim* who argued against the purported self-evident status of natures was al-Bāqillānī (d. 1013). A near contemporary of Avicenna, and for much of his adult life a resident of Baghdad, he was also one of the first to systematize and popularize the newly emerging Ash'arite *kalām*, which took a more traditionally Islamic approach to theological and philosophical issues. Al-Bāqillānī observed:

> Concerning what [the philosophers] are in such a stir, namely that they know by sense perception and necessarily that burning occurs from fire's heat and intoxication from excessive drink, it is tremendous ignorance. That is because that which we observe and perceive sensibly when one drinks and the fire comes into contact is only a change of the body's state from what it was, namely, one's being intoxicated or burnt, no more. As for the knowledge that this newly occurring state is from the action of whatever, [such a causal relation] is not observed; rather it is something grasped through rigorous inquiry and examination.[51]

In other words, although we observe the constant conjunction of two types of events—whether fire's contacting cotton and the cotton's burning or intoxication following excessive drinking—one does not observe the causal connection or mechanism that explains such regularities. Based solely on sense perception, one could equally explain the regularity of our observations by appealing to a custom or habit on the part of God to bring about one type of event on the occasion of another type of event. For example, it might be that when fire is placed in contact with cotton, God, not the fire, causes the burning of the cotton. Both interpretations—whether the natural causation of *falsafa* or the occasionalism of *kalām*—are underdetermined, should one appeal solely to sense perception.

A second in the chain of *kalām* arguments against Aristotelian natures was intended to show that in fact natures taken alone could not be causally efficacious. Again let us consider an argument derived from al-Bāqallānī.[52] We observe around us the temporal succession of various and different events. If this temporal succession of events is due solely to natures, then the nature might be either eternal or temporal. Now since nature does not act by choice but always acts in the same way, if it were eternal, then from all eternity there would have been the same actions and the same events. Thus one could not explain the variety and differences of temporal events. If the natures that cause the temporal succession of various and different events are themselves temporal, that is to say, various and different natures arise and so produce various and different events, then there must be a cause for the temporal origination of those new natures. Consequently one can again ask about the origination of the new nature: "Is it caused by a nature and if so is that nature eternal or temporal?" Here one finds oneself once again facing the initial question. Clearly, then, if every cause acts through a nature, one is on the road to infinite regress. The adherents of *kalām* denied the possibility of an infinite series absolutely, whether an infinite series extending into the past or an infinite series of presently existing natural causes. Thus the purported series of natural causes must terminate with God. Of course, one could say that God acts through a finite series of intermediary natural causes, but why complicate matters when the earlier argument had shown that there is no empirical reason for assuming causal relations between various observable events? Simplicity, then, would suggest that one needs only a single cause. According to this account it is God, rather than the natures of things, that causally determines everything in the world at every instant. The origins of Islamic occasionalism—the view that reserves all causality for God and God alone—may well have had its origins in *kalām* critiques of Aristotelian natures.[53]

CONCLUSION

In Islamic occasionalism, one sees an extreme response to a question that first arose in the late Hellenistic world and then influenced discussions of the understanding of nature in the medieval Arabic-speaking world: "What is God's causal relation to the natural world?" Among those working in the *falsafa* tradition, the answer initially seems to have been that God created natures and matter, impressing the one into the other. Here God is the efficient cause of the natural world's existence. Subsequent thinkers, most notably Avicenna, relegated the task of impressing natures into prepared matter to an immaterial substance below God, identified with the Active Intellect or Giver of Forms. For certain later Muslim philosophers, such as Averroës, God apparently stands to the world only as its final cause, not its efficient cause, and so the issue of making natures that come from without and are then subsequently impressed into matter fell by the wayside. Indeed, when one turns to the Latin West and its reception of Arabic philosophy, in a real sense it was Averroës who led the way on this point—not in restricting God's role to final causation alone (for many Latin scholastics saw God as both final and efficient cause), but in rejecting the need for a separate substance (al-Fārābī's and Avicenna's Active Intellect or Giver of Forms) to explain how natures are impressed in matter. Thus, at least by the time of Thomas Aquinas (1225–1274 CE), Avicenna's Giver of Forms seemed to play no significant role in physics, and the Active Intellect had come to be identified with an internal cognitive faculty belonging individually to each human. In short, the Active Intellect was no longer considered a separate substance, as philosophers working in the Islamic world had commonly held.

In stark contrast to those working in the *falsafa* tradition, practitioners of *kalām* developed an occasionalistic outlook on the world, which simply did away with the intermediacy of natures and made God the direct efficient cause of all actions in the world. While none of the theological treatises of the *mutakallimūn*, in which they themselves laid out these arguments, made it into Latin translations, their thought was nonetheless known to Latin scholastics indirectly. Moses Maimonides (1135–1204 CE), for instance, mentioned *kalām* positions in his *Guide of the Perplexed*,[54] and Averroës incorporated into his *The Incoherence of the Incoherence* virtually the whole of al-Ghazālī's *The Incoherence of the Philosophers*, in which al-Ghazālī approvingly mentioned *kalām* theories.[55] Both Averroës' and Maimonides' works were in turn translated into Latin and played significant roles in the development of Latin scholasticism. Despite the relatively early presence of *kalām* theories available in Latin (even if at one remove), it would be difficult to trace direct lines of influence to similar views in Europe, such as seventeenth-century occasionalism and David Hume's criticism of causation in the early modern period.[56] Still, there are certain marked notes of agreement between the two groups. For example, both linked occasionalism with their anti-Aristotelian polemics, and attacked the causal theory that underwrote Aristotelian science.[57] Ironically, while in the West these very points of similarity were seen as part of a scientific outlook that helped bring about the European scientific revolution, their success in the Islamic context, with the accompanying critique of natures taken as causal principles, has been seen (rightly or wrongly) as contributing to the decline of Islamic science.

NOTES

1. Aristotle, *Physics* II.1.192b21–23.

2. Ibid; Arabic in *Arisṭuṭālīs, aṭ-Ṭabīʿī*, ed. Abdurrahman Badawi, 2 vols. (Cairo: General Egyptian Book Organization, 1964–65).

3. Al-Kindī, *Rasāʾil al-Kindī al-Falsafīya*, ed. Muhammad Abu Rida, (Cairo: Dār al-Fikr al-ʿArabī, 1953), 1:165.

4. Al-Fārābī, *Falsafat Arisṭuṭālīs*, ed. Muhsin Mahdi (Beirut: Dār Majallat Shiʿr, 1961), 89; English translation in *Philosophy of Plato and Aristotle*, trans. Muhsin Mahdi (Ithaca, NY: Cornell University Press, 2002), 128.

5. *ʿUyūn al-masāʾil*, in *Alfārābī's Philosophische Abhandlungen*, ed. Friedrich Deiterici (Leiden: E. J. Brill, 1890), 60.

6. Yaḥyá ibn ʿAdī, *Maqālāt Yaḥyá ibn ʿAdī al-falsafīya*, ed. Sahban Khalifat (Amman: Publications of the University of Jordan, 1988), 269–70.

7. In Aristotle, *Arisṭuṭālīs, aṭ-Ṭabīʿī ad* 192b21–23.

8. Avicenna, *The Physics of* The Healing, ed. and trans. Jon McGinnis (Provo, UT: Brigham Young University Press, 2009), I.5 (3), 39.

9. Ibn Bājja, *Sharḥ as-Samāʿ aṭ-ṭabīʿī Arisṭuṭālīs*, ed. Majid Fakhry (Beirut: Dār an-Nahār li-n-Nashr, 1973), 24.

10. Ibn Ṭufayl, *Hayy Ben Yaqdhân, roman philosophique d'Ibn Thofaīl*, ed. Léon Gauthier (Beirut: Imprimerie Catholique, 1936), 85, English translation in *Hayy Ibn Yaqzān*, trans. Lenn Goodman (New York: Twayne Publishers, 1972), 132.

11. Averroës, *al-Jawāmiʿ fī l-falsafa Kitāb as-samāʿ aṭ-ṭabīʿī*, ed. Josep Puig (Madrid: Instituto Hispano-Arabe de Cultura, Consejo Superior de Investigaciones Científicas, 1983), 19–20; Long Commentary on the *Physics* (Venice: *apud Junctas*, 1562), 48C–49K (repr. Frankfurt: Minerva, 1962).

12. For general accounts of theories of causation in the Islamic middle ages see Taneli Kukkonen, "Causality and Cosmology, the Arabic Debate," in Eevan Martikainen, ed., *Infinity, Causality and Determinism: Cosmological Enterprises and their Preconditions* (Frankfurt: Peter Lang, 2002), 19–43; and Thérèse-Anne Druart, "Metaphysics," in *The Cambridge Companion to Arabic Philosophy*, ed. Peter Adamson and Richard C. Taylor (Cambridge: Cambridge University Press, 2005), 327–48.

13. Al-Kindī, *Fī l-ibāna ʿan anna ṭabīʿat al-falak mukhālafa li-ṭabāʾiʿ al-ʿanāṣir al-arbaʿa*, in al-Kindī, *Rasāʾil*, 2:40 (henceforth *Ṭabīʿat al-falak*) (emphasis added).

14. Ar-Rāzī, *Maqāla fīmā baʿd aṭ-ṭabīʿa*, in *Rasāʾil Falsafīya*, ed. Paul Kraus (Cairo: Universitatis Fouadi I Litterarum Facultatis Publicationum, 1939), 116 (repr. Frankfurt:

Institut für Geschichte der arabisch-islamischen Wissenschaften, 1999) (emphasis added). This text, while I believe it is an authentic work by ar-Rāzī, is almost certainly not his *Maqāla fīmā ba'd aṭ-ṭabī'a*, but from his *Fī l-madkhal ilá l-'ilm aṭ-ṭabī'ī* [Introduction to Physics], which also went under the title *Sam' al-kiyān* [Auscultatio physica].

15. Aristotle, *Physics* II.1.193a9–b21.

16. See for example *Physics* VIII.1, where Aristotle argued that the motion of the heavens could not have begun at some first moment of time; *De caelo* I.11–12, where he argued that the heavens must be ungenerated and necessary; and finally *Metaphysics* Z.8, where he argued that form and matter absolutely do not come to be, but only particular instances of forms in matter come to be.

17. See Aristotle, *Physics* VIII.5 and *Metaphysics* Λ.7.

18. For the reception of Aristotle's *Physics* (with a particular emphasis on the role of the Neoplatonist John Philoponus's *Physics* commentary) see Paul Lettinck, *Aristotle's Physics and Its Reception in the Arabic World* (Leiden: E. J. Brill, 1994).

19. For discussions of Ammonius's influence on philosophy in the Islamic Middle Ages, see Robert Wisnovsky, *Avicenna's Metaphysics in Context* (Ithaca, NY: Cornell University Press, 2003), part 1; and Amos Bertolacci, "Between Ammonius and Avicenna: al-Fārābī's treatise *On the Goals of Aristotle's* Metaphysics," in *The Reception of Aristotle's* Metaphysics *in Avicenna's* Kitāb al-Šifā': *A Milestone of Western Metaphysical Thought* (Leiden: E. J. Brill, 2006), 65–110.

20. For a discussion of Aristotle's account of mixtures and various classical, medieval, and contemporary interpretations of Aristotle's account, see Rega Wood and Michael Weisberg, "Interpreting Aristotle on Mixture: Problems about Elemental Composition from Philoponus to Cooper," *Studies in History and Philosophy of Science* 35 (2004): 681–706.

21. For a discussion of ar-Rāzī's intellectual influence in the medieval Islamic world, see Lenn E. Goodman, "Muḥammad ibn Zakariyyā' al-Rāzī" in *History of Islamic Philosophy*, ed. Seyyed Hossein Nasr and Oliver Leaman (London and New York: Routledge 1996), 198–215.

22. See, ar-Rāzī, *Kitāb ash-shukūk 'alá Jālīnūs* [Doubts Concerning Galen], ed. Mahdi Muhaqqiq (Tehran: al-Ma'had al-'alī al-'ālamī li-l-fikr wa-l-ḥadāra l-Islāmīya, 1993).

23. For a general discussion of the relation between Aristotle's physics and Ptolemy's astronomy and the challenges it presented to Arabic-speaking philosophers and astronomers, see George Saliba, "Aristotelian Cosmology and Arabic Astronomy," in *De Zénon d'Élée à Poincaré, Recueil d'études en homage à Roshdi Rashed*, ed. Régis Morelon and Ahmad Hasnawi (Leuven: Peeters, 2004), 251–68.

24. Al-Kindī, *Ṭabī'at al-falak*, 40–44.

25. Al-Kindī, *Fī ibāna 'an al-'illa l-fā'ila l-qarība lil-kawn wa-l-fasād*, in al-Kindī, *Rasā'il*, 1:226.

26. See Abraham D. Stone, "Avicenna's Theory of Primary Mixture," *Arabic Sciences and Philosophy* 18 (2008): 99–119, for a discussion of the Greek origins of this problem and Avicenna's theory of the Giver of Forms as a solution to it.

27. Al-Fārābī, *Mabādi' al-mawjūdāt*, in Fauzi Najjar, ed., *as-Siyāsa al-madanīya al-mulaqqab bi-mabādi' al-mawjūdāt* (Beirut: Imprimerie Catholique, 1964), 36–37.

28. See al-Fārābī, *Falsafat Arisṭuṭālīs*, 129–130 and *Mabādi' al-mawjūdāt*, 54–55.

29. See Avicenna, *al-Kawn wa-l-fasād*, ed. Mahmud Qasim (Cairo: General

Egyptian Book Organization, 1969), XIV; and *The Metaphysics of* The Healing, ed. and trans. Michael E. Marmura (Provo, UT: Brigham Young University Press, 2005), IX.5, for discussions of the role of the "Giver of Forms" (wāhib al-ṣuwar) in the processes of generation and corruption.

30. Avicenna, *al-Af'āl wa-l-infi'ālāt*, ed. Mahmud Qasim (Cairo: General Egyptian Book Organization, 1969), II.1, 256.

31. The most extensive discussion of Avicenna theory of the "Giver of Forms" is Jules L. Janssens, "The Notions of *Wāhib al-Ṣuwar* (Giver of Forms) and *Wāhib al-'aql* (Bestower of Intelligence) in Ibn Sīnā," in *Intellect et Imagination dans la Philosophie Médiévale*, ed. Maria Cândida Pacheco and José Francisco Meirinhos (Turnhout, Belgium: Brepols, 2006), 1:551–62.

32. There is even reason to believe that the great critic of *falsafa*, al-Ghazālī, may have incorporated Avicenna's theory into *kalām*, albeit with significant modifications; see Jon McGinnis, "Occasionalism, Natural Causation and Science in al-Ghazālī," in *Arabic Theology, Arabic Philosophy, from the Many to the One: Essays in Celebration of Richard M. Frank*, ed. James E. Montgomery (Leuven: Peeters, 2006), 441–63.

33. It is worth noting that Illuminationist philosophy seems not to have made it into Latin and so had no real influence on Latin philosophy and science. For a discussion of Arabic optical theories that would be influential on the Latin *perspectiva* tradition, see David C. Lindberg, *Theories of Vision from al-Kindi to Kepler* (Chicago: University of Chicago Press, 1976).

34. Arabic in Suhrawardī, *The Philosophy of Illumination*, ed. and trans. John Walbridge and Hossein Ziai (Provo, UT: Brigham Young University Press, 1999), 129.

35. Ibn Bājja, *al-Wuqūf 'alá l-'aql al-fa''āl*, in *Rasā'il ibn Bājja al-ilāhīya*, ed. Majid Fakhry (Beirut: Dār an-Nahār li-n-Nashr, 1968), 107; French translation in Thérèse-Anne Druart, "Le Traité d'Avempace sur 'Les choses au moyen desquelles on peut connaître l'intellect agent,'" *Bulletin de Philosophie Médiévale* 22 (1980): 73–77.

36. Ibn Bājja, *Sharḥ as-Samā' aṭ-Ṭabī'ī*, 23.

37. Averroës, *Talkhīṣ kitāb al-usṭuqissāt li-Jālīnūs*, ed. George Anawati and Sa'id Zayed (Cairo: The General Egyptian Book Organization, 1987), 55.

38. See Averroës' middle commentary on *Generation and Corruption, Commentarium medium in Aristotelis de generatione et corruptione libros*, ed. Francis Fobes (Cambridge, MA: The Medieval Academy of America, 1956), book II, para. 56, 146–48; English translation in *Averroes on Aristotle's* De Generatione et Corruptione, *Middle Commentary and Epitome*, trans. Samuel Kurland (Cambridge, MA: The Medieval Academy of America, 1958), 101–2.

39. Ibn Bājja, *Sharḥ as-Samā' aṭ-ṭabī'ī*, 27; cf. Averroës' long commentary on the *Metaphysics* Z.9 (comment 31), *Tafsīr mā ba'd aṭ-ṭabī'īyāt*, ed. Maurice Bouyges, 3 vols. (Beirut: Imprimerie Catholique, 1938–1952), 878–886.

40. Averroës, *Tafsīr mā ba'd aṭ-ṭabī'īyāt*, 885.

41. See, e.g., his *Decisive Treatise*, trans. Charles E. Butterworth (Provo, UT: Brigham Young University Press, 2001), 16 (21).

42. See Averroës' long commentary on the *Metaphysics* Λ.6 (comments 29–36), *Tafsīr mā ba'd aṭ-ṭabī'īyāt*; English translation in Charles Genequand, *Ibn Rushd's Metaphysics* (Leiden: E. J. Brill, 1986), 134–50.

43. See Dimitri Gutas, *Greek Thought, Arabic Culture: The Graeco-Arabic Translation*

Movement in Baghdad and Early 'Abbāsid Society (2nd-4th/8th-10th centuries) (London: Routledge, 1998), 72–73.

44. See F. E. Peters, *Aristoteles Arabus: The Oriental Translations of Commentaries on the Aristotelian Corpus* (Leiden: E. J. Brill, 1968), 35–36.

45. For at least one historical account of this complaint, see David S. Margoliouth, "The Discussion between Abū Bishr Mattā and Abū Sa'īd as-Sīrāfī on the Merits of Logic and Grammar," *Journal for the Royal Asiatic Society* (1905): 79–129.

46. *Yaẓharu*, literally, "[the thing] becomes apparent."

47. Arabic in Richard M. Frank, *Beings and their Attributes, the Teaching of the Basrian School of the Mu'tazila in the Classical Period* (Albany: State University of New York Press, 1978), 80n3.

48. Al-Ghazālī, *Iqti'ād fī l-i'tiqād*, ed. 'Abdallah Muhammad al-Khalili (Beirut: Dār al-kutub al'ilmīya, 2004), 59; English translation in Michael E. Marmura, "Ghazali's Chapter on Divine Power in the Iqti'ād," *Arabic Science and Philosophy* 4 (1994): 312.

49. Dissenters included ar-Rāzī in his *Maqāla fīmā ba'd aṭ-ṭabī'a*, 116, who maintained natures are neither immediately perceptible, even if their purported actions might be, nor is their existence a first principle of the intellect; and Avicenna in his *The Physics of* The Healing, I.5 (4), 40, who argued that although the natural philosopher must accept the existence of natures as one of his first principles, that existence could be demonstrated in first philosophy.

50. Aristotle, *Physics* II.1.193a3–6.

51. Al-Bāqillānī, *Tamhīd*, ed. Richard McCarthy (Beirut: Librairie Orientale, 1957), 43 (77); al-Ghazālī provides a similar argument in his *The Incoherence of the Philosophers*, ed. and trans. by Michael E. Marmura (Provo, UT: Brigham Young University Press, 1997), 171 (discussion 17).

52. Al-Bāqillānī, *Tamhīd*, 34–37 (59–65).

53. For discussions of Islamic occasionalism, see Majid Fakhry, *Islamic Occasionalism, and Its Critique by Averroës and Aquinas* (London: George Allen & Unwin, Ltd, 1958).

54. See Moses Maimonides, *Guide of the Perplexed*, part I, chapter 73.

55. See Averroës, *Destructio Destructionum Philosophia Algazelis*, ed. Beatrice H. Zedler (Milwaukee: Marquette University Press, 1961).

56. One might also note the atomism of Nicholas of Autrecourt in the Middle Ages. Unfortunately, *kalām* atomism was not discussed in this study. A brief survey of it is available in the Stanford Encyclopedia of Philosophy (http://plato.stanford.edu/entries/arabic-islamic-natural/), and a more extensive study is offered by Alnoor Dhanani, *The Physical Theory of Kalām, Atoms, Space, and Void in Barisan Mu'tazilī Cosmology* (Leiden: E. J. Brill, 1994); and again in Abdelhamid I. Sabra, "*Kalām* Atomism as an Alternative Philosophy to Hellenizing *Falsafa*," in *Arabic Theology, Arabic Philosophy, from the Many to the One*, 199–272. As for a discussion of *kalām* influences on Nicholas of Autrecourt, see Zénon Kaluza, "Nicolas d'Autrécourt et la tradition de la philosophie grecque et arabe," in *Perspectives arabes et médiévales sur la tradition scientifique et philosophique grecque*, ed. Ahmad Hasnawai, Abdelali Elamrani-Jamal, and Maroun Aouad (Leuven: Peeters, 1997), 365–93.

57. It is worth noting, however, that unlike later European occasionalists, Muslim theologians did not altogether deny real qualities, nor did they adopt a mechanical

outlook to replace Aristotelian forms as a causal explanation. So, for example, they would affirm that the wine is red because of the accident of redness in it; however, God is the cause of the redness in the wine as well of our perceiving the wine as red. For a discussion of causal explanations in medieval Latin and early modern thought, see Steven Nadler, "Doctrines of Explanation in Late Scholasticism and in the Mechanical Philosophy," in *The Cambridge History of Seventeenth-Century Philosophy*, ed. Daniel Garber and Michael Ayers (Cambridge: Cambridge University Press, 1998), 513–52.

CHAPTER 4

Natural Knowledge in the Latin Middle Ages

Michael H. Shank

In the rough millennium between Boëthius (d. 524 CE) and Regiomontanus (d. 1476 CE), the inquiry into nature in Europe changed drastically. Starting from a modest Roman inheritance, the intellectuals of medieval Europe built new conceptual, methodological, and institutional frameworks for their study of nature.[1] Accordingly, this chapter seeks to characterize the emergence of several medieval approaches to nature in a series of transformations that proceeded at an accelerating pace to the fifteenth century. Today, the English word "science" and its counterparts in other vernaculars denote primarily the systematic inquiry into nature, an activity that is firmly anchored in the world's universities. This partnership traces its ancestry to the later Middle Ages.

The long process that culminated in these transformations had humble beginnings. Between the fifth and tenth centuries, the western part of the former Roman Empire found itself literally in a postcolonial situation. Formerly occupied peoples with varying degrees of access to the language, infrastructure, and culture of their Roman colonizers survived new waves of Germanic, Hun, and Norse occupiers. The descendants of these peoples eventually coexisted and mingled, building a distinctive political and intellectual world by selection, amalgamation, and innovation. Intellectually, they eventually came to terms with the legacies of Greco-Roman culture and the Judeo-Christian scriptures in Latin, the language imposed by Rome and subsequently promoted by the Church of Rome. The most momentous developments of the early medieval era were anonymous and technical rather than theoretical: the diffusion of agricultural improvements in particular transformed the economy and the society.[2] Building on compendia of late-Roman learning about nature, the scholars of this era likewise succeeded in extending their understanding beyond what they had inherited. They clearly desired to know more. By the eleventh and twelfth centuries, larger numbers of scholars built on these early efforts by translating and absorbing a massive library of natural-philosophical, medical, astrological, and mathematical works primarily from the Arabic.

The thirteenth century marks a turning point in the histories of world culture and of the scientific enterprise. The emerging universities institutionalized learning, giving the teaching, study, and advancement of theoretical approaches to natural knowledge a permanent home. The ups and downs of the intervening history notwithstanding, the partnership of the scientific enterprise with the university still thrives today. The medieval universities not only created a quasi-autonomous space for natural philosophy and the mathematical sciences in the faculties of arts, but also institutionalized law, medicine, and theology. Masters in the last two faculties often continued to develop the natural philosophy and the mathematical and analytical tools of their earlier education.

The era also witnessed both an increasing specialization of the inquiry into nature, and crossovers among specialties. In the early Middle Ages, the object under consideration largely determined the type of inquiry to which it belonged. Scholars in the later Middle Ages, however, would recognize that the same object of inquiry could be approached from several points of view, with different questions and methods. The moon, for example, fell under the purview of astronomy (motions and positions), natural philosophy (composition), optics (phases), and medicine (its relation to times for bloodletting). The number of *scientiae* proliferated, and new approaches found their way into, and between, established categories and curricula. Thanks to the universities, these transformations in the study of nature also found a home in the larger culture. The basic concepts, language, and assumptions of natural philosophy, medicine, and the mathematical sciences became familiar among the growing literate elites that the universities produced and also among the ones they served.

LATE ANTIQUITY AND THE POSTCOLONIAL EARLY MIDDLE AGES

To understand the medieval achievement, it is crucial to appreciate an oft-forgotten fact: at the height of the Roman Empire, the Latin language provided limited access to Greek medicine, natural philosophy, and the

Hellenistic mathematical sciences, which only the best-educated Romans could read in the original. The others relied on encyclopedias, handbooks, and summaries for a second- or thirdhand acquaintance with such material. In the end, Roman civilization produced only a handful of original Latin writings focused primarily on nature and related topics.[3]

Self-consciously practical, the Romans were aware of, but lacked interest in, Greek theoretical contributions (in marked contrast to Greek religion and literature). Some Romans bragged about their innocence of philosophy, including natural philosophy; others were suspicious of both the Greeks' speculations and their medicine. Cicero saw his fellows as concerned with specific calculations and measurement,[4] not with abstract Greek mathematics, a neglect symbolized by the overgrown tomb of Archimedes in Syracuse. When necessary, however, the Romans did tap Greek expertise. To reform the calendar, Julius Caesar turned to the Alexandrian Sosigenes, while medicine in Rome was a stereotypically Greek endeavor that Cato, Juvenal, and Pliny targeted with their barbs.[5] It was in the Roman Empire of the second century that Ptolemy and Galen produced the most impressive works in, respectively, the mathematical sciences and medicine—in Greek, not Latin.

This state of affairs shaped the cultural predicament of the learned in the former Roman colonies once the central administration of the empire fizzled. Judged by Roman rather than Alexandrian standards, the traditional notion of a precipitous medieval decline in the inquiry into nature seems grossly exaggerated, even more so if the vantage points are the banks of the Thames, the Seine, the Rhine, or the Danube.

The Latin writers from the first century BCE to the ages of Augustine (d. 430 CE) and Boëthius did little original inquiry into nature, but they performed a crucial service by summarizing parts of the Greek heritage or by translating parts of it into Latin. Their budding technical vocabulary and categories of thought would frame Latin-language approaches to nature until the twelfth century. These were only the first fruits, however. Only after scholars embraced and translated the high culture of Islamic civilization did readers of Latin for the first time appropriate the bulk of Greek works about nature and develop their own more adequate technical vocabulary.[6]

THE ORGANIZATION OF KNOWLEDGE

Central to the later medieval understandings of nature were its reorganizations of the classical relationships among the disciplines. Roman writers usually treated knowledge of nature in terms borrowed from the Greek. Vitruvius (d. 25 BCE) used the terms *physica* (natural [philosophy]) and *physicus* (natural [philosopher]) to denote the theory and the theoreticians of nature. While *philosophia* usually encompassed the acquisition and evaluation of all theoretical knowledge, Vitruvius specifically associated the word with the inquiry into nature: "*philosophia* explains the nature of things, which in Greek is called *physiologia*."[7]

Some intellectuals in late Roman antiquity thought of philosophy in loftier terms—as a goal to which lesser disciplines could lead. Augustine (d. 430 CE) wrote *De ordine*, a treatise on the seven liberal arts or, as he called them, "disciplines," that is, "sciences" (Latin translated *epistēmē*, the Attic Greek word for the highest form of knowledge and still the modern Greek word for "science," as both *scientia* and *disciplina*). After passing through the verbal disciplines (grammar, logic, and rhetoric), the student seeking enlightenment ascended in abstraction from musical theory through astronomy and geometry, presumably to the pinnacle of arithmetic, the simplest discipline, which required no sense perception. The mathematical disciplines were thus the highest rungs on the ladder to philosophy, leading eventually to the true Neoplatonic goal, "the One."[8]

All this no doubt seems very abstract and remote from the study of nature. But for Augustine and his Neoplatonist contemporaries, the four mathematical disciplines did deal with the natural world. Music and astronomy treated the ratios of harmony and the celestial spheres, while number and "continuous quantity" (the subjects of arithmetic and geometry) were among nature's fundamental elements. After converting to Christianity, Augustine touted these disciplines as handmaidens (*ancillae*) useful for understanding scripture. His later grappling with creation shows more esteem for the problem of understanding the natural world than did many of his philosophical contemporaries.[9] His conviction that the Creator had implanted the seeds of natural causality in all things makes his brand of Christianity look downright empirical among the ethereal Neoplatonisms, Gnosticisms, and mystery religions of his contemporaries.[10]

In Martianus Capella's famous "pagan" encyclopedia, *The Marriage of Philology and Mercury* (ca. 400 CE), the liberal arts also served as handmaidens, this time to Eloquence (*Philologia*). Throughout the early Middle Ages in particular, this work offered a valuable elementary introduction to the four mathematical disciplines.[11] The Roman valuation of eloquence inspired many treatises on grammar and rhetoric, leaving Latin introductions to the other liberal arts or to philosophy in shorter supply. No one felt more keenly than Boëthius the urgency of addressing this defect, but by then it was almost too late. Hoping to translate all of Plato and Aristotle into Latin, he completed several of Aristotle's logical works and Euclid's *Elements*, fragments of which survived into the early Middle Ages. He coined *quadrivium* (the fourfold way) for the Platonic mathematical disciplines of arithmetic, geometry, music, and astronomy, and wrote introductions to them: "whoever neglects these has lost the whole teaching of philosophy."[12]

Boëthius's advice came too late for Rome, but not for posterity, to which he also passed on Aristotle's distinction among the three theoretical parts of philosophy, ascending from "physics" (*physica*) through mathematics to theology.[13] The latter term meant not the revealed tenets of Christianity, but rather the subject matter of Aristotle's *Metaphysics*, or what Augustine called "natural theology" (*theologia naturalis*), that is, rational inquiry into the first cause and being in general.[14] Boëthius identified the "mathematical part" of philosophy with the *quadrivium*, and the "natural part" (*physica*) with natural philosophy, theoretically the study of all change or motion in nature. In late antiquity and the later Middle Ages, natural philosophy effectively came to mean Aristotle's works on the subject, excluding the change-related topics he had omitted (for example, medicine and alchemy).[15] Before the translation of Aristotle into Latin in the twelfth century, however, the *quadrivium* passed for the main areas of learning about nature.[16]

Early medieval scholars preserved definitions and organized summaries but, in Grant's words, "it apparently did not occur to them to pose broad questions about nature and its operations . . . they were not issue oriented."[17] A growing number of exceptions, however, went beyond the encyclopedic tradition, raised questions, or examined specimens. The concerns of Gregory of Tours (d. 594 CE) about the proper celebration of Easter drew him into calendaric work, including some observations of the stars.[18] New research has shown that scholars in the Carolingian era went beyond the calendar, for example, grappling with discrepancies between inherited sources, improving on them, and reviving planetary astronomy, which had no liturgical or calendaric significance.[19] Books of Anglo-Saxon plant remedies displayed a keen knowledge of local flora and an organization useful for identifying plants in the field.[20] Other developments in cosmology and medicine should cause us to pause much longer than we can here.[21] These endeavors, which fell mostly outside the *quadrivium*, were beginning to leave the older taxonomy behind.

From the tenth century onward, the contacts of Latin scholars with Islamic civilization began to change the intellectual landscape. In his cathedral school teaching, Gerbert of Aurillac (d. 1003 CE) used armillary spheres and astrolabes that he had brought back from Spain. These, and related devices such as the abacus, would typify the hands-on *quadrivium* teaching of the eleventh century. The same era witnessed unprecedented developments in the rationalization of law,[22] which motivated the search for more aids to systematic reasoning. Without having an axiomatic deductive system, scholars used hints from Boëthius effectively to construct the criteria for one: "the fact that he described his procedure as that of 'mathematics and other sciences' [*in mathematica . . . ceterisque disciplinis*] led medieval authors before the introduction of Aristotle's *Posterior Analytics* to try to develop a general theory of scientific method from it."[23] Such efforts are signs of the scholars' intense craving for knowledge and for the tools necessary to acquire and organize it.

TWELFTH-CENTURY *PHYSICA*

The enthusiasm of the Latins for translating Arabic works culminated between 1140 and 1180 and stimulated wide-ranging ferment in most categories of learning. In the early twelfth century, Adelard of Bath's (d. 1152) contacts with Islamic civilization led him to think of his own culture as impoverished.[24] Arabic offered access to not only Greek mathematics, natural philosophy, medicine, and logic in translation but also many original products of Islamic civilization in these fields. Nature became an obsession that transpires in the writings, often poems, of the masters associated with the cathedral schools (Bernard Sylvestris's *Cosmographia*, Alan of Lille's *Plaint of Nature*), in the sculpture that decorated the cathedrals themselves, and in law, where natural law took on unprecedented importance.[25]

While lopsided developments in logic were upsetting the traditional balance of the *trivium*, the liberal arts also came under fire from a different direction. In the 1120s, William of Conches's *Philosophia* proposed to "dissolve the marriage of Mercury and Philology." His attack on Martianus Capella's book was both rhetorical and substantive. Beyond his disdain for eloquence, William proposed a clarification of the roles of philosophy and *physica* that was inspired by the new medical and natural philosophical learning associated with Constantine the African and medical teaching in Salerno.[26] As William emphasized, the *philosophus* and the *physicus* each consider the same objects from the created world. The former focuses on the *fact* of their existence (treated as necessary), while the latter considers the *explanation* of their existence (treated as *probabilis*, that is, plausible). Although the first chapter of Genesis asserts the existence of various beings (the *fact*), it does *not explain how* they came into being. William insisted that, in fulfilling his mandate of supplying the missing explanation, the *physicus* was doing nothing contrary to Genesis.[27]

William of Conches's contemporaries understood what he meant by *physica*: "investigating the causes of things in their effects and the effects from the causes."[28] Not everyone liked it, however. One opponent attacked his "philosophical, or better, physical" approach to the Biblical account of creation.[29] Even John of Salisbury thought that some of "the *physici* . . . exalt beyond measure the power of nature" and went too far in disputing about the soul, the resurrection, and creation—to the detriment of the faith. Others disliked the *physicus*'s skepticism about miracles.[30] The times were changing. Augustine had valued the liberal arts in order to understand scripture (as did his followers). For William, however, scripture did not explain how the world came to be. This lacuna called for *physica*, a causal natural philosophy. He was not shy in castigating his critics: "Ignorant themselves of the forces of nature and wanting to have company in their ignorance, they do not want people to look into anything; they want us to believe like peasants and not ask the reason behind things."[31]

In the Roman and early medieval worlds, natural philosophy had been a taxonomic rubric more than a full-blown body of knowledge. By the early twelfth century, natural philosophy emerged as a program for the causal explanation of nature and developed some content. To twelfth-century eyes, Plato's *Timaeus* was particularly exciting: his use of Pythagorean ratios, triangles, and Platonic solids to explain the origins of the cosmos made him the natural philosopher of the era. The pump was primed: between 1140 and 1180, a massive influx of works recently translated from the Arabic filled the category of natural philosophy with so much new content that it was overflowing.

Meanwhile, drawing on his own exposure to Arabic learning, Adelard of Bath was adding new depth to the *quadrivium*, to which contemporaries paid scant attention.[32] Adelard's early-twelfth-century translation of Euclid's *Elements* replaced the fragments of the Boethian translation. Along with the new vision of a causal natural philosophy, scholars now had access to the archetype of an axiomatic, deductive, mathematical system, and to the crucial prerequisite for understanding astronomy, optics, and mechanics.

NEW CLASSIFICATIONS OF LEARNING: FROM *ARTES* TO *SCIENTIAE*

Early on, Adelard of Bath had advocated reliance on reason rather than authority (understood as bookish learning and as revelation), and had insisted on letting nature do the teaching. But he too depended heavily on the textual tradition and advanced it.[33] The transformations in knowledge that he envisioned required many new books. Their proliferation helps to explain the enthusiasm for classifications of learning, an odd activity best understood as an attempt to cope with the flood of new materials and of specialized schools "of grammar, of arts, of divinity, of civil or canon law, of medicine, and more."[34]

Reports of ancient books in Spain led scholars like Gerard of Cremona to Toledo, where he learned Arabic and translated dozens of works into Latin. His translation of Ptolemy's *Almagest*—the work that drew him to Spain—would do for astronomy what Adelard's Euclid had done for geometry a few decades earlier. Two-thirds of the newly translated treatises belonged to the nonmusical mathematical sciences.[35] The remainder included generous portions of Aristotelian works in logic and especially natural philosophy. By the early thirteenth century, almost all of the Aristotelian corpus was available in Latin. With this new material so heavily focused on the natural world, it is scarcely surprising that *scientia* increasingly developed naturalistic connotations as well.

New knowledge was bursting the seams of the traditional categories used to circumscribe it. In the later twelfth century, *ars* ("art") took on the additional meaning of "method" (that is, a system of rules) while *disciplina* could both refer to "theory" (*doctrina speculativa*) and serve as synonym for *ars* and *scientia*.[36] Whereas *physica* continued to designate the study of nature generally (from natural philosophy to medicine), the expression *scientia naturalis* and its cognates (*scientia naturae, scientia de natura*) surfaced with increasing frequency in the later twelfth century (Daniel Morley) and especially after the thirteenth (Robert Kilwardby, Pietro d'Abano, and others). Kilwardby defined *naturalis scientia* as the branch of theoretical knowledge pertaining to "knowledge of mobile/changeable body in respect of its mobility/changeability" (that is, the change or motion, not the body, is the focus).[37]

The translation of al-Fārābī's *De scientiis* had prepared this transition by dividing "natural science" (*scientia naturalis*) into eight parts, each identical to the title of an Aristotelian natural philosophical book: *Physics, On the Heavens, On Generation and Corruption, Meteorology, On Animals, On the Soul, On Plants* and *On Minerals*.[38] This list would soon swell the "arts" curriculum at Paris in the mid-thirteenth century, and many other universities besides.

The association of the *trivium* with the "verbal sciences" (*scientiae sermocinales*) and of the *quadrivium* with the "real sciences" (*scientiae reales*, that is, devoted to things, *res*), mediated the transition from the language of the liberal arts to that of the sciences. The category "real sciences" underwent extensive transformations. This new, more flexible division could encompass all three of Aristotle's theoretical sciences/philosophies, with a multiplicity of subdivisions under *physica* and *mathematica* in particular.[39] Moreover, the category of *scientiae reales* permitted the inclusion of such non-liberal arts as medicine, law, metaphysics/theology, as well as the practical *scientiae* of ethics, politics, and economics. In the late twelfth century, Hugh of St. Victor, who wrote *Practica geometriae*, had already classified the mechanical arts under philosophy. Although traditionally opposed to the liberal arts for being "adulterine," they nevertheless contained knowledge worthy of being called *scientia*. Twelfth-century interest in these areas surfaces in the new coinages *ingeniarius* and *ingeniator* to designate builders of war machines.[40]

The influx of new material generated complaints about the old classification. As one early-thirteenth-century author noted, the liberal arts omitted natural philosophy, metaphysics, and poetry.[41] For Aquinas at midcentury, "the seven liberal arts do not adequately divide theoretical philosophy,"[42] even as Robert Kilwardby's lengthy *Origin of the Sciences* discussed forty different specialized *scientiae*. Kilwardby took Aristotle's division of all propositions into ethical, physical, and logical (in the *Topics*) as his framework.[43] Here, the category "physical" encompasses all three divisions of the speculative sciences (theology/metaphysics, *physica*, and mathematics), all understood as divine (that is, its objects are not created by man). Dante's *Convivio*, aimed at readers of the Tuscan vernacular, organized the "sciences" (*scienze*) according to the structure of the cosmos. After matching each of the liberal arts, from grammar through astronomy/astrology, to its planetary sphere, Dante associated the three remaining spheres with the "theoretical philosophies," from *scienza naturale* (or *fisica*) to theology or "divine science."[44]

UNIVERSITIES AS INSTITUTIONAL CLASSIFICATIONS OF LEARNING

These theoretical schemata are informative, but they must take a back seat to practice. Already before the translation movement, several towns were renowned for the specialization of their learning: medicine in Salerno, law in Bologna, the arts and theology in Paris, for example. During the thirteenth century, not only these specialties but also the bulk of the translated Greco-Arabic material would find their places in a new type of institution. The *studium generale*, or university, made knowledge its main business on an unprecedented scale. The model caught on quickly. New universities multiplied in the fourteenth and fifteenth centuries, numbering some sixty by 1500.[45] It is difficult to exaggerate their importance as institutional homes for the pursuit of natural philosophy, medicine, and the mathematical sciences, as well as the crucial ancillary disciplines of logic and mathematics.

The organization of the universities embodied a new classification of learning initially subdivided according to two predominant models. The northern European one, patterned on Paris, consisted of a faculty of arts and three higher faculties (medicine, law, and theology). The northern Italian pattern associated with Bologna had grown out of an eleventh-century school of law. In the mid-thirteenth century, Bologna put law in one faculty, the "arts" and medicine in another, adding theology in the fourteenth century.[46]

By naming their introductory and largest faculty for the "arts," the universities institutionalized a terminology that was almost as out-of-date then as it is today (bachelor of arts, and so forth). By the mid-1250s, the faculty of arts at Paris required Aristotle's *libri naturales*, which fit neither the *trivium* nor the *quadrivium*. The disciplines of the *trivium* were taught, but the many new subspecialties of logic eventually outnumbered grammar and rhetoric. Obsolete or not, the "liberal arts" would remain a cliché of academic speeches for centuries.[47]

While thirteenth-century masters were sometimes called *artistae*, they thought of themselves generically as philosophers, *philosophi*, particularly at Paris, where some deliberately set themselves against the theologians. Tellingly, the bishop of Paris's Condemnation of 1277 included the propositions: "That the only wise men of the world are philosophers" and "There is no higher life than the philosophical life." When the masters changed hats, they adopted more specific designations for themselves and their colleagues: *logicus*, *physicus*, *perspectivus*, *astronomus*, and the like.[48] Some of them clearly specialized as authors or teachers, especially in the mathematical sciences and medicine. In Paris, John of Sacrobosco's popular textbooks were exclusively astronomical and mathematical, while questions about, and texts of, the "intermediate sciences" are often grouped in academic manuscripts, suggesting that the latter were studied, and probably taught, together.[49] In Vienna, John of Gmunden (d. 1442) chose to specialize in teaching mathematics and astronomy for two decades. By the early fifteenth century, Bologna and Kraków each created a first chair, then a second, dedicated to astronomy/astrology. From the thirteenth and fourteenth centuries, specialized treatises on anatomy and the dissections that accompanied them (Paris, Bologna) preceded the establishment of new chairs in anatomy and eventually surgery.[50]

Symptoms of specialization also transpired in genres. The chief format for inquiry in natural philosophy was the *quaestio* ("whether local motion can produce heat," "whether elements persist formally in a mixed body," et cetera).[51] Technical astronomical and optical work was often organized in treatises, rather than in the "questions" more familiar in natural philosophy and cosmology. Even elementary works in this tradition were treatises (for example, Sacrobosco's *Tractatus de sphera* and Langenstein's *De reprobatione ecentricorum et epicyclorum*).

Masters who practiced natural philosophy, the mathematical sciences, or medicine typically had a strong sense of their methodological autonomy. Their explanatory naturalism therefore did not disappear when they went on in theology. Already in the twelfth century, Gilbert of Poitiers had pointed to the different meanings of some words when used in theology and natural philosophy. In the thirteenth century, both the theologian Albertus Magnus and the master of arts Siger of Brabant claimed the right and the duty of ignoring miracles and the faith when they were reasoning as *physici* (natural philosophers). John Blund (d. 1248) saw no problem in disagreeing with saints about the natural world. In matters of faith, saints were mouthpieces of the Holy Spirit, but in "natural matters" (*de rebus physicis*), they spoke as men, and could therefore be deceived.[52] In the fourteenth century, the theologian Nicole Oresme discussed many wonders that, he argued, either had or would eventually have physical explanations: "There is no reason to take recourse to the heavens, the last refuge of the weak, or to demons, or to our glorious God as if he would produce these effects directly, more so than those effects whose causes we believe are well known to us." Among the learned, the belief in the natural necessity of secondary causes was wide-ranging, from the use of mechanical models in astronomy to the striking naturalism of the rabbi and astronomer Levi ben Gerson, who excluded miracles above the sphere of the moon when interpreting scripture. Conclusions about the natural world were cordoned off from miracles by such expressions "according to physics," "speaking naturally," or "according to the light of natural reason."[53] The whole point of these activities was to solve problems only with criteria accessible the senses and the intellect. Only in such an environment could the arguments of pagans like Aristotle and Ptolemy and infidels like Averroës and Ibn al-Haytham command such respect.

The bounds of naturalism were nevertheless contested. At the behest of conservative theologians, the bishop of Paris's Condemnation of 1277 took issue, among other things, with the determinism of many propositions circulating in the arts faculty, as well as with the general uppitiness of the masters of arts. At stake here and elsewhere was the place of the new learning in the hierarchy of knowledge. Although immersed in the Aristotelian revival and very attuned to mathematics, Roger Bacon also sympathized with an Augustinian hierarchy of knowledge. Unlike William of Conches, he believed that scripture contained all wisdom and that the various branches of philosophy served the faith as handmaidens, a nostalgic perspective that surfaces now and again into the late fourteenth century.[54] That role was fading, however: members of the faculties of arts thought and behaved not as handmaidens, but as autonomous agents.

Well they might have. Since attrition rates hovered around 80 to 90 percent between matriculation and the master's degree, most students never entered a higher faculty, let alone theology. Even without a degree, their sojourn in the "arts" faculty nevertheless advanced their careers. The analytical skills and advanced literacy they gained signified competence.[55] The universities thus diffused the inquiry into nature throughout the culture, for they exposed vast numbers of students, who would run the growing princely, ecclesiastical, and town bureaucracies, to disciplines and texts that treated the operations and structures of the natural world. Courts did not want to be left out of this trend: King Charles V of France, for example, commissioned several translations of university texts into French. Some 250 of the 1,239 manuscripts in his library belonged to the *quadrivium*, the second largest category; 40 to 60 each were natural philosophical and medical or surgical.[56]

The logic and natural philosophy of the "lower" faculty shaped the higher faculties profoundly, for the "step" (*gradus* = degree, step) structure of university curricula required an advanced arts degree for admission to a higher faculty. The questions that theological and medical masters raised, respectively, about the scientific status of theology and the natural philosophical status of medicine show strikingly how much they had internalized the criteria of the arts faculty and used them to analyze and to justify their own knowledge claims.

THE RANGE OF *SCIENTIA*

By the early thirteenth century, the translation of Aristotle's works, the impressive Arabic commentaries on them, and much original Arabic medical, natural philosophical, and mathematical material heightened the interest of questions about the scientific status of specific areas of inquiry. The central treatises of Aristotle's natural philosophy—the *Physics*, *On the Soul*, and *On the Heavens*—provided concrete examples not of dogmatic expositions but of natural knowledge in the making. These were reasoned arguments that critically evaluated multiple competing possibilities before choosing a final position. Meanwhile Aristotle's treatise on scientific demonstration (the *Posterior Analytics*), along with the interest in Euclid and Ptolemy, focused much attention on standards of knowledge and deductive methods in particular.

This new material transformed the way intellectuals thought about the knowledge of nature. After some initial opposition, it was hard for them not to call most of it *scientia*, and they did.[57] In literary contexts, *scientia* generically meant "knowledge." In philosophical contexts, *scientia* often meant "knowledge attained by demonstration," following the *Posterior Analytics'* ideal for the presentation of ascertained, mostly natural, knowledge (Aristotle's examples came largely from mathematics, natural philosophy, and the mathematical sciences). Indeed, to denote such demonstration, the twelfth century coined the adjective *scientificus* ("knowledge-producing"), which Dante brought into the Tuscan vernacular (*scientifico*).[58] Just as often, however, *scientia* was accompanied by specific qualifiers denoting one discipline (*scientia stellarum, scientia de motu, scientia medicine, scientia dialecticae, scientia de plantis, scientia de ingeniis*, and so forth). In such situations, the term designated a coherent body of specialized knowledge, without necessarily presupposing a deductive or demonstrative framework (for example, "Is there a science concerning generation and corruption?"). At Bologna, official documents used *scientia* as a synonym for faculty (*scientia medicine*).[59] Although the term *scientia* thus extended from general connotations to field-specific bodies of knowledge analogous to what we now call sciences, its range caused little confusion in context.

For Kilwardby and others, *scientia* alone could be a synonym for *physica* ("natural philosophy" or "the study of nature"), reflecting the subject's importance and omnipresence in the faculty of arts. For a younger contemporary of his, philosophy was divided into the mechanical and liberal sciences (not "arts"); the latter were further divided into "natural philosophy broadly understood," moral philosophy, and rational philosophy (that is, the disciplines of the *trivium*). The category "natural philosophy broadly understood" in turn encompassed metaphysics, the "mathematical sciences" (the *quadrivium*), and natural philosophy in the strict sense, a schema that Bonaventure also used.[60] As these schemata suggest, there were many medieval views of the matter.

The wide applicability of the term *scientia* was surely one reason for the emphasis on classification: the proliferation of new species demanded a fresh taxonomy. These later-medieval discussions often disguised methodological debates. Which tools and methods were appropriate to a particular inquiry? And what was their relation to the tools of related *scientiae*?

Not least, the deductive ideal of demonstration raised important related questions about certainty in specific *scientiae*: Where did the foundation of certainty lie? And how could one get from there to the various sciences?

FROM THE *QUADRIVIUM* TO THE INTERMEDIATE SCIENCES (*SCIENTIAE MEDIAE*)

Aristotle had already raised such disciplinary questions by calling astronomy, optics, and harmonics "the more physical of the mathematical sciences" (*Physics* 2.2). As one of the first Latin commentators on Aristotle's *Physics* and *Posterior Analytics*, Robert Grosseteste (d. 1253) explored the gray areas in these works, notably the relations of mathematics and natural philosophy. He articulated a framework for their interaction that would resonate into the later Middle Ages. He held that, for disembodied intellects, theology offered greater certainty than mathematics and *physica*. For us here below, however, mathematics yielded greater certainty, an echo of Ptolemy's *Almagest*. But Grosseteste had also read Aristotle. He therefore treated disciplines like astronomy, music/harmonics, and *perspectiva* not simply as mathematics, but as composite subjects: "purely mathematical subjects to which are added natural 'accidents'" (nonessential properties). He used the expression "subalternate sciences" (*scientiae subalternatae*) to convey Aristotle's view: astronomy and optics/*perspectiva* were sciences subalternated to geometry, harmonics/music to arithmetic.[61]

What does this mean? A science is subalternate to another when it borrows some principles from that other (the "subalternating" science). Take a planetary sphere (from astronomy) or a light source (from *perspectiva*). Although each is a physical entity, aspects of each—the sphere and the rectilinear ray, respectively—can be abstracted for geometrical treatment. Since the principles that govern spheres and straight lines come from geometry (not astronomy or optics), geometry is the subalternating science in each case.

It is crucial to emphasize that Grosseteste gave the mathematical sciences a significant causal role. Echoing Aristotle (*Posterior Analytics* 1.13), he argued that when the natural philosopher (*physicus*) and the "perspectivist" (*perspectivus*) each considers the rainbow, the former knows only the fact (*quia—that* it is) while the latter knows its cause (*propter quid—why* it is).[62] The practitioner of a subalternate science that draws on mathematics thus possesses causal knowledge (the most valued kind); the practitioner of *physica* (natural philosophy) in this case does not. The combination of mathematics and natural philosophy thus gave optics a powerful explanatory boost, a thesis that Roger Bacon would also defend passionately.[63] If one was interested in causal explanations, there was nothing inferior about the "subalternate" mathematical sciences.

There were many schemas of subalternation.[64] Grosseteste's analysis of Aristotelian subalternation effectively split up the old *quadrivium* between mathematics (arithmetic and geometry) and the subalternate sciences of astronomy and musical theory wedged between mathematics and natural philosophy. Aquinas (d. 1274) called this category the "intermediate sciences" (*scientiae mediae*), those that "apply mathematical principles to natural things."[65]

Were these sciences more mathematical or physical? There was no clear consensus. In one passage about astronomy, Bacon cast his lot with the *naturales mathematici* (physical mathematicians) against the *puri mathematici* (pure mathematicians) who neglected the physical. While some loosely lumped the intermediate sciences with natural philosophy, others called them "intermediate mathematics" (*mathematica media*).[66]

THE EXPANSION OF THE "INTERMEDIATE SCIENCES"

Following the translations, older classification schemes were cracking. The Roman *quadrivium* had been oblivious to optics, which was already sophisticated after Euclid and Ptolemy and became vastly more so after Ibn al-Haytham. A composite thirteenth-century list of "intermediate sciences" would include newer disciplines like the "science of machines" (*scientia de ingeniis*) or the "science of weights" (*scientia de ponderibus*).[67] Aquinas had specifically excluded the motion of corruptible bodies from mathematical treatment (and the intermediate sciences) while conceding exceptional status to astronomy, whose incorruptible entities move but are treated mathematically.[68] This was not the final word, however. Recent successes in the classic intermediate sciences were hard to ignore. Theodoric of Freiberg's mathematical explanation of refractions and reflection in raindrops beautifully illustrated the power of geometry to explain the rainbow, which had been a natural philosophical problem for Aristotle. Successes such as these stimulated some fourteenth-century natural philosophers to theorize about problems of motion in mathematical terms. The approaches of these *calculatores* associated with early-fourteenth-century Merton College, Oxford, spread to mid-fourteenth-century Paris, then to the late-fourteenth-century German universities, and to fifteenth-century Italy.[69]

The two most famous of these approaches deserve mention. The "latitude of forms" quantified changes in qualities (for example, degrees of heat, speed, whiteness, and so forth) and analyzed them as a function of time. A second approach, "the proportions of speeds in motions" associated with Thomas Bradwardine, criticized Aristotle's treatment of speeds in relation to forces and resistances and used proportion theory to develop original solutions to this and related problems. Both brought mathematical tools to bear on natural philosophical problems.[70] In their own lists of "intermediate sciences," Henry of Langenstein (d. 1397) included the "latitude of forms" and Regiomontanus (d. 1476) "the proportions of speeds in motions." At the interface between mathematics and natural philosophy, the intermediate sciences were clearly a growing category.[71] These are but two of the new technical "analytical languages"[72] that characterized the era. Scholars created new logical and logico-mathematical techniques that they put to use in many facets of natural philosophy and even theology. They transformed the character of natural philosophy and theology so deeply that one historian sees in just one of them "a new intellectual factor . . . a typical change of mentalities."[73]

Not everyone appreciated these innovations or their diffusion. The sixteenth-century Aristotelian natural philosopher Pietro Pomponazzi, for example, criticized the *Liber calculationum* of Richard Swineshead, the Oxford *calculator*, for its excess of mathematics in natural philosophy, and disapproved of this new hybrid.[74] Galileo's analyses of free fall and projectile motion would eventually show just how pregnant such hybrids could become.

FIRST PRINCIPLES OF NATURAL KNOWLEDGE

Grosseteste's analysis of the intermediate sciences had presupposed that natural knowledge emerges, at least in part, from a deductive framework. Any conclusion must properly be able to trace its roots to reliable first principles. Those who worried about the ultimate source of principles necessary to ground all the sciences faced several options.

For many, especially thinkers in the Augustinian tradition, Christian theology seemed the best repository of the highest principles. As first cause of everything and guarantor of all knowledge, God would surely ground first principles securely. But others insisted on distinguishing Christian theology from a metaphysics or "theology" of first principles that, like Aristotle's, was accessible to all rational thinkers, regardless of creed. This distinction was effectively built into the structure of the universities. Masters of arts who were not studying theology could not treat (revealed) theology, but they routinely commented on the "theology" in

Aristotle's *Metaphysics*, a standard work in the faculty of arts. A third approach sought the foundations of specific sciences not in a higher science, but in experience. Indeed, a well-documented bevy of thinkers at Oxford treated the principles of natural philosophy in precisely this way.[75] Although schematic, this spectrum of approaches illustrates the very different ways in which medieval thinkers who all admired the *Posterior Analytics* sought to ground their natural knowledge.

Indeed, attitudes toward Aristotle's *Posterior Analytics* warrant a closer look. This work clearly set out a goal of universal and necessary natural knowledge that found a receptive audience in the early thirteenth century: "demonstrative knowledge" certainly counted as scientific. But was *only* knowledge of this sort scientific? As Eileen Serene has emphasized, the enduring enthusiasm for the Aristotelian *ideal* of demonstration gives a false impression of unanimity among late medieval thinkers. Their views of truth, necessity, and certainty—the key components of demonstration—varied widely.[76] Hard thinking about the problems that Aristotle had raised (and others that his obscure points generated) introduced innovations, doubts, and emendations.

These developments show what an oversimplification it is to identify natural-scientific knowledge in the later Middle Ages exclusively with Aristotle's ideal of demonstration, or with causal knowledge, or with universal propositions, or with knowledge of the "essential natures of things." In fact, late medieval thinkers took some remarkably un-Aristotelian positions. Kilwardby already recognized that many cognitions appropriately called *scientia* were probable rather than certain, especially in natural philosophy. To be sure, in his hierarchy of true and certain knowledge, metaphysics and mathematics tied for first place, as the dignity of the former's subject matter vied with the certainty of the latter's mode of demonstration. *Physica* came next, followed by *ethica*; *mechanica* occupied the lowest level of certainty. Kilwardby conceded that "natural philosophy [*physica*] does not always conclude from necessary [propositions] . . . it often concludes by means of probable [propositions] that, in truth, are false." For him, *physica* could attain certainty not in all, but in many contingent things, for the latter display great regularity (unlike the more erratic arts).[77]

As ideals, certainty and universality did not disappear, of course, but it is crucial to understand that not everyone thought they were required for scientific knowledge. Bacon held that "the certitude of a particular experience of singulars is more perfect and greater than that of a universal experience." John Duns Scotus (d. 1308) held that experience could make first principles evident. But he also believed that events (including natural ones) occur because many *particular* causes act together. In addition, the unaided human mind could have direct, certain, and evident knowledge of *singular* objects. Such views seriously undermined the status of the universal-oriented demonstrative syllogism as the sole instrument of scientific knowledge; they were, in Marrone's words, "dynamite for established Aristotelianism."[78]

The creativity of the fourteenth century also extended logic beyond Aristotle's syllogistic to include particular claims derived from experience and sense perception. It was possible to gain demonstrative knowledge not simply of universal propositions but also of particular ones. William of Ockham (d. 1349) discussed a new type of syllogism, the expository syllogism, designed expressly to produce nonuniversal conclusions. Even though it has received little attention, "this may well be termed a revolutionary innovation."[79]

Nor did all medieval natural philosophers seek knowledge of the "essential nature" of things. John Buridan (d. after 1358) gave that job to the metaphysician, not the natural philosopher: "no science besides metaphysics has to consider the essential nature of a thing 'simply' . . . the *physicus* need not know what a man is or what an ass is 'simply'; he may describe such things by means of some changes or some operations."[80] Specialization had its advantages. After giving the metaphysician his due, the natural philosopher could study change without addressing ultimate reality. It is not a coincidence that Buridan was also the leading logician of the later Middle Ages, who showed that Aristotle's syllogistic was in fact a subset of "consequentiae," a more general logic of implication (p implies q) whose variables are not terms but propositions.[81] By the mid-fourteenth century, medieval natural philosophy had, and used, tools other than the syllogism.

PRACTICE, PRACTITIONERS, AND BOUNDARIES

The emerging universities had enthusiastically embraced the new natural philosophy, medicine, and (on a smaller scale) the mathematical sciences. Even at Paris, the site of most early opposition to it, Aristotelian natural philosophy had become a central curricular requirement by the mid-thirteenth century. To an unprecedented degree, the study of nature thus acquired a secure home, not in one institution, but in a proliferating *species* of institution. As universities multiplied and their enrollments rose in the fourteenth and fifteenth centuries, their curricula exposed several hundred thousand individuals to the study of nature, making it an ineradicable part of the European cultural landscape.[82]

From a quantitative point of view natural philosophy was more than the heart of the university's approach to the study of nature; it also strongly influenced such related endeavors as academic medicine and theology and even the philosophical courses in the *studia* of some mendicant orders.[83] Its required texts were more numerous, and its lecture sequences were often longer than those in the mathematical sciences, employing more masters, generating more commentaries, and enrolling more students. The universities thus set the tone for the way later medieval masters practiced and advanced natural philosophy and, to a lesser extent, the mathematical sciences as well.

How, then, was natural philosophy practiced? As its main task was the critical scrutiny of assumptions, arguments, and conclusions about all kinds of change, its cast was largely, but not exclusively, verbal. This scrutiny, which was logic-intensive, took place both in oral disputations[84] and in written lectures, questions, commentaries, and expositions. Disputations, which centered on a particular question for example, "whether the world is eternal"), were required of both students and masters. Student disputations took place almost daily in some instances, before peers who were penalized for failing to participate. The more sophisticated masters' disputations were usually weekly events, which younger students were also required to audit for two years.

In most fields, the disputation and the disputed question served both to teach and to advance the state of knowledge. (As their medium, the mathematical sciences often used the treatise rather than the question, although exceptions do exist). Students witnessed the analysis of problems from several points of view, with objections and responses, and encountered genuine disagreements among masters, who sometimes changed their minds.[85] Although it was later stereotyped as empty haggling, masters valued the disputation highly as the road to truth. It earned the praise of Jewish observers of the university, some of whom adopted its approach in their own writings.[86] This oral culture left its mark on the texts of natural philosophy, often framed as disputed questions that spar with actual or imagined opponents.

In this era, much "science was *livresque*" (bookish), "not just set down in books; it was largely carried out in books."[87] For Ockham at least, the verbal cast of this enterprise was a principled position: "properly speaking, *naturalis scientia* is not about corruptible and generable things, nor about natural substances, nor about moveable things, for such things are subjects or predicates in no conclusion known by *naturalis scientia*."[88] Rather, natural knowledge consisted of *propositions*, which in turn consisted of words signifying things, concepts, and/or other words. It was the careful scrutiny of propositions using rigorous logical analysis that led Ockham to conclude (among other things) that motion was not itself a thing, but a name for the behavior of a thing.[89]

Murdoch's expression "natural philosophy without nature" nicely makes the point that many medieval natural philosophers were more interested in conceptual analysis than the direct observation of nature or experimentation.[90] A key feature of the enterprise in the fourteenth century was the exploration of hypothetical scenarios, sometimes with surprising results. After exploring the hypothetical rotation of Earth, Nicole Oresme confessed that he could find no compelling rational or empirical evidence either to prove or to disprove it. He also analyzed the behavior of a stone falling through a hypothetical tunnel along the diameter of Earth, concluding that the stone would oscillate with a pendulum-like motion from one side to the other.[91]

If much medieval natural philosophy, like most theory today, was bookish, not all of it was—to say nothing of medicine and the mathematical sciences. Some books testify to their authors' wrestling with nature through observation and experimentation. In the "science of plants," the "science of stones," and the "science of animals" (considered by most as belonging to natural philosophy), observation played a nontrivial role. Knowledge acquired from local observation appears in Hildegard of Bingen's writings on fungi, birds, and animals from the Rhineland. Albertus Magnus systematically observed the bat and experimented on the behavior of ants whose antennae he had removed—arguably the first since Theophrastus to take up Aristotle's "causally oriented study of the 'animals and plants around us.'" A hunter who spent hours observing raptors and their prey in their habitats, Holy Roman Emperor Frederick II (d. 1250) criticized Aristotle in his detailed *On the Art of Hunting with Birds*. He also studied avian anatomy and determined experimentally that vultures find carrion by sight, not smell.[92] Clearly, Aristotle was not treated as infallible. His positions often framed the box, but fourteenth-century natural philosophers also thought outside it. They criticized, and proposed alternatives to, his views on many topics, from new rules of motion and explanations of the acceleration of falling bodies to discussions of the possible rotation of Earth, which he had held to be impossible.[93]

The intermediate sciences relied on the book, but also went beyond it. While most *astronomi/astrologi* could make predictions using only astronomical tables in a windowless room, the minority who worried about the foundations of the "science of the stars" did appeal to the sky. The new, late-thirteenth-century *Alfonsine Tables* professes a rhetoric of long-term observation in Spain, even though few new observations shaped them.[94]

But in 1292 in Paris, William of St. Cloud measured the precession of the equinoxes, leading him to reject the variable rate of the Arabic theory of trepidation. Once the *Alfonsine Tables* entered Paris, Johannes de Muris both modified them and tested their accuracy (he built a giant quadrant with a radius of more than four meters). He and nine colleagues observed the solar eclipse in 1337 with "several good astrolabes."[95] The *Provençal* rabbi and *savant* Levi ben Gerson (d. 1344) initiated in Europe interior meridian-line solar observations, determined the solstice, invented the "Jacob's staff" to facilitate longitudinal measurements, and used his own observations to criticize Ptolemy's lunar theory, to which he created an alternative.[96] Johannes Regiomontanus (d. 1476) also criticized the failure of existing tables to predict planetary positions accurately and of Ptolemaic theory to predict planetary sizes correctly.[97]

The literature on experiment in the Middle Ages is growing. A widely diffused late-thirteenth-century alchemical treatise, pseudo-Geber's *Summa perfectionis*, involved procedures that were fully experimental, since they were used as tools of analysis, discovery, and identification. Significantly, its approach to theory was empirical rather than metaphysical, drawing on corpuscular themes in Aristotle's work.[98] Theodoric of Freiberg used artificial apparatus (probably a spherical urine glass) to model the path of sunlight in the raindrop, and geometry to explain the primary and secondary rainbows. Peter of Maricourt devised a host of experiments to produce the first systematic study of the magnet.[99] Anatomy, once tied to craft and under the control of barber-surgeons, also became a subject in late-medieval medical faculties, as the practice of dissection acquired pedagogical utility.[100]

Whereas "lifers" such as Siger of Brabant (d. ca. 1280), John Buridan, and John of Gmunden spent entire careers teaching in the arts faculty, most masters of arts taught but a few years, others for a decade or more.[101] The few who entered medicine or theology had the greatest opportunity to extend their study of the "natural books" or the mathematical sciences at the university. Like today's graduate assistants, they survived for years by teaching in the arts faculty, while studying in the higher faculty.[102] Many kept abreast of natural philosophy, remained engrossed by its questions, and practiced its methodological autonomy from theology. When Aquinas commented once again on Aristotle's *Physics* shortly before his death, it was presumably because the subject still fascinated him, not because his theology needed a handmaiden in extremis.

A few years later, driven by conservative theologians and enacted by the bishop of Paris, the Condemnation of 1277 sought to assert the dominance of theology over philosophy by challenging some limits on philosophical reasoning; in effect, it also sealed the existing methodological separation of the faculties of arts and of theology at Paris. It forbade masters of arts who were not also theologians from making theological pronouncements. Natural philosophy was about the ordinary course of nature and human reason; it was not about God, not even when the small minority of masters of arts who became theologians engaged in it.[103] In effect, the interface between natural philosophy and theology was a one-way valve. Long trained in the faculties of arts, the theologians knew their natural philosophy as well as anyone and they used it.[104] In this way, natural philosophical issues, assumptions, and techniques deeply influenced many academic discussions in theology, medicine, and even law.[105]

THE CONTESTED PLACE OF THEOLOGY IN THE HIERARCHY OF KNOWLEDGE

The oft-repeated claim that, in the Middle Ages, theology was queen of the sciences is widely misunderstood and flatly false without substantial qualifications. Everyone boasted of the superiority of his own discipline, of course. When the Parisian theologians did so, particularly in their struggles of the 1260s and 1270s, they were staking a claim, not necessarily settling a dispute.[106] Whereas everyone conceded the supreme dignity of the *object* of theology (God), not everyone conceded the superiority of theology as a way of knowing. Indeed, theologians themselves often asked explicitly "whether theology is a science," and their answers varied.

Both the question and some answers to it illuminate the extent to which the philosophical criterion of natural reason set the terms of the debate in other faculties. Like *scientia, theologia* has various context-dependent meanings and some ambiguity. It was Aristotle—no Christian—who had first ranked *theologia* ("first philosophy" or metaphysics) as the highest of the sciences because it dealt with the first principles of being. It was this discipline that Aquinas also called *"scientiarum rectrix et regulatrix"* (the ruler and governess of the sciences), and he did so while commenting on Aristotle's *Metaphysics*. In short, he was explaining a pagan philosopher, not outlining a quintessentially medieval Christian view.[107]

The ambiguity of "theology"—metaphysics versus the principles of Christian faith—had already raised the hackles of some thirteenth-century theologians. For Alexander of Hales, "First Philosophy [= metaphysics] . . . is the theology of the philosophers, and concerns the cause of causes"—not to be confused with Christian theology, as Albertus Magnus agreed.[108] From the late thirteenth century to around 1340, when theologians asked "whether theology is a science," they meant: Do the premises, arguments, and conclusions of the new, systematic, logic-intensive discipline of Christian theology meet the criteria of Aristotle's *Posterior Analytics*?[109] Not even the supreme dignity of its object exempted theology from methodological scrutiny in the university. While Aquinas argued that theology was a science, Henry of Ghent and Ockham denied that it was.[110] For Ockham, the principles of a science must be better known than its conclusions. But the principles of theology are the articles of faith, which, as he repeatedly noted, are so far from self-evident that they appear "false to all, or to the majority, or to the wisest."[111] Their being known only to God or the blessed disqualified them as premises in a demonstration. Theology was therefore not a science.

Was Aquinas's theology therefore a pseudoscience? For Ockham, science just is true, so the notion of false (= pseudo) science is, strictly speaking, a contradiction.[112] Kilwardby nevertheless closely approximates the concept of a pseudoscience. In the science of the stars, he distinguishes three parts: a mathematical part called astronomy and two astrological parts, the first of which is natural while the other is "quasi-natural—not truly natural, but mendacious and superstitious."[113]

Wherever one drew its boundaries, the science of the stars offered theology ever more serious competition. Figures from Bernard Sylvester in the twelfth century, through Pietro d'Abano in the fourteenth, to Johannes Regiomontanus and Francesco Capuano da Manfredonia in the fifteenth all placed astronomy/astrology at the apex of the sciences.[114] By the fifteenth century, astronomy/astrology was much in demand in courts and cities as well as in universities, some of which even founded new chairs in the subject (such as Bologna and Kraków).

CONCLUSION

In the early Middle Ages, scholars did their best to reconstruct a picture of natural knowledge from the Latin remnants of classical natural philosophy, mathematical science, and medicine. They soon developed a thirst for more. The many twelfth-century translations of new Greek and Arabic texts drastically increased the available material. By the mid-thirteenth century, Aristotle's "books about nature," supported by his works on logic and method, had become the dominant means of studying nature in Europe. Natural philosophy had found a permanent home at the heart of the new universities, where masters of arts cultivated them and brought them to bear on their work in the higher faculties. Medicine and such mathematical and mixed sciences as optics, astronomy, and the science of weights also flourished in or near the universities, among the minority of scholars who specialized in these subjects.

The thirteenth century assimilated this material so successfully that innovative trends had appeared already at midcentury. It was during the fourteenth century, however, that masters at Oxford and Paris in particular extended natural philosophy into new territory. In criticizing Aristotle's views on motion, they brought proportion theory, geometry, and theoretical measurement to bear on the problem of change and other areas of natural philosophy. Their work expanded the scope of the intermediate sciences into territory that would have surprised not only Aristotle, but also their thirteenth-century predecessors. In the later fourteenth and fifteenth centuries, new universities sprang up in the Holy Roman Empire and in eastern Europe. The masters who spearheaded this institutional growth drew on recent scientific developments in the older universities and diffused them widely. Masters of arts who took advanced degrees in medicine and theology brought their learning to bear on their new studies. By the fifteenth century, the spread of the universities and their growing enrolments had rooted the study of nature permanently in European culture.

NOTES

I thank Ann Blair, Joan Cadden, Victor Hilts, David C. Lindberg, Lynn Nyhart, Evgeny Zaytsev, and my fellow editors for comments and criticisms, and the Institute for Research in the Humanities at the University of Wisconsin-Madison for its incomparable support. Uncredited translations are my own.

1. During this period, the meaning of *natura* also was far from invariant; see Tullio Gregory, "L'idea di natura nella filosofia medievale prima dell' ingresso della fisica di Aristotele: il secolo XII," in *La filosofia della natura nel Medioevo* (Milan: Società editrice Vita e pensiero, 1966), 27–65; Jacques Chiffoleau, "*Contra naturam*. Pour une approche casuistique et procédurale de la nature médiévale," *Micrologus* 4 (1996): 265–312; and Lawrence Roberts, ed., *Approaches to Nature in the Middle Ages* (Binghamton: NY, 1982).

2. Brian Stock, "Science, Technology, and Economic Progress in the Early Middle Ages," in *Science in the Middle Ages*, ed. David C. Lindberg (Chicago: University of Chicago Press, 1978), 1–51, esp. 23–32.

3. William Stahl, *Roman Science: Origins, Development, and Influence to the Later Middle Ages* (Madison: University of Wisconsin Press, 1962), 66, 71, 79, 96.

4. Cicero, *Tusculan Disputations*, I.ii.5.

5. Heinrich von Staden, "Liminal Perils: Early Roman Receptions of Greek Medi-

cine," in *Tradition, Transmission, Transformation: Proceedings of Two Conferences on Pre-Modern Science held at the University of Oklahoma*, ed. F. Jamil Ragep, Sally Ragep, and Steven Livesey (Leiden: Brill, 1996), 369–418.

6. For astronomy, see Emmanuel Poulle, "Le vocabulaire de l'astronomie planétaire du XIIe au XIVe siècle," *La diffusione delle scienze islamiche nel medio evo europeo* (Rome: Accademia Nazionale dei Lincei, 1987), 193–212, esp. 195.

7. Vitruvius, *On Architecture*, bk. 8, ch. preface, sec. 1.II, 132; bk. 8, ch. preface, sec. 4.II, 136. He contrasts *physici* and *philosophi* in bk. 7, ch. preface, sec. 2.

8. Ilsetraut Hadot, *Arts libéraux et philosophie dans la pensée antique* (Paris: Études Augustiniennes, 1984), 86–87, 122; Danuta R. Shanzer, "Augustine's Disciplines: *Silent diutius Musae Varronis?*" in *Augustine and the Disciplines: From Cassiciacum to* Confessions, ed. Karla Pollmann and Mark Vessey (Oxford: Oxford University Press, 2005), 69–112.

9. R. A. Markus, "Marius Victorinus and Augustine," in *The Cambridge History of Later Greek and Early Medieval Philosophy*, ed. A. H. Armstrong (Cambridge: Cambridge University Press, 1970), 395.

10. David C. Lindberg, "Science and the Early Christian Church," *Isis* 74 (1983): 509–30, esp. 526–27.

11. William H. Stahl and Richard Johnson, *Martianus Capella and the Seven Liberal Arts* (New York: Columbia University Press, 1971), esp. 21–25.

12. Stahl, *Roman Science*, 199; Wesley Stevens, "Marginalia in the Latin Euclid," in *Scientia in margine: Études sur les marginalia dans les manuscrits scientifiques du moyen âge à la Renaissance*, ed. Danielle Jacquart and Charles Burnett (Geneva: Droz, 2005), 117–37; George Molland, "The Quadrivium in the Universities: Four Questions," in Scientia *und* ars *im Hoch- und Spätmittelalter*, ed. Ingrid Craemer-Ruegenberg and Andreas Speer (Berlin: De Gruyter, 1994), 66–78, esp. 66; Stahl and Johnson, *Martianus Capella*, 92.

13. Boëthius, *De trinitate*, bk. 2, Loeb 8–9, 88.

14. Augustine, *The City of God*, trans. by David Wiesen (Cambridge, MA: Harvard University Press, 1968), esp. bk. 8, ch. 1 (vol. 3, 3–7) defines theology as "a Greek word by which we understand thought and speech about the divine" (translation emended). In addition to "natural" theology, he discusses elsewhere (bk. 8, ch. 5 [vol. 3, 23]) such pagan varieties as "mythical" (*fabulosa*) and "political" (*civilis*) theology.

15. Edward Grant, *A History of Natural Philosophy from the Ancient World to the Nineteenth Century* (Cambridge: Cambridge University Press, 2007), 42–43.

16. Philippe Delhaye, "La place des arts libéraux dans les programmes scolaires du XIIIe siècle," in *Arts libéraux et philosophie au moyen-âge* (Montreal/Paris, 1969), 172; Ralph McInerny, "Beyond the Liberal Arts," in *The Seven Liberal Arts in the Middle Ages*, ed. David Wagner (Bloomington: Indiana University Press, 1983), 250.

17. Grant, *A History of Natural Philosophy*, 105.

18. Stephen McCluskey, "Gregory and Tours, Monastic Timekeeping, and Early Christian Attitudes to Astronomy," *Isis* 81 (1990): 8–22; Stephen C. McCluskey, *Astronomies and Cultures in Early Medieval Europe* (Cambridge: Cambridge University Press, 1998).

19. Bruce Eastwood, "The Astronomy of Pliny, Martianus Capella and Isidore of Seville," in *Science in Western and Eastern Civilization in Carolingian Times*, ed. Paul Butzer and Dietrich Lohrmann (Basel: Birkhäuser, 1993), 161–80, esp. 177; Bruce East-

wood, *Ordering the Heavens: Roman Astronomy and Cosmology in the Carolingian Renaissance* (Leiden: Brill, 2007), esp. ch. 1.

20. Linda Ehrsam Voigts, "Anglo Saxon Plant Remedies and the Anglo-Saxons," *Isis* 70 (1979): 250–68, esp. 266–68.

21. Monica Green, ed. and trans., *The Trotula: A Medieval Compendium of Women's Medicine* (Philadelphia: University of Pennsylvania Press, 2001); Margaret Gibson, "The Continuity of Learning circa 850–circa 1050" and "The Study of the *Timaeus* in the XIth and XIIth Centuries," in Margaret Gibson, *"Artes" and Bible in the Medieval West* (Aldershot: Variorum, 1993).

22. Guy Beaujouan, "The Transformation of the Quadrivium," in *Renaissance and Renewal in the Twelfth Century*, ed. Robert Benson and Giles Constable (Cambridge, MA: Harvard University Press, 1982), 463–87, esp. 464; Harold Berman, "The Origin of Western Legal Science," in *Law and Revolution: The Formation of the Western Legal Tradition* (Cambridge, MA: Harvard University Press, 1983), 120–64.

23. Charles Lohr, "The Pseudo-Aristotelian *Liber de causis* and Latin Theories of Science in the Twelfth and Thirteenth Centuries," in *Pseudo-Aristotle in the Middle Ages: The* Theology *and Other Texts*, ed. Charles Burnett (London: The Warburg Institute, 1986), 54.

24. Adelard of Bath, *Conversations with his nephew; On the same and the different; Questions on natural science; and, On birds*, ed. and trans. Charles Burnett (Cambridge: Cambridge University Press, 1998), 2–3.

25. Marie-Dominique Chenu, *Nature, Man, and Society in the Twelfth Century*, ed. Jerome Taylor and Lester K. Little (Chicago: University of Chicago Press, 1968), 4–11; Berman, *Law and Revolution*, 144–47.

26. Joan Cadden, *Meanings of Sex Difference in the Middle Ages: Medicine, Science, and Culture* (Cambridge: Cambridge University Press, 1993), 54–70.

27. Joan Cadden, "Science and Rhetoric in the Middle Ages: The Natural Philosophy of William of Conches," *Journal of the History of Ideas* 56 (1995): 1–25, esp. 4–7; Alexander Fidora and Andreas Neiderberger, "Philosophie und Physik zwischen notwendigem und hypothetischem Wissen: Zur wissenschaftstheoretischen Bestimmung der Physik in der *Philosophia* des Wilhelm von Conches," *Early Science and Medicine* 6 (2001): 22–34, esp. 25–26, 31–32; Peter Dronke, "Scientific Speculations," *A History of Twelfth-Century Western Philosophy*, ed. Peter Dronke (Cambridge: Cambridge University Press, 1988), 154.

28. Hugh of St. Victor, *Didascalicon*, II.30.

29. Tullio Gregory, "La nouvelle idée de nature et de savoir scientifique au XIIe siècle," in *The Cultural Context of Medieval Learning*, ed. John E. Murdoch and Edith Sylla (Dordrecht: D. Reidel 1975), 198.

30. Brian Lawn, *The Rise and Decline of the Scholastic 'Quaestio Disputata' with Special Emphasis on its Use in the Teaching of Medicine and Science* (Leiden: Brill, 1993), 22–23; Chenu, *Nature, Man, and Society*, 10–14.

31. Edward Peters, "*Libertas Inquirendi* and the *Vitium Curiositatis* in Medieval Thought," in *The Concept of Freedom in the Middle Ages: Islam, Byzantium and the West*, ed. George Makdisi, Dominique Sourdel, and Janine Sourdel-Thomine (Paris: Les Belles Lettres, 1985), 89–98, on 92.

32. Alison Drew, "The *De eodem et diverso*," in *Adelard of Bath: An English Scientist*

and Arabist of the Early Twelfth Century, ed. Charles Burnett (London: Warburg Institute, 1987), 21.

33. Margaret Gibson, "Adelard of Bath," in Burnett, *Adelard of Bath*, 10–11; Adelard of Bath, *Conversations with his Nephew*, 226–27.

34. Jacques Verger, "The First French Universities and the Institutionalization of Learning: Faculties, Curricula, Degrees," in *Learning Institutionalized: Teaching in the Medieval University*, ed. John Van Engen (Notre Dame: University of Notre Dame Press, 2000), 5–19, esp. 6–7; Serge Lusignan, *Parler vulgairement: Les intellectuels et la langue française aux XIIIe et XIVe siècles*, 2nd ed. (Paris: Vrin, 1986), 52.

35. Olaf Pedersen, "Du *quadrivium* à la physique. Quelques aperçus de l'évolution scientifique au moyen âge," in *Artes liberales von der antiken Bildung zur Wissenschaft des Mittelalters*, ed. Josef Koch (Leiden: Brill, 1959), 107–23, esp. 112.

36. John of Salisbury equated *ars* with the Greek *methodos*; Jeannine Quillet, "De l'art des conjectures à la science divine selon Nicolas de Cues," in Craemer-Ruegenberg and Speer, Scientia *und* ars, 95; David Luscombe, "Scientia and Disciplina in the Correspondence of Peter Abelard and Heloïse," in *"Scientia" und "Disciplina": Wissenstheorie und Wissenschaftspraxis im 12. und 13. Jahrhundert*, ed. Rainer Berndt, Matthias Lutz-Bachmann, and Ralf M. W. Stammberger (Berlin: Akademie Verlag, 2002), 79–89, esp. 83.

37. Burnett, *Adelard of Bath*, 143–44; Pietro d'Abano, *Lucidator dubitabilium astronomiae*, ed. Graziella Federici-Vescovini (Padua: Programma e 1 + 1 Editore, 1988), 105 and passim. Morley calls Hesiod a *"naturalis scientie professor"* (Brian Stock, *Myth and Science*, 4); d'Abano preferred *scientia naturalis* to *physica*. Nancy Siraisi, *Arts and Sciences at Padua: The Studium of Padua before 1359* (Toronto: Pontifical Institute of Mediaeval Studies, 1973), ch. 4: Scientia Naturalis et Metaphysica, 109–142, esp. 111; Robert Kilwardby, *De ortu scientiarum*, ed. Albert Judy (London: The British Academy, 1976), 17.

38. Al-Farabi, *Catalogo de las ciencias*, ed. and trans. by Ángel Gonzales Palencia (Madrid: Consejo naciónál de las Investigaciones scientificas, 1953), 161; Marshall Clagett, "Some General Aspects of Physics in the Middle Ages," *Isis* 39 (1948): 31–34.

39. Jacob Hans Josef Schneider, "Scientia Sermocinalis/Realis: Anmerkungen zum Wissenschaftsbegriff im Mittelalter und der Neuzeit," *Archiv für Begriffsgeschichte* 35 (1992): 58–60, 80.

40. Elspeth Whitney, *Paradise Restored: The Mechanical Arts from Antiquity through the Thirteenth Century* (Philadelphia: American Philosophical Society, 1990), 75–127; Beaujouan, "Transformation of the Quadrivium," 463–64; Quillet, "De l'art des conjectures," 95–106, esp. 95; Dietrich Lohrmann, "Les marges dans les manuscrits d'ingénieur," in *Scientia in margine*, 217–240, esp. 218.

41. Claude Lafleur, *"Scientia* et *ars* dans les introductions à la philosophie des maîtres ès arts de l'université de Paris au XIIIe siècle," in Craemer-Ruegenberg and Speer, Scientia *und* ars, 47–49.

42. Armand Maurer, *St. Thomas Aquinas: Division and Methods of the Sciences* (Toronto: Pontifical Institute of Mediaeval Studies, 1963), 11.

43. Kilwardby, *De ortu scientiarum*, 194–95; Olga Weijers, "L'appellation des disciplines dans les classifications des sciences aux XIIe et XIIIe siècles," *Archivum Latinitatis Mediae Aevi* 46/47 (1988): 42–43.

44. Dante, *Convivio*, III.xiii (xiv), 8; *Le opere di Dante: Testo critico della Società Dant-*

esca Italiana, 2nd ed. (Florence: Società Dantesca Italiana 1960), 185–89, esp. 186; also Tiziana Suarez-Nani, "Dante Alighieri ou la convergence des arts et des sciences," in Craemer-Ruegenberg and Speer, Scientia *und* ars, 126–42, esp. 132–33.

45. Jacques Verger, *Men of Learning in Europe at the End of the Middle Ages*, trans. by L. Neal and S. Rendall (Notre Dame, IN: University of Notre Dame Press, 2000), 50–51.

46. Paul O. Kristeller, "Philosophy and Medicine in Medieval and Renaissance Italy," *Organism, Medicine, and Metaphysics: Essays in Honor of Hans Jonas*, ed. Stuart F. Spicker (Dordrecht: Reidel, 1978), 29–40, esp. 33.

47. Paul Czartoryski, "La notion d'université et l'idée de science à l'université de Cracovie dans la première moitié du XVe siècle," *Mediaevalia philosophica polonorum* 14 (1970): 25–27.

48. Edward Grant, *God and Reason in the Middle Ages* (Cambridge: Cambridge University Press, 2001), 215; Alain de Libera, "Faculté des arts ou faculté de philosophie? Sur l'idée de philosophie et l'idéal philosophique au XIIIe siècle," in *L'enseignement des disciplines à la Faculté des arts (Paris et Oxford, XIIIe- XVe siècles)*, ed. Olga Weijers and Louis Holtz (Turnhourt: Brepols, 1997), 429–44, esp. 439.

49. Guy Beaujouan, "Le quadrivium et la faculté des arts," in Weijers and Holtz, *L'enseignement des disciplines*, 191–92; Danielle Jacquart, "Rapport de la Table ronde: Les disciplines du quadrivium," in Weijers and Holtz, *L'enseignement des disciplines*, 239–47, esp. 242–43, 247.

50. Nancy Siraisi, *Medieval and Early Renaissance Medicine: An Introduction to Knowledge and Practice* (Chicago: University of Chicago Press, 1990), 86–97.

51. Grant, *A History of Natural Philosophy*, 236–38.

52. John Marenbon, "Gilbert of Poitiers and the Porretans on Mathematics in the Division of the Sciences," in *"Scientia" und "Disciplina": Wissenstheorie und Wissenschaftspraxis im 12. Und 13. Jahrhundert*, ed. Rainer Berndt, Matthias Lutz-Bachmann and Ralf M. W. Stammberger (Berlin: Akademie Verlag, 2002), 37–69, esp. 56; Daniel Callus, "The Function of the Philosopher in Thirteenth-Century Oxford," in *Beiträge zum Berufsbewusstsein des mittelalterlichen Menschen*, ed. Paul Wilpert (Berlin: De Gruyter, 1964), 152–62, esp. 159.

53. Luca Bianchi, "Loquens ut naturalis," in *Le verità dissonanti: Aristotele a la fine del medioevo*, ed. Luca Bianchi and Eugenio Randi (Bari: Laterza, 1990), 33–56; Bert Hansen, *Nicole Oresme and the Marvels of Nature: A Study of his* De causis mirabilium *with Critical Edition, Translation, and Commentary* (Toronto: Pontifical Institute of Mediaeval Studies, 1985), 136–37; Michael H. Shank, "Mechanical Thinking in European Astronomy (13th–15th Centuries)," in *Mechanics and Cosmology in the Medieval and Early Modern Period*, ed. Massimo Bucciantini, Michele Camerota, and Sophie Roux (Florence: Leo Olschki, 2007), 3–27, esp. 27; Bernard R. Goldstein, "Galileo's Account of Astronomical Miracles in the Bible: a Confusion of Sources," *Nuncius* 5 (1990): 9.

54. David C. Lindberg, "Science as Handmaiden: Roger Bacon and the Patristic Tradition," *Isis* 78 (1987): 518–36, esp.527–30; Nicholas Steneck, "A Late Medieval *Arbor Scientiarum*," *Speculum* 50 (1975): 245–69, esp. 257–63.

55. Jacques Paquet and Jozef Ijsewijn, eds., *The Universities in the Late Middle Ages* (Leuven: Leuven University Press, 1978), 13n20.

56. Lusignan, *Parler vulgairement*, 134–37.

57. See the excellent overview by Steven J. Livesey, "Scientia," in *Medieval Science,*

Technology, and Medicine: An Encyclopedia, ed. Thomas Glick, Steven J. Livesey, Faith Wallis (New York: Routledge, 2005), 455–58.

58. Giacinta Spinosa, "Neologismi aristotelici et neoplatonici nelle teorie medievali della conoscenza: *alteratio, alteritas, sensitivus, scientificus, cognoscitivus*," in *Aux origines du lexique philosophique européen: L'influence de la* latinitas, ed. Jacqueline Hamesse (Louvain-la-Neuve: Fédération internationale des instituts d'études médiévales, 1997), 181–220, esp. 209–14.

59. Aegidius Aurelianus, *Quaestiones super De generatione et corruptione*, ed. Ladislaw Kuksewicz (Amsterdam: B. R. Grüner Publishing Co., 1993), 3; Alfonso Maierù, *University Training in Medieval Europe*, ed. and trans. Doreen Pryds (Leiden: Brill, 1994), 76–80.

60. Kilwardby, *De ortu scientiarum:* "The Greeks call nature *phisim* and body *phisicum* and the knowledge/science *phisicam*" (17); Weijers, "L'appellation des disciplines," 52; Claude Lafleur, "Les texts didascaliques," in Weijers and Holtz, *L'enseignement des disciplines*, 345–72, esp. 354; Emma T. Healy, *St. Bonaventure: De reductione artium ad theologiam: A Commentary with an Introduction and Translation* (St. Bonaventure, NY: The Franciscan Institute, 1939), 4.

61. Grosseteste, "Commentary on the *Physics*," cited in W. R. Laird, "Robert Grosseteste on the Subalternate Sciences," *Traditio* 43 (1987): 147–69, esp. 150, 153, 155; Steven Livesey, "The Subalternation of the Sciences," in *Science and Theology in the Fourteenth Century* (Leiden: Brill, 1989), 20–53.

62. Cited in William Wallace, *Causality and Scientific Explanation* (Ann Arbor: University of Michigan Press, 1972–74), 1: 220n35.

63. David C. Lindberg, "Roger Bacon and the Origins of Perspectiva in the West," in *Mathematics and its Applications to Science and Natural Philosophy in the Middle Ages*, ed. Edward Grant and John E. Murdoch (Cambridge: Cambridge University Press, 1987), 249–68, esp. 258–59.

64. Avicenna subalternated medicine to natural philosophy, on a par with the mathematical sciences; G. C. Anawati, "Ibn Sina," *Dictionary of Scientific Biography*, vol. 15, on 497–98.

65. Thomas Aquinas, *The Division and Methods of the Sciences: Question V and VI of his Commentary on the* De Trinitate *of Boethius*, trans. Armand Maurer, 3rd rev. ed. (Toronto: Pontifical Institute of Mediaeval Studies, 1963), 37–38, 45.

66. Roger Bacon, *Communia naturalia*, II, pt. 5, ch. 17, in *Opera hactenus inedita Rogeri Baconi*, fasc. 4, ed. Robert Steele (Oxford: Clarendon Press, 1909), 444; Jean Gagné, "Du *quadrivium* aux *scientiae mediae*," in *Arts libéraux*, 984–86; Frank Hentschel and Martin Pickavé, "*Quaestiones mathematicales*. Eine Textgattung der Pariser Artistenfakultät im frühen 14. Jahrhundert," in Aertsen, Emery, and Speer, *Nach der Verurteilung von 1277: Philosophie und Theologie an der Universität von Paris im letzten Viertel des 13. Jahrhunderts* (Berlin: Walter De Gruyter, 2001), 618–34, esp. 632–33nn 64, 71, 73. One fourteenth-century physician added *perspectiva* to the quadrivium; Graziella Federici-Vescovini, "L'inserimento della 'perspectiva' tra le arti del Quadrivio," in *Arts libéraux*, 969–74, esp. 974n30.

67. Gagné, "Du *quadrivium* aux *scientiae mediae*," in *Arts libéraux*, 978–79; Ernest A. Moody and Marshall Clagett, eds., *The Medieval Science of Weights (Scientia de ponderibus): Treatises ascribed to Euclid, Archimedes, Thabit ibn Qurra, Jordanus de Nemore and Blasius of Parma* (Madison: University of Wisconsin Press, 1952), 150–51.

68. Aquinas, *The Division and Methods of the Sciences*, 38–39.

69. Edith D. Sylla, "The Fate of the Oxford Calculatory Tradition," in *L'Homme et son univers au moyen âge*, ed. Christian Wenin (Louvain-la-Neuve: Editions de l'Institut Supérieur de Philosophie, 1986), 992–98; Lawn, "The Diffusion of Mertonian Ideas," in *Rise and Decline of the Scholastic 'Quaestio Disputata,'* 53–65.

70. John E. Murdoch and Edith Dudley Sylla, "The Science of Motion," in Lindberg, *Science in the Middle Ages*, 224–41; and Marshall Clagett, *The Science of Mechanics in the Middle Ages* (Madison: University of Wisconsin Press, 1959).

71. Steneck, "A Late Medieval *Arbor Scientiarum*," 254, and Joannes Regiomontanus, *Opera collectanea*, ed. Felix Schmeidler (Osnabrück: O. Zeller Verlag, 1972), 46; Amos Funkenstein, *Theology and the Scientific Imagination* (Princeton, NJ: Princeton University Press, 1986), 303–17. In the manuscript of his *De Revolutionibus*, Copernicus also included geodesy/surveying and mechanics in the list, omitted in print, that precedes book 1, chapter 1; Copernicus, *De Revolutionibus orbium coelestium*, trans A. M. Duncan (1976; Norwalk: The Easton Press, 1993), 35.

72. John E. Murdoch, "Philosophy and the Enterprise of Science in the Later Middle Ages," in *The Interaction between Science and Philosophy*, ed. Yehuda Elkana (Atlantic Highlands: Humanities Press, 1974), 51–74, esp. 58–71.

73. Alain de Libéra, "Le développement de nouveaux instruments conceptuels et leur utilisation dans la philosophie de la nature au XIVe siècle," in *Knowledge and the Sciences in Medieval Philosophy*, Proceedings of the 8th International Congress of Medieval Philosophy, ed. Simo Knuuttila et al. (Helsinki: s.n., 1990), 158–97, esp. 172.

74. John E. Murdoch and Edith Sylla, "Swineshead, Richard," *Dictionary of Scientific Biography* (New York, 1970–90), 13:209. Francis Bacon later complained that natural philosophy had been made subservient to mathematics and medicine (*Novum organum*, LXXX); Grant, *A History of Natural Philosophy*, 305–6.

75. Edith Dudley Sylla, "The *a posteriori* Foundations of Natural Science: Some Medieval Commentaries on Aristotle's *Physics*, Book I, chapters 1 and 2," *Synthèse* 40 (1979): 147–87.

76. Eileen Serene, "Demonstrative Science," in *The Cambridge History of Later Medieval Philosophy*, ed. Norman Kretzmann, Anthony Kenny, and Jan Pinborg (Cambridge: Cambridge University Press, 1982), 496.

77. Kilwardby, *De ortu scientiarum*, 136–37.

78. George Molland, "Roger Bacon's De laudibus mathematicae: A preliminary study," in *Texts and Contexts in Ancient and Medieval Science: Studies on the Occasion of John E. Murdoch's Seventieth Birthday*, ed. Edith Sylla and Michael McVaugh (Leiden: Brill, 1997), 68–83, esp. 71; Serene, "Demonstrative Science," 512–13; Steven P. Marrone, "Concepts of Science among Parisian Theologians in the Thirteenth Century," in *Knowledge and the Sciences in Medieval Philosophy*, vol. 3, ed. Reijo Työrinoja, Anja Inkeri Lehtinen, and Dagfinn Føllesdal (Helsinki: s.n., 1990), 124–33, esp. 130–31.

79. Ockham defines it as "two singular premises arranged in the third figure, which, however, can yield a singular, a particular, or an indefinite conclusion, but not a universal one . . ." William of Ockham, *Summa logicae*, ed. G. Gál (St. Bonaventure, NY: Franciscan Institute, 1974), 3: 1, 16; I. M. Bochenski, *A History of Formal Logic*, trans. by Ivo Thomas, 2nd ed. (New York: Chelsea Publications, 1970), 232–33.

80. Serene, "Demonstrative Science," 516 (whose translation I modified slightly).

81. Hendrik Lagerlund, "The Systematization of Modal Syllogistics," in *Modal Syllogistics in the Middle Ages* (Leiden: Brill, 2000), 30–160; Jack Zupko, *John Buridan: Portrait of a Fourteenth-Century Arts Master* (Notre Dame, IN: University of Notre Dame Press, 2003), 59–77.

82. Edward Grant, "Science and the Medieval University," in *Rebirth, Reform, and Resilience: Universities in Transition, 1300–1700*, ed. James Kittelson and Pamela Transue (Columbus: Ohio State University Press, 1984), 68–102.

83. Maierù, *University Training in Medieval Europe*, 12–14; Grant, "Science and the Medieval University," 84–85.

84. Edith Sylla, "The Oxford Calculators," in Kretzmann, Kenny, and Pinborg, *The Cambridge History of Later Medieval Philosophy*, 542–48.

85. Serge Lusignan, "L'enseignement des arts dans les collèges parisiens au Moyen Âge," in Weijers and Holtz, *L'enseignement des disciplines*, 43–49, esp. 47; Olga Weijers, "La 'disputatio,'" in Weijers and Holtz, *L'enseignement des disciplines*, 393–404, esp. on 393–94, 401–2.

86. Michael Shank, *"Unless You Believe, You Shall Not Understand," Logic, University, and Society in Late Medieval Vienna* (Princeton, NJ: Princeton University Press, 1988), 151–52; Collette Sirat, "L'enseignement des disciplines dans le monde hébreu," in Weijers and Holtz, *L'enseignement des disciplines*, 495–509, esp. 504–8.

87. John E. Murdoch, *Album of Science: Antiquity and the Middle Ages* (New York: Scribner's Sons, 1984), 3.

88. William of Ockham, *Expositio super VIII libros Physicorum*, ed. V. Richter and G. Leibold (St. Bonaventure, NY: Franciscan Institute, 1986), paragraph 30.

89. Max Lejbowicz, "Les disciplines du *quadrivium:* l'astronomie," in Weijers and Holtz, *L'enseignement des disciplines*, 195–211, esp. 199–200.

90. John E. Murdoch, "Natural Philosophy without Nature," in Roberts, *Approaches to Nature in the Middle Ages*, 171–213.

91. Grant, *A History of Natural Philosophy*, 215–43; Nicole Oresme, *Le Livre du ciel et du monde*, ed. Albert Menut and Alexander Denomy (Madison: University of Wisconsin Press, 1968), 144–45, 520–39.

92. Kenneth F. Kitchell and Irven M. Resnick, "Hildegard as Medieval 'Zoologist,'" in *Hildegard of Bingen: A Book of Essays*, ed. Maud Burnett McInerny (New York: Garland, 1998), 25–52; Danielle Jacquart, "L'observation dans les sciences de la nature au moyen âge: limites et possibilités," *Micrologus* 4 (1996): 55–75; quotation about Albertus Magnus from James Lennox, "The Disappearance of Aristotle's Biology: A Hellenistic Mystery," *Apeiron* 27 (1994): 7–24, 23; R. James Long, "The Reception and Interpretation of the Pseudo-Aristotelian *De Plantis* at Oxford in the Thirteenth Century," in Työrinoja, Lehtinen, and Føllesdal, *Knowledge and the Sciences in Medieval Philosophy*, 3:111–23, esp. 121; Bernard Ribémont and Geneviève Sodigné-Costes, "Botanique médiévale: tradition, observation, imaginaire. L'exemple de l'encyclopédisme," in *Le moyen âge et la science: approches de quelques disciplines et personnalités scientifiques médiévales*, ed. Bernard Ribémont (Paris: Klincksieck, 1991), 153–69.

93. Grant, *A History of Natural Philosophy*, 192–99.

94. "Judah, son of Moses, son of Mosca, and Rabbi Isaac ibn Sid say: The science of astrology is a subject that cannot be investigated without observations. Yet, the observations made by the experts in this discipline cannot be completed by a single

man," in José Chabás and Bernard R. Goldstein, Archimedes, eds., *The Alfonsine Tables of Toledo* (Dordrecht: Kluwer Academic Publishers, 2003), 8: 136, 141–43.

95. Lynn Thorndike, "Query no. 14," *Isis* 14 (1930): 421; Pierre Duhem, *Le système du monde: Histoire des doctrines cosmologiques de Platon à Copernic* (Paris: Hermann, 1913–59), 4:16–17; Lejbowicz, "Les disciplines du *quadrivium:* l'astronomie," 195–211, esp. 209–11; Beaujouan, "Le *quadrivium* et la Faculté des arts," 194.

96. Bernard Goldstein, "Theory and Observation in Medieval Astronomy," *Isis* 63 (1972): 39–47, esp. 47; Goldstein, *The Astronomy of Levi ben Gerson (1288–1344)* (New York: Springer Verlag, 1985),7–8, 13–14; Goldstein, "Before the Sun in the Church," *Journal for the History of Astronomy* 32 (2001): 73–77; and Jacquart, "L'observation dans les sciences de la nature au moyen âge," 57–59.

97. Noel M. Swerdlow, "Regiomontanus on the Critical Problems of Astronomy," in *Nature, Experiment and the Sciences*, ed. Trevor Levere and William Shea (Boston: Kluwer, 1990), 165–95, esp. 173.

98. William R. Newman, "Art, Nature, and Experiment among Some Aristotelian Alchemists," in Sylla and McVaugh, *Texts and Contexts in Ancient and Medieval Science*, 305–17, esp. on 305–12; Newman, "The Place of Alchemy in the Current Literature on Experiment," in *Experimental Essays—Versuch zum Experiment*, ed. Michael Heidelberger and Friedrich Steinle (Baden-Baden: Nomos Gesellschaft, 1998), 9–33, esp. 23–25; Newman, "The Medieval Tradition of Alchemical Corpuscular Theory" (23–44), and "Erastus and the Critique of Chemical Analysis" (45–64) in *Atoms and Alchemy: Chymistry and the Experimental Origins of the Scientific Revolution* (Chicago: University of Chicago Press, 2006).

99. Sabetai Unguru, "Experiment in Medieval Optics," in *Physics, Cosmology, and Astronomy, 1300–1700: Tension and Accommodation*, ed. Sabetai Unguru (Dordrecht: D. Reidel: 1991), 163–81; Danielle Jacquard, "L'observation dans les sciences de la nature au moyen âge," esp. 75; Guy Beaujouan, "La prise de conscience de l'aptitude à innover (le tournant du milieu du XIIIe siècle)," in Ribémont, *Le moyen âge et la science*, 5–14, esp. on 8.

100. Katharine Park, *Secrets of Women: Gender, Generation, and the Origins of Human Disssection* (Cambridge, MA: Zone Books, 2006).

101. William J. Courtenay, "The Arts Faculty at Paris in 1329," in Weijers and Holtz, *L'enseignement des disciplines*, 55–69, esp. 62.

102. John Marenbon, "The Theoretical and Practical Autonomy of Philosophy as a Discipline in the Middle Ages: Latin Philosophy, 1250–1350," in *Knowledge and the Sciences in Medieval Philosophy*, vol. 1, ed. Monika Asztalos, John E. Murdoch, and Ilkka Niiniluoto (Helsinki,(s.n., 1990), 262–74.

103. Grant, "The Relations between Natural Philosophy and Theology," in *A History of Natural Philosophy*, 239–73.

104. Alain de Libéra, "Faculté des arts ou faculté de philosophie?," 440–44.

105. Paul Czartoryski, "La notion d'université et l'idée de science," 29.

106. Verger, "The First French Universities," esp. 10–11, 14–15; John Baldwin, "Theology, the Queen of the Faculties," in *The Scholastic Culture of the Middle Ages, 1000–1300* (Lexington, MA: Heath, 1971), 79–97.

107. W. D. Ross, ed., *Aristotle's Metaphysics* (1924; Oxford: Oxford University Press, 1966), 1:350–51; Jonathan Barnes, ed., *The Complete Works of Aristotle* (Princeton,

NJ: Princeton University Press, 1984), 2:1619–20; Thomas Aquinas, prologue to the *Commentary on Aristotle's Metaphysics*, trans. John P. Rowan (Chicago: Henry Regnery Company, 1961), 1: 1–2.

108. Thomas Aquinas, *Summa theologiae*, introduction, q. 1, ch. 1, quoted in Marie-Dominique Chenu, *La théologie comme science au XIIIe siècle*, 3rd ed. (Paris: Vrin, 1957), 94; Henri Hugonnard-Roche, "La classification des sciences de Gundissalinus et l'influence d'Avicenne," *Études sur Avicenne* (Paris: Les Belles Lettres, 1984) 89–90.

109. Amos Funkenstein, "Scholasticism, Skepticism, and Secular Theology," in *Skepticism from the Renaissance to the Enlightenment*, ed. Richard H. Popkin and Charles B. Schmitt (Wiesbaden: O. Harrassowitz, 1987), 49; Chenu, *La théologie comme science au XIIIe siècle*; Shank, "On Paralogisms in Trinitarian Doctrine (1): The Early Fourteenth Century," in *"Unless You Believe, You Shall Not Understand,"* 57–86, 135–36.

110. Stephen F. Brown, "Late Thirteenth Century Theology: 'Scientia' pushed to its Limits," in Berndt, Lutz-Bachmann, and Stammberger, *"Scientia" und "Disciplina,"* 249–60, esp.252–53.

111. Ernest A. Moody, *The Logic of William of Ockham* (1935; New York: Russell and Russell, 1965), 211.

112. John Duns Scotus, *Philosophical Writings: A Selection*, trans. by Alan Wolter (Indianapolis: Hackett, 1987), 20, 171.

113. Kilwardby, *De ortu scientiarum*, 32–33, 43: "this science (= astrologia) is in part mendacious and fictitious . . ."

114. Lindberg, "Roger Bacon and the Patristic Tradition," 531–32; Christine Silvi, *Science médiévale et vérité: Étude linguistique de l'expression du vrai dans le discours scientifique en langue vulgaire* (Paris: Honoré Champion, 2003), 453; Stock, *Myth and Science*, 27–28; Joannes Regiomontanus, "Oratio . . . in praelectione Al-fragani," in Schmeidler, *Opera collectanea*, 43–53; Michael H. Shank, "Setting Up Copernicus? Francesco Capuano da Manfredonia's *Expositio* on the *Sphere* of Sacrobosco," *Early Science and Medicine* 14 (2009): 290–315, esp. 296; Graziella Federici-Vescovini, "La place privilégiée de l'astronomie–astrologie dans l'encyclopédie des sciences théoriques de Pierre d'Abano," in Työrinoja, Lehtinen, and Føllesdal, *Knowledge and the Sciences in Medieval Philosophy*, 3:42–51, esp. 46–48; Mieczyław Markowski, "Von den mittelalterlichen Ansätzen eines Wandels zum kopernikanischen Umbruch im Wissenschaftsverständnis," in Craemer-Ruegenberg and Speer, Scientia *und* ars, 79–94, esp. 92–93.

CHAPTER 5

Natural History

Peter Harrison

The term "natural history" has a quaint and old-fashioned air about it, no doubt partly on account of its mental association with museums, partly because it refers to a discipline that, from about the end of the nineteenth century, seems to have receded into the background to make way for the contemporary biological sciences. On reflection, the discipline is also designated by a decidedly odd combination of words that imply a subject matter—nature—for which a chronological approach of the kind suggested by "history," seems entirely inappropriate. The strangeness of "natural history" is only compounded when we consider the ways in which the discipline was regarded by its practitioners in previous ages. Wolfgang Franz, writing in the seventeenth century on natural history, describes his enterprise in these terms:

> The history of Brutes, which by some is not unjustly called Ζωογραφια [zoographia], or a Description of living creatures, is that part of Physicks which treateth of Brute beasts. We may properly call this one part of Physicks, because it treateth of the nature of things; for Physicks is either Physicks properly so called, or you may divide it into Metaphysicks, and Mathematicks. Physicks properly so called comprehendeth under it the nature of Meteors, Metals, Plants, Stars, the four Elements, men, and Brutes. Some would have the consideration of Brutes be brought under Medicine, which I think belongeth more properly to Philosophy.[1]

Franz's observations convey to the modern-day reader something of the unfamiliar boundaries of the disciplines as they were conceived in the early-modern period, and of their perplexing relationships with other areas of study. Franz speaks of a history of animals sometimes referred to as "zoographia" and which is part of "physicks." He also allows that there are those who believe that the history of animals ought to be included in medicine but states that it more properly belongs in "philosophy." This attempt to situate the history of animals serves as a salient reminder of the alien character of the disciplines as they were conceived a little over three hundred years ago. Franz took it for granted that readers of his history of animals would recognize it as part of the genre of "natural history," an area of scholarship that included the additional categories of histories of plants and stones. "History," like "physics" and "philosophy," thus had a somewhat broader meaning than that which we are familiar.

A reading of other natural histories from this period brings additional surprises: justifications for the study of nature were markedly different from those of our own era, the practice of natural history called for procedures utterly unlike those that we commonly associate with the natural sciences, and the very object of study—"nature"—was conceptualized differently. Finally, if we consider the career of natural history, from its revival in the sixteenth century to its decline in the nineteenth, in each of these aspects—its relation to other disciplines, its motivations and justifications, its methods, its subject matter—the enterprise of natural history underwent significant change. In short, the natural history of these earlier historical periods bears only a remote resemblance to any currently practiced discipline.

THE REVIVAL OF ANCIENT LEARNING

A typical Renaissance natural history dealt with one of the three categories: animal, vegetable, or mineral. Further divisions were often observed within these broad categories. Four-footed animals were usually distinguished from the rest of the animal kingdom. Some writers separated viviparous (bearing live young) from oviparous (egg-laying) quadrupeds; some treated fish and "aquatic animals" together. "Insects" was a rather broad category that encompassed most of the invertebrate species, including the "imperfect" creatures, those thought to generate spontaneously. Plants, "God's vegetable creatures" as one author called them, were the other major concern of natural historians. Histories of plants were more consciously focused on the provisions of cures than were works on animals, and for this reason were more closely aligned with the practice of medicine or "physic." Indeed, during the Middle Ages, the chief repository of knowledge about nature was not natural history, but medicine.[2] In addition to living things, "stones" or "minerals" were also the subjects of natural histories along with specific geographical locations, both familiar and exotic, aspects of animal and human psychology, and even human productions.

The content and style of works of natural history is best conveyed through consideration of a typical example, such as *The Historie of Foure-footed Beastes* (1607) by English clergyman and naturalist Edward Topsell (1572–1625?). In the preface to the work, Topsell states that it is his intention to provide the reader with descriptions of "the true and lively nature of every Beast, with a discourse of their severall Names, Conditions, Kindes, Vertues (both naturall and medicinal), Countries of their breed, their love and hate to Mankinde, and the wonderfull worke of God in their Creation, Preservations, and Destruction."[3] Topsell's thirty-page account of the lion, for example, commences with a general introduction accompanied by an illustration, before proceeding to an account of the "severall names of Lions" in ancient and modern languages, etymological speculations on those names, the kinds of lions, their geographical distributions, their color, epithets, voice, rage, food and eating, cruelty, hatred, terrors, lust, adultery, and respect for their elders. There follows a selection of stories about lions, usually to do with their relations with humans—Androcles and the lion, unfortunate men devoured by lions, men who have overcome lions, men transfigured into lions. Topsell then gives an account of the mental characteristics of lions, their understanding, their anger and its signs, epigrams of poets concerning the rage of lions. Specific practical advice concerning the hunting and taming of lions is provided next, along with the "uses" of their parts. References to statues of lions, the constellation of the lion, scriptural allusions to the lion follow. The entry concludes with a separate section on "the medicines of the Lyon."[4]

The reader is thus presented with what seems to be an odd combination of fact, folklore and fantasy, literary allusions, poetry, and pharmacology. While some information appears in a form that might pass as "zoology" today, the intention of the author seems to be to provide a comprehensive account of this animal not as it appears in nature, but as it is represented in texts. Topsell's work is primarily one of scholarship rather than observation. The claims it makes are authenticated through appeals to textual authorities, the vast majority of which are ancient sources.

Topsell thus announces on his title page that "the story of euery Beast is amplified with Narrations out of Scriptures, Fathers, Phylosophers, Physicians, and Poets."[5]

Natural historians of this period made few claims to originality. On the contrary, for most of its practitioners, natural history was an attempt to reconstruct an ancient body of knowledge. Ulisse Aldrovandi (1522–1605), professor of medicine ("physic") and curator of the botanical gardens at Bologna, saw as his task the revival of a natural history that had been "buried for so many hundreds of years in the gloom of ignorance and silence."[6] Conrad Gesner (1516–1565), the foremost natural historian of his time and chief source for Topsell's *Historie*, also spoke about "renewing old and ancient things which were forgotten." The task of the natural historian, he suggested, was "to restore things from Death, or ruine which were sould thereto and to restore the names of things, and things by their names."[7] The assumption of humanist scholars was that a more-or-less complete knowledge of nature could be found in ancient texts: Aristotle (384–322 BCE) on animals, his disciple Theophrastus (ca. 371–286 BCE) on plants and stones, and the physician Dioscorides (ca. 40–80 CE) on plants. To these must be added the thirty-seven books of *Naturalis Historia* of the Latin encyclopedist Pliny the Elder (23/4–79 CE).[8] The expression "natural history" itself can be traced back to the Greeks, for whom *historia* carried the broad sense of "an enquiry into something remarkable."[9] Aristotle's works on animals had been known in Latin translation from the thirteenth century, and Pliny had been available in piecemeal form in the two distinct generations of medieval encyclopedias—those of Isidore of Seville (ca. 560–636 CE) and the Venerable Bede (ca. 673–735 CE), and the relatively more recent works of Vincent of Beauvais (ca.1194–1264 CE), Bartholomew Anglicus (fl. 1220–1240 CE), and Albert the Great (d. 1280 CE).[10]

Claims about the existence of a coherent ancient science that had once been universally revered need to be treated with some caution. There is a sense in which ancient natural history was actually the construction of Renaissance scholars who sought to legitimate their own activities by associating them with an ancient practice.[11] In fact, Renaissance works drew upon an enormous number of ancient sources that share few common disciplinary characteristics. In addition to Aristotle and Pliny, whose writings are separated by several centuries and already display disparate methods and motives, natural historians drew upon the allegorizing collections of animal lore to be found in the medieval bestiaries and books of birds, which for the most part were based on the *Physiologus*, a work written in the early Christian period.[12] These works were actually aids to the allegorical interpretation of scripture and acted as a repository of the symbolic moral and theological meanings of the creatures. Numerous other sources—poets, philosophers, theologians, and historians—were also utilized. Topsell's entry "Lyon" is fairly typical in that it relies heavily upon Aristotle and Pliny, but also makes use of a variety of other authorities: the *Physiologus* and a biblical commentary of Ambrose of Milan from the Patristic period, the thirteenth-century *De animalibus* of Albert the Great (itself a commentary on Aristotle), more recent authors such as Girolamo Cardano (1501–1576), and others. A further indication of the scale of the more ambitious of these projects can be gauged from the range and number of Conrad Gesner's sources. The preface to Gesner's five-volume *Historiae animalium* (1551–1587) provides the reader with an impressive inventory of sources that is testament to the author's monumental energy and linguistic facility: 4 Hebrew texts, 82 Greek authors, no fewer than 175 Latin authors, and several works by contemporary German, Italian, and French naturalists. Gesner's work epitomizes the scholarly approach of the Renaissance humanists to natural history.

Part of the task of the Renaissance natural historian was to correct errors of transmission through philological research and the comparison of extant texts. Through such labors these scholars were able to excise many of the errors and conflations that had crept into works on plants and animals.[13] In addition to these grammatical labors, humanist scholars could also harmonize accounts in the various sources under consideration. In his *Castigationes plinanae* (1493–1493), a work devoted to correcting the errors of Pliny, humanist scholar Ermolao Barbaro reduced the life span of the elephant, given by Pliny as two to three hundred years, to one hundred and twenty, the figure that Aristotle had provided.[14] Renaissance natural historians were thus not mere copyists, as some of their critics alleged, and although they dealt primarily with texts their efforts led to considerable improvements in knowledge of the natural world. Inevitably however, the enterprise of natural history moved beyond text-based activities. One reason for this lay in the problem faced by scholars in identifying unknown species from written accounts and unfamiliar names. On encountering an unknown Greek name in a text, a scholar could simply transliterate the name, as many medieval Arab translators had done, or attempt to find a known species that matched the description. The unsatisfactory practice of transliteration had led to the very conflations and confusions that the humanists sought to avoid, and hence the most satisfactory way to resolve the difficulty was to go to the field to identify the specimen that answered to the description. Thus scholars whose original interests had been solely philological found themselves developing an interest in actual animals and plants.

Correct identification was particularly crucial for plant species, for in order to prescribe cures physicians needed to be able to identify from the Latin nomenclature the appropriate local species. English divine and naturalist William Turner (1508?–1568), observing that existing herbals were "al full of unlearned cacographees and falselye naminge of herbes," set himself the task of correcting such deficiencies.[15] His *Libellus de re herbaria novus* (1538) was a lexicon of herbs that listed Latin names of herbs in alphabetical order along with their variants, with English and Greek equivalents. Turner had resorted to identifying entries in the Greek texts through fieldwork, and his herbal not only included the local names but was the first herbal in England to include local species previously unmentioned in Greek and Latin texts. When Turner shifted his attention to the study of birds, he encountered many of the same problems. Very few local species had received mention in the classical sources, and fieldwork was called for to remedy the deficiencies. Similar programs were undertaken on the Continent by Conrad Gesner, Hieronymous Bock (1498–1554), and Leonard Fuchs (1501–1566).[16] Such scholars found themselves not merely reproducing the descriptions of the ancients but imitating their methods.

The need to supplement the ancient catalogs of creatures was made all the more necessary by the voyages of discovery. The animals and plants of the New World issued a serious challenge to the monopoly over natural history held by the ancients. Explorer Amerigo Vespucci observed of the fauna of the new world, "Pliny did not touch upon a thousandth part of the species of parrots and other birds and the animals too which exist in those same regions."[17] This may have been something of an exaggeration, but nonetheless, improvements in navigation had resulted in a vast explosion of zoological data. While the hundreds of new species of animals provided delights for the curious, of more immediate practical import were new plants that promised new cures. In *Joyfvll Nevves out of the Newe Founde Worlde* (1577), Nicholas Monardes set forth "the rare and singular vertues of diuerse and sundrie Hearbes, Trees, Oyles, Plantes, and Stones, with their applications as well for Physic as Chirurgerie."[18] Works such as these, when combined with the publications of local botanists, point to a dramatic growth in botanical knowledge during this period. It is estimated that between 1550 and 1700, the number of known plants increased at least fourfold.[19] In the seventeenth century, the invention of the microscope also opened up new vistas, bringing to light innumerable creatures of unimaginable variety. Had Aristotle been alive in the seventeenth century, one author observed, he would had to have written a new history of animals, for "the first Tome of Zoography is still wanting."[20]

Early-modern naturalists thus found themselves increasingly torn between two apparently different kinds of activity, the one philological and grammatical, the other to do with matters of empirical fact. The new orientation of natural history called for a set of observational procedures of equal rigor that could supplement the more traditional text-based approaches. Eventually, the new observational approach was to supplant the authority of the ancients. By the beginning of the seventeenth century the idea of rehabilitating an ancient natural history was overtaken by calls for the complete reformation of the discipline.

THE REFORMATION OF NATURAL HISTORY

Natural histories of the seventeenth century show an increasing tendency to be critical of written sources, to stress firsthand observation, to set out specific criteria for accepting eyewitness accounts—in short, to move natural history away from a preoccupation with documents to a direct engagement with nature. While this process might originally have been motivated by a need to supplement the writings of the ancients, the late sixteenth century witnessed a growing reaction against the authority of the ancients, fuelled by a general impulse to reform the spheres of religion and learning. The Protestant Reformation had successfully challenged the authority of the Roman Church and its de facto philosophical saint, Aristotle. The German reformer Martin Luther (1483–1546), who had instituted reforms within the Church, had argued that the universities, "where only that blind, heathen teacher Aristotle rules," also stood in need of "a good, thorough reformation."[21] Luther's sentiments were echoed by educational reformers. Swiss preacher of medical reforms, Paracelsus (1493–1541), argued for the supremacy of the *Codex Naturae*, the book of nature, over against the books of Galen, Avicenna, and Aristotle. Inquiring minds, he insisted, ought to turn to that library of books that "God himself wrote, made and bound."[22] To know God's book, moreover, scholars needed to leave their cloistered libraries and enter the world of things. In England, Francis Bacon (1561–1626) allowed that there might have been a time in antiquity "when natural science was perhaps more flourishing" but nevertheless insisted that "new discoveries must be sought from the light of nature, not fetched back out of the darkness of antiquity."[23] This was particularly true in natural history, he argued, where "we see there hath not been that choice and judgment used as ought to have been; as may appear in writings of Plinius, Cardanus, Albertus, and divers other of the Arabians, being taught with much fabulous matter, a great part not only untried, but notoriously untrue."[24]

None of this meant the end of resort to written texts. Rather, the new engagement with nature brought with it a more critical and selective approach to the use of such authorities. Nehemiah Grew (1641–1712), one of the fathers of plant anatomy, announced in his catalog of the Royal Society's natural history collection that "I have made the Quotations, not to prove things well known . . . as if *Aristotle* must be brought to prove a Man hath ten Toes." Instead, authorities were cited "To be my Warrant, in matters less credible."[25] The same care is taken in the *Ornithology* (1678) of John Ray and Francis Willoughby, who state that in the composition of the work they have not "scraped together whatever of this nature is any where extant, but have used choice, and inserted only such particulars as our selves can warrant upon our own knowledge and experience, or whereof we have assurance by the testimony of good authors or sufficient Witnesses."[26] Ray and Willoughby proposed three legitimate sources for natural histories—personal warrant, good authors, and sufficient witnesses. The importance of the first had long since been stressed by Bacon, whose principle it was to "admit nothing but on the faith of eyes, or at least of careful and severe examination; so that nothing is exaggerated for wonder's sake." Only thus could natural history be purged of "fables and superstitions and follies."[27] While the bookish orientation of Renaissance natural historians may seem somewhat remote from these new observational practices, there are significant points of continuity. The humanist practices of excerpting, compiling, classifying, and sifting of textual evidence are not unconnected to the business of weighing and assessing testimonies and dealing with discrete matters of fact. Hence, the mental habits involved in the compilation of encyclopedic works of natural history could be, and undoubtedly were, reoriented toward the study of the empirical world.[28] But while both forms of natural history needed criteria for distinguishing good authors and reliable witnesses, such criteria became increasingly important during the seventeenth century.

New tests for the trustworthiness of observers stressed social status, education and training, personal virtues, and institutional settings. It has been suggested that in early-modern England credibility was associated with one's "gentlemanly" status.[29] It is also likely that standards for the reliability of testimony were imported from the legal context of the courtroom into newer scientific institutions, while the practice of taking medical histories also provided a precedent for the new factually based natural histories.[30] The Parisian Royal Academy of Sciences, founded in 1666, stressed the importance of both social status and institutional setting in legitimating observation claims. Relations issued under the auspices of the society, it was claimed, "do contain only Matters of Fact, that have been verified by a whole Society, composed of Men which have Eyes to see these sorts of things." These observers were "indowed with the Spirit of Philosophy and Patience," unlike the purveyors of travelers' tales such as merchants and soldiers, who were more likely to report curiosities and exaggerated novelties.[31] Similar claims about the perspicacity of the Fellows of the Royal Society of London were made by its first historian, Thomas Sprat (1635–1713).[32]

The reform of natural philosophy was not merely to be accomplished through turning from books to the natural world and establishing quality assurance mechanisms for witnesses. The very notion of "nature" and the conditions under which it was to be observed also underwent a significant transformation. In his proposals for a new style of natural history, Francis Bacon thought that the natural historian ought to be concerned with three aspects of nature: nature observed in its natural course, nature "erring and varying," and nature "altered and wrought."[33] For the second category Bacon hoped for a "particular natural history of all prodigies and monstrous births of nature; of everything in short that is in nature new, rare, and unusual."[34] The third category included not merely human productions but what Bacon referred to as "experimental history," in which nature was diverted from its normal course by experimentation. The scope of natural history was broadened to incorporate the unnatural and the nonnatural. This was a significant departure from the Aristotelian view that the science of nature was to be founded on generalizations made from common sense observations of nature in its normal course. The Baconian scheme was premised upon the notion that nature would only reveal its operations when put to the test, and forced out of its "natural" state. The inclusion of the "preternatural" among the legitimate objects of a natural history also ran counter to the Aristotelian tradition, for it encouraged an interest in the exceptional, and inhibited the premature formulation of general explanatory rules.

Bacon made a second claim about the natural world that signals a major shift in the understanding of the religious significance of nature. Nature, he said, provides evidence of the attributes of the Deity, but does not reflect his image.[35] The notion that the world, in some sense, bore the image of the Deity lay at the heart of the medieval emblematic understanding of the world. According to that view, all of the creatures in the world are natural symbols or mirrors of the transcendent. Nature could thus be studied as a text for its emblematic, theological meanings. This was the premise of the *Physiologus* and the bestiaries, of medieval allegorical readings of scripture, and of the emblematic components of Renaissance natural histories. Bacon insisted that the book of nature could not be read in this fashion and articulated the modern view that allegories are not about eternal ontological relations but linguistic conventions.[36] This new view of the natural world is evident in the things that seventeenth-century natural histories begin to exclude.[37] Nehemiah Grew, for example, declared that his descriptions of objects in the natural history collection of the Royal Society would be innocent of the traditional emblematic associations: "After the Descriptions; instead of medling with Mystick, Mythologick, or Hieroglyphick matter; or relating Stories of Men who were great Riders, or Women that were bold and feared not horses; as some others have done: I thought it much more proper, To remarque some of the Uses and Reasons of Things."[38] These omissions represent a monumental change in the way in which nature was conceived. The creatures have not been placed in the world to represent or communicate divine truths. They are there to be used. The Baconian ideal is thus realized in Grew's focus on a rational knowledge of natural objects that will facilitate their material exploitation.[39]

THE JUSTIFICATIONS AND USES OF NATURAL HISTORY

Early-modern works of natural history were often prefaced by lengthy claims about the dignity and worth of the discipline. It was frequently claimed that in antiquity natural history had been practiced or sponsored by individuals of the highest standing. These included the biblical patriarchs, the wise King Solomon, and Alexander the Great. No doubt much of this had to do with attracting wealthy or highly placed patrons, the support of whom was necessary for any large-scale project. In his *Herball or Generall Historie of Plantes* (1597), a work thoughtfully dedicated to Sir William Cecil, Lord High Treasurer of England, John Gerard (1545–1612) thought it worth informing his Lordship that the study of plants in the past had enjoyed the support of "noble Princes" such as Solomon and Alexander the Great, who had "joyned this study with their most important matters of state."[40] In Italy, the Medici popes and princes had self-consciously modeled themselves on the classical rulers, enthusiastically sponsoring collections of natural objects and facts about nature, and the related activity of the translation of ancient Greek texts.[41] With a slight tone of reproach, one English promoter of natural history pointed to the generous support for the study of nature in other countries: "I see examples of this munificence in our age to give me comfort: Ferdinand the Emperor and Cosimus Medices, Prince of Toscane are herein registred for the furthering this science of plants."[42] John Tradescant the Younger (1608–1662), in the catalog of his father's famous natural history collection, wrote that because this collection of rarities offered more "variety than any one place known in Europe could afford" the publication of its catalog would be "an honour to our Nation."[43] The promotion of natural history could thus become a matter of national pride.

Natural history was also claimed to be a religious duty, sanctioned by scripture. Adam's cultivation of the Garden of Eden and his naming of the beasts, the confusion of languages at the Tower of Babel, the collection and housing of animals in Noah's Ark—each of these narratives provided important justification for the practice of natural history and informed its methods. When Gesner spoke of restoring old and ancient things now forgotten, he was referring not only to the writing of the Greek philosophers but to an even more ancient Adamic science. In urging the recovery of "the names of things, and things by their names," he was alluding to Adam's naming of beasts in paradise, an episode generally taken as indicative both of Adam's perfect knowledge of the natures of living things, and of the existence of a natural language in which names express true natures. Edward Topsell thus explained that Adam had been an accomplished naturalist and for this reason natural history was "Divine, and necessary for all men to know."[44] The notion of an Adamic science also motivated Bacon and the Baconians in the Royal Society of London. According to Bacon, "the first acts which man performed in Paradise consisted of the two summary parts of knowledge; the view of the creatures, and the imposition of names."[45] Bishop Thomas Sprat, in his apologetic history of the Royal Society of London, went even further, agreeing that the pursuit of natural history and philosophy were religious duties, because they had been Adam's sole vocation.[46] Adam's knowledge, moreover, underwrote his dominion over the creation. The study of nature, in Bacon's project, was to effect "a restitution and reinvesting (in great part) of man to the sovereignty and power . . . which he had in his first state of creation."[47] The idea of an Adamic natural history persisted until well into the eighteenth century, when Carl Linnaeus (1707–1778), on the frontispiece of the 1760 edition of *Systema naturae*, depicted himself as Adam simultaneously naming the animals and writing the *Systema*.[48]

The cultivation of formal botanic or physick gardens was also justified as the partial reestablishment of an original Edenic order. Such gardens came to be regarded as living books, and as representative of Adam's encyclopedic knowledge. In the hundred years from the middle of the sixteenth century, these gardens sprang up all over Europe and were regarded as "living catalogues of plants, living catalogues of creation."[49] The human imposition of order on these living objects also embodied the restoration of lost dominion. In the eighteenth century, Carl Linnaeus, author of the classificatory system of binomial nomenclature, reordered the botanic gardens in Uppsala according to his new system, planting out the separate species in a determined order, representing the fixed patterns that God had used in the Creation. Linnaeus also thought that the domestication of the wilderness was a means of restoring it to its paradisal state.[50] Thus from the sixteenth to the eighteenth century, the encyclopedic garden was regarded as a living pharmacopoeia, "a kind of surrogate Bible," and a recreation of the garden of Eden.[51]

If botanic and physick gardens were regarded as reproductions of the original Eden, museums came to be thought of as recreations of Noah's Ark. Thus the collection of "Nature's admirable works" that belonged to the Englishman John Tradescant the Elder (d. 1638) and provided the basis for Oxford's Ashmolean Museum was popularly known as "Tradescant's Ark." The Old Testament patriarch was regarded as a vital link in the transmission of ancient knowledge from Adam to Moses, author of the Pentateuch. In order to fashion a vessel that would provide suitable accommodation for all the animals, and to stock it with appropriate provisions for them, Noah would have needed to have been quite an acute naturalist.[52] Early-modern custodians of natural history collections could regard themselves as latter-day Noahs, bringing together in a systematic fashion the whole realm of living things from the far-flung reaches of the Earth.[53]

Biblical sanctions for the pursuit of natural history were accompanied by a number of claims for the direct usefulness of the activity. At the beginning of his *Historiae animalium,* Conrad Gesner listed three benefits of natural history: animals and plants provide cures and other practical benefits, they teach us important moral lessons, and they are emblems of divine truths and point to the wisdom of the Deity.[54] The profitableness of histories of plants in the provision of cures was not seriously doubted by anyone, in spite of the inroads made by the novel chemical medicines of the Paracelsians and Helmontians. It was for this reason that natural history was closely associated with medicine. One seventeenth-century English physician wrote that "the word Physician, derived from the Greek φυσιχϖς, is plainly and fully rendred by the word *Naturalist,* (that is) one well vers'd in the full extent of Natural and natural things."[55] Only a precise knowledge of nature, the writer claimed, would enable the physician to prepare and apply appropriate cures. While animals were also used in the preparation of cures, they were used far less frequently than plants.

Hence the justification for the study of animals had to be sought elsewhere, and in particular, in the religious and moral sphere.

One traditional reason for being acquainted with the characteristics of animals was that it assisted in the interpretation of scripture. Wolfgang Franz informed readers of his *Animalium historia sacra* (1653) that the treatise should be of use and benefit "not only to physicians, but also to all scholars, and more especially to Divines," for "many places in Scripture cannot be interpreted without the knowledge of Animals."[56] Plants could also serve as aids to biblical exegesis. Ralph Austen (d. 1676) spoke of "the frequent use of *Similitudes* betweene the Church of God and Fruit-trees, and betweene our Saviour and Fruit-trees," allusions best understood through familiarity with these trees and their fruits.[57] With the gradual demise of allegorical readings of scripture and the gradual fading of the emblematic view of the natural world, the idea that each creature was a divinely instituted symbol became increasingly unsustainable. However, it was strongly argued that animals still taught important moral and theological lessons. The ancients—Aelian, Plutarch, and Pliny in particular—had sought moral lessons in the characteristics of animals, so it is not surprising that their latter-day imitators followed suit. Gesner wrote that "there want not instructions out of beasts, by imitation of whose examples, the lives and manners of men are to be framed to another and better practise."[58] English experimental philosopher Robert Boyle (1627–1691) thought that animals were both "doctors of divinity" and "teachers of ethics."[59] All creatures were thought to serve both moral and practical uses. Dogs, for example, were models of love and faithfulness, elephants of meekness, bees of justice, turtles of chastity.[60] While animals were no longer symbols of eternal verities they were still living object lessons in morality, and indeed claims that natural history could play an important role in the moral formation of those who studied it were not uncommon even in the nineteenth century.[61]

Finally, and perhaps most important of all, natural histories came to be thought of as vital components of natural theology. It was thought that truths about God could be discovered through the study of either of his "two books"—the book of scripture and the book of nature. Natural theology was concerned with the latter. Topsell claimed that nature is "that Chronicale which was made by God himself, every living beast a word, every kind being a sentence, and all of them together, a large History."[62] John Johnston, in his *Nature of Four-Footed Beasts*, spoke likewise of "Natures book, wherein we may behold the supreme power," going so far as to say that God "is comprehended under the title of natural history."[63] Somewhat confusingly, these works often contain an explicit denial that they are concerned with "divinity." In this context, "divinity" refers to revealed theology, which was concerned with more specific truths of particular religious traditions. Natural theology, or theology derived from nature, by way of contrast, was thought to be concerned with universal features of all religious traditions. Thus John Ray's assertion that he published the *Ornithology* "To the illustration of Gods glory, by exciting men to take notice of, and admire his infinite power and wisdom" is entirely consistent with his avowed omission of discussions of "divinity."[64] The singular benefit of natural theology as developed from the study of nature was that it gave rise to universal religious truths of the kind likely to lead to religious harmony rather than division. This was in keeping with the Enlightenment preference for a universal and rational religion.

From about the middle of the seventeenth century a new hybrid genre of natural theology emerged, called "physico-theology," the primary purpose of which was to lay bare the evidence of divine wisdom and goodness as it was manifested in the natural world. The advent of physico-theology ushered in the golden age of natural history, which flourished in the eighteenth century on account of its theological credentials. Notable works in this genre are John Ray's classic *The Wisdom of God Manifested in the Works of Creation* (1691) and William Derham's *Physico-Theology: or A Demonstration of the Being and Attributes of God from the Works of Creation* (1711–1712). One hundred years after the inauguration of the lectures the impulse that had inspired them was still strong. William Paley's *Natural Theology* (1802) was perhaps the most popular work on the topic ever written, and it was followed by the Bridgewater Treatises (1833–1840) a group of works devoted to the subject of "the power, Wisdom, and Goodness of God, as manifested in the Creation." These treatises, written by authoritative figures and informed by the latest findings in the natural sciences, were runaway bestsellers.[65]

While these theologically oriented natural histories were particularly popular in England, the vogue for physico-theology also became widespread throughout Europe. Bernard Fontenelle, for many years secretary of the Parisian Academy of Sciences, sanctified the activities of that august body with his claim that the divine wisdom was most plainly seen in the "mechanism of the animals" and that "true physics can be elevated into a kind of theology."[66] His sentiments were reflected in the writings of Dominique Réverend and Noël Antoine Pluche, both members of the Catholic clergy who wrote classics of physico-theology that were subsequently translated into English.[67] Dutch naturalist Bernard Nieuwentijt produced *L'existence de Dieu, démontrée par les merveilles de la nature* (1725), devoted to the standard theme of natural theology.[68] The structure and behavior of insects was again a popular topic. Friedrich Lesser, the Lutheran pastor of Nordhausen, Thuringia, produced a complete treatise on the subject entitled *Théologie des Insectes, ou Demonstration des Perfections de Dieu dans tout ce qui concerne les Insectes* (1742).[69] Swedish taxomonist Carl Linnaeus announced that the world was like "a museum" in which God had arranged all the living creatures as "admirable proofs of his wisdom and power."[70] For many in the early-modern period, the pursuit of natural history was an intrinsically theological activity.

HISTORY, NATURAL HISTORY, AND NATURAL PHILOSOPHY

"As all things have their revolutions, so hath naturall History the same chance. It was held for a goddess . . . but now a dayes it is so despised, that it is of no esteem at all; this matter needs no proving."[71] John Johnston's pessimistic assessment of the status of natural history, issued in 1631, was in some respects not far from the mark. In spite of the lofty rhetoric stressing its manifold uses, natural history never received the kind of recognition reserved for its more highly regarded relation, natural philosophy. William Wallace has observed of the Renaissance curriculum that "little or no attention" was paid to Aristotle's natural history.[72] According to Gilbert Jacchaeus, professor of philosophy and medicine at Leiden, natural history was not taught in the universities because it did not involve demonstrative science, it was not difficult enough to require the assistance of a teacher, and there was insufficient time to include it in the two- or three-year philosophy course.[73] The situation improved little during the course of the sixteenth and seventeenth centuries. In his *Herbal*, John Gerard rued the fact that "that which sometime was the study of great Phylosophers and mightie Princes, is now neglected."[74] Histories of animals, too, were "in a great measure neglected by English men," according to John Ray.[75] The modest status of the discipline was reflected in the number of university chairs devoted to its profession. John Johnston expressed doubts about whether natural history was "entirely professed in any University or School (except Bolognia where Aldrovandus was)."[76] In spite of its much-vaunted theological usefulness and its popular appeal, natural history was slow to make inroads into the traditional university curriculum.

The fortunes of natural history in this respect were very much determined by its relationship with bordering disciplines. On the one hand, as natural *history*, it was one of the historical disciplines.[77] On the other hand, as *natural* history, its subject matter was nature. The hybrid character of the enterprise conspired against the attainment of higher status, for it lacked both the human relevance of civil history and the explanatory power of natural philosophy. The English philosopher Thomas Hobbes (1588–1679) gives some sense of the relative position of natural history amongst the disciplines: "The register of knowledge of fact is called history. Whereof there be two sorts: one called natural history . . . The other, is civil history; which is the history of the voluntary actions of men in commonwealths. The registers of science, are such books as contain the demonstrations of consequences of one affirmation, to another; and are commonly called books of philosophy." Philosophy is itself divided into a number of categories by Hobbes, the first of which concerns natural bodies, and which is natural philosophy.[78] The difference between natural history and natural philosophy is clear from Hobbes's distinction between the realm of facts and the realm of the sciences. Natural history, as concerned with mere description, belongs to the former. Natural philosophy, on the other hand, is a science concerned with causal explanation. In this understanding of things, one of the prospects for raising the status of natural history would be to provide it with a role related to the more elevated goals of natural philosophy. This was more or less the position of Aristotle, who had understood his *History of Animals* as prolegomena to philosophical zoology that would deal with universal judgments rather than particular and accidental facts.[79] Of more immediate importance, this was precisely the vision of natural history that Francis Bacon set out at the beginning of the seventeenth century and that, along with its association with natural theology, provided a source of legitimacy for natural history for the next two hundred years.

Bacon's plan for the renovation of the sciences had involved a new representation of their relations, and his "tree of knowledge" was of utmost importance for understandings of the connections between various branches of knowledge throughout the seventeenth and eighteenth centuries. In the sciences concerning nature, pride of place was given to the history of nature, with Bacon asserting that "the foundation must be laid in natural history."[80] This natural history was to be a reformed enterprise with more strict conditions for the collection of evidence and a proper arrangement of materials with a view to the service they were to provide for natural philosophy. Bacon credits Pliny with having been the only "person who ever undertook a Natural History according to the dignity of it" but adds that "he was far from carrying out his undertaking in a manner worthy of the conception."[81] Bacon's conclusion was thus that "all the natural history we have, whether in the mode of inquiry or in the matter collected, is quite unfit for the end which I have mentioned, namely, the Foundation of [natural] Philosophy."[82] However, the facts of a history of nature, including the preternatural and experimental, collected with appropriate safeguards to ensure their accuracy, would "give light to the discovery of causes and supply a suckling philosophy with its first food."[83] Bacon thus envisaged the disciplines of natural philosophy forming a pyramid: Natural history at the bottom, physic in the middle, metaphysics close to the top, and at the pinnacle "the summary law of nature" that God used in the Creation.[84] Bacon's view of the relations of the disciplines was adopted by the Royal Society of London and was reproduced in eighteenth-century maps of knowledge in such works as John Harris's *Lexicon Technicum* (1704), Ephraim Chambers's *Cyclopedia* (1728), and in the famous frontispiece of the *Encyclopédie* (1751). The Baconian vision of the methods of the sciences, along with his conception of their interrelation, persisted until the middle of the nineteenth century.[85]

In spite of this attempted elevation of the role of natural history, because it dealt with description rather than causal explanation, many continued to regard it as something less than genuine "science" or "philosophy." The discovery of causes and general explanatory principles lay in the domain of philosophy. English Aristotelian John Sergeant (1622–1707) thus dismissed Baconian methods as "utterly Incompetent or Unable to beget Science." This was because, in his view, it was "meerly Historical, and Narrative of Particular Observations" from which it was impossible to deduce universal conclusions.[86] It was against such long-standing prejudices that Wolfgang Franz, with whose description of the scope of natural history this chapter began, was to insist that natural history should really be regarded as "belonging more properly to philosophy." More realistically, perhaps, it could be contended that the important thing about such natural histories was that they contained true observations, even if these provided only a partial account of the relevant phenomena. In his *Mémoires pour servir à l'histoire naturelle des Animaux* (1671–1676), Claude Perrault (1613–1688) of the Parisian Academy of the Sciences observed that some natural histories were memoirs "confined to the Narrative of some particular Acts, of which the Writer has certain knowledge." These accounts were "only the parts" of a history, yet were to be preferred on account of their "certainty and truth."[87]

Certain and true such relations might have been, but together they could hardly be said to comprise a coherent body of knowledge. The danger of this approach was that natural history might become disparate collections of data ungoverned by any theoretical conceptions or ultimate end. Perrault defended this method by claiming that "things of this Nature might be put into *Mémoires*, which are as it were *Magazines*, wherein are lockt up all sorts of things, to be made use of in times of need."[88] Such "memoirs" were common genres for the production of natural history in the seventeenth and eighteenth centuries: witness Robert Boyle's *Memoirs for the Natural History of Humane Blood* (1683) and Robert Plot's *The Natural History of Oxfordshire* (1677), subtitled "being an essay toward the natural history of England." The issue for natural historians was exactly how much factual information was required, and how long the business of philosophical speculation should be postponed. Perrault observed that "it is impossible to Philosophize without making some general Propositions which ought to be grounded on the knowledg of all particular things, whereof Universal notions are composed; and that we still have a long time to work, before we can be instructed in all the particulars necessary for this End."[89] The danger of the Baconian model of natural history was that it could paradoxically lead to the triumph of those very impulses that Bacon had sought to stifle: the "over-curious diligence in observing the variety of things." During the eighteenth century, critics of the Royal Society frequently alleged that its members were more concerned with "the collection of diverting and amusing specimens" than with "the sort of programmatic and socially influential enterprise that Bacon had hoped to promote."[90]

NATURAL HISTORY AND THE HISTORY OF NATURE

The Baconian vision of natural history was of collections of facts unencumbered by theoretical preconceptions. This was commendable in principle, but had the almost inevitable consequence that some kind of ordering principle would move into the domain of natural history in order to fill the explanatory vacuum. To a degree, this was why natural theology came to provide an important governing principle of natural history for much of the eighteenth and nineteenth centuries. However, natural theology was by no means the sole systematizing principle operating in natural history during this period. Prior to the close of the seventeenth century the study of nature had been "historical" only in the special sense that it dealt with observed facts. From this time natural history also began to include elements that were "historical" in the familiar sense, that is, to do with questions of origins and change over time. Natural history, in short, became chronological.

One impetus for this change arose out of the tendency to read the Bible, and Old Testament narratives in particular, in an historical or literal, rather than allegorical, sense. Works dealing with the natural history of

Earth, for example, took as their point of departure the observations of the "sacred historian" Moses, presumed author of the first five books of scripture. In 1695, John Woodward (1665–1728), professor of physic at Gresham College, London, published his *Essay towards a Natural History of the Earth* (1695). In this work he resorts to Moses's account of the formation of Earth—only in the latter's capacity "as an Historian" he assures us—even though he is engaged in the writing of a "physical discourse."[91] In doing so, Woodward was following the lead of other authors who had also sought to combine natural and sacred history, the most famous example of which was *Telluris Theoria Sacra* (*Sacred Theory of the Earth*) of Thomas Burnet (1635?–1715), which appeared in 1681. Burnet's work sets forth "the original of the earth, and all of the general changes which it hath already undergone or is to undergo till the consummation of all things" and represents an interesting reversal of the usual relation of natural history and natural philosophy, for it is an application of Descartes' natural philosophy to an aspect of natural history.[92] It was the first in a series of similar works that attempted to combine the natural history of Earth with sacred chronology. These include the *Essay* of Woodward, *A New Theory of the Earth* (1696) by Newton's successor in the Lucasian Chair, William Whiston, and Thomas Robinson's *New Observations of the Natural History of this World of Matter, and of this World of Life* (1696).

A number of these works also involved speculations about the origin and development of living things. The biblical narratives of the Fall and Flood suggested mutations both of Earth and of its human inhabitants. In the seventeenth century, Burnet and Whiston had already explained how these events in sacred history had wrought changes in the bodies of both humans and animals. During the eighteenth century natural historians took up the story. For them, the diversity of the human race called for some account of the changes wrought on successive generations of human beings that had altered them from the unity of their original created perfection. These changes were usually accounted for in terms of degeneration from an original ideal. The biblical account of the Fall meshed neatly with a predominant renaissance theme that stressed the perpetual degeneration of all natural things. Added to this was the long-standing Aristotelian principle that what is stable, fixed, and immutable is superior to that which is subject to vicissitude. The diversity of human customs and manners, of modes of divine worship, of languages—all bore witness to the deviation of Adam's race from an originally perfect physiognomy, robust health, adherence to a single universal language and the one true religion. Climatic change, together with the dispersion of living things over the face of Earth—both of which could be directly linked to human transgressions—were the most commonly cited causes of the degeneration of animals and plants. By a happy coincidence the recently discovered continent of America seemed to provide a living laboratory in which would be confirmed the deleterious effects of climate and dispersion.

These themes converged in the theories of organic degeneration of George-Louis Leclerc, comte de Buffon (1707–1788). Buffon is best known for his *Histoire naturelle, générale et particulière* (1749–1804), probably the most ambitious work of natural history ever produced. The groundwork for this vast opus commenced during Buffon's employment as keeper of the Royal Gardens (*Jardin du Roi*, now the *Jardin des Plantes*) and emerged out of a project to catalog the royal collections of natural history. Buffon transformed this modest task into a massive undertaking to provide a comprehensive account of the whole of nature. A total of fifty volumes were planned, of which thirty-six appeared during Buffon's lifetime. The immense scope of the project, combined with the fact that its compilation was essentially the work of a single mind, meant that Buffon was able to draw connections between the various subdisciplines of natural history and to make universal claims about living things in general. Buffon thus introduced an element of explanation into natural history, seeking a single principle that could account for change and diversity in living things. He also encouraged other natural historians to move beyond description, to link facts together, to draw generalizations, and thus "to arrive at a higher form of knowledge."[93]

According to Buffon, the diversity of living things proceeds from "change, deterioration, degeneration." The specific agents of change include climate, diet, domesticity, transportation, and migration.[94] Some of Buffon's compatriots posited similar agents of change. Pierre-Louis Maupertuis could thus claim that inhabitants of glacial regions were "deformed," while Corneille de Pauw argued along similar lines that all of the native species of America were "degenerate."[95] But Buffon went further, entertaining the possibility that structurally similar creatures might be the descendants of a common ancestor.[96] Buffon did not develop these ideas in any great detail, however, and ultimately conceded—whether motivated by piety or prudence is not clear—that "it is certain from revelation that . . . every species emerged fully formed from the hands of the creator."[97] Buffon's ideas show how nature could become the subject of a history in the chronological sense. But he differs from earlier historians of Earth by ignoring the kinds of historical records that earlier writers had referred to, including sacred scripture. For Buffon, it is "the archives of nature" that are to be consulted.[98] This new conception signals another important shift in the study of nature away from static metaphors for the natural world—book, mirror, garden, museum—to a dynamic one. The archive is a record of historical change. Buffon's approach thus contrasted with both the enterprise of physico-theologians such as Ray, Derham, and Pluche, who typically sought to catalog remarkable instances of creaturely adaptations that would serve as specific examples of divine ingenuity and benevolence, and those historians of Earth who sought evidence for their theories in scripture. These developments did not necessarily entail a departure from theistic readings of nature, but rather the introduction of a new understanding of the principle of change in the natural world. That the divine purposes were being worked out in human history was never seriously held in doubt prior to the nineteenth century. The attempt to link chronological or sacred history with natural history brought with it the quest for evidence of a single divine plan being worked out in the realm of nature. Authors of the composite natural/sacred histories of Earth had already grasped this principle. William Whiston insisted that "the State of the *Natural* is always accommodated to that of the *Moral* World."[99] A similar principle was asserted by German philosopher Gottfried Wilhelm Leibniz (1646–1716), who claimed that a harmony existed between "the physical realm of nature and the moral realm of grace," arguing that "the laws of motion and the changes of bodies serve the laws of justice and control."[100]

The rudiments of such conceptions were already evident in the seventeenth-century speculations about the chain of being and the justice of God. This notion of the great chain of being was an ancient conception according to which all existing things were graded on a scale, beginning from nonbeing, through imperfect creatures, higher animals, human beings, angels, to God. A number of thinkers had reasoned, however, that being fixed to a certain position on the ladder of being would limit the happiness of creatures and that God would have created orders of being capable of improvement beyond their original station. Some speculated that this would simply entail the resurrection of animals and plants at the last judgment and their restoration to perfection in which they had existed prior to their Fall. Others, drawing upon the Neoplatonic tradition, thought that the same outcome could be achieved by allowing for the transmigration of souls through the various levels of the ladder of being. Lady Anne Conway (1631–1679) claimed that God's justice was expressed in the transmutation of things "out of one *Species* into another." The tendency of such transmutations is toward perfection: "the nature of all creatures . . . to increase, and infinitely advance towards a farther Perfection."[101] Leibniz, who was acquainted with Conway's ideas, also spoke of "the transformation of animated bodies," suggesting that the substances that make up living things manifest "the law of the continuous progression of its own workings."[102]

These theologically informed notions of biological progression were to inspire the developmental schemes of Swiss naturalist Charles Bonnet (1720–1793), who was the first individual to use the term "evolution" in the context of natural history. In *Considérations sur les corps organisés* (1762), Bonnet rehearses the idea that all the creatures that have ever lived, and ever shall live, have existed in microscopic form, each encased within its parent, from the time of the creation. The next phase of the argument appeared in *La Palingénésie philosophique* (1770), where Bonnet argues that this present world is but one phase of a cosmic succession of worlds. The history of the cosmos is punctuated by geological catastrophes that bring each world to a close, and initiate the next. The biblical Flood was the most recent such terrestrial event.[103] The world before the Flood was inhabited by creatures quite different from our own, yet that exist in the present world in a more advanced form. Each human soul is encased in a number of germs—one for each recreation of the world. As successive worlds die and are born again, human souls progress through the various germs, in time attaining the highest form of perfection.[104] Animals too, are promoted through the revolutions of the world. Bonnet imagines "a continual progress, more or less gradual, of all species towards a higher perfection." All creatures on the scale of being "will be continually changing in a constant and determined order."[105] This was natural history on a grand cosmic scale, at its most theoretically and metaphysically sophisticated.

The enterprises of Buffon and Bonnet are somewhat removed from the ideals of Bacon, insofar as they attempt to explain the diverse phenomena of nature according to a single principle. These later writers strived to bring lawful explanation into natural history and thus make it more like natural philosophy. The introduction of these grand organizational principles into natural history was among a number of factors that enhanced its public stature. "The science of natural history," wrote one observer in 1765, "is more cultivated than it ever has been . . . at present natural history occupies the public more than experimental physics or any other science."[106] Another consideration in the changing status of natural history was the popularity of natural theology. As we have seen, less dynamic accounts of natural history than those of Buffon and Bonnet could also be organized around a single principle—that of design. Many eighteenth- and nineteenth-century descriptive natural histories are really natural theologies governed by the principle of divine forethought. Peter Roget (1779–1869), physiologist, secretary of the Royal Society, and creator of the famous thesaurus, declared in his Bridgewater Treatise that with the single principle of design, he had been able to marshal the facts of natural history into a "methodized order," to unite them into "comprehensive generalizations" and to establish their "mutual connections." He was thus able to bring "a unity of design, and that scientific form, which are generally wanting in books professedly treating of Natural Theology" while at the same time, cherishing the hope that his work "might prove a useful introduction to the study of Natural History."[107]

It was but a small step from these pious efforts to more heterodox accounts of the later nineteenth century.[108] In his *Vestiges of the Natural History of Creation* (1844), Robert Chambers thus spoke of a divine principle that permeated the natural world. God was "present in all things," and all things conformed to an "original divine conception." A single law governed both natural orders: "The inorganic has one final and comprehensive law—Gravitation. The organic, the other great department of mundane things, rests in like manner on one law, and that is—Development."[109] Ironically, just as Newton's assertion that gravity is the constant and lawful activity of the Deity had led to the naturalistic assertion that God was no longer necessary to explain the phenomena of attraction, so too with the claim that a single divine principle pervaded the organic world. Charles Darwin (1809–82), perhaps the most famous naturalist of them all, also claimed that a single selective principle lay beneath the origin of the diverse species. This principle did not seem to require direct divine activity, yet in a sense represents the culmination of the whole theological tradition of natural history in its quest for a single explanatory principle for the diversity of living things. Thus in the frontispiece of the sixth edition of the work for which he is most famous, *On the Origin of Species* (originally published in 1859), Darwin could quote from the Bridgewater Treatise of William Whewell (1794–1866): "with regard to the material world . . . we can perceive that events are brought about not by isolated interpositions of Divine power, exerted in each particular case, but by the establishment of general laws." In another, perhaps unintentionally ironic gesture, Darwin cites in the same place Francis Bacon on the study of the book of nature. In fact Darwin's *Origin of Species* signaled the end of the dominance of Bacon's influential philosophy of science. Darwin had come to the recognition that the Baconian ideal of natural history would mean the indefinite postponement of theorizing and that no general laws of absolute certainty would emerge from the inductive processes as Bacon had hoped. Darwin proposed instead the formulation of hypotheses that would guide the collection of facts. Natural selection was thus proposed not as a certain law, but as the most probable explanation of the disparate facts related to the origins of species: "The line of argument often pursued throughout my theory is to establish a point as a probability by induction, and to apply it as a hypothesis to other points, and see whether it will solve them."[110] Darwin's break with Baconianism led to the charge that his theory of evolution by natural selection was "unscientific"—a claim that was true if Bacon's standards were to be upheld.[111] The acceptance of Darwin's theory thus depended on a revision of the criteria for what constituted a proper use of the inductive method.

Darwin refashioned the methods of natural history by modifying Baconian notions of induction, while at the same time enabling a new conception of the sciences that included the study of the natural world without recourse to theistic explanations. Admittedly, natural selection was, from the first, a controversial mechanism for biological change. But it is significant that from the time of the publication of *On the Origin of Species*, proposed alternative mechanisms were themselves almost invariably nontheistic.[112] The first generation of Darwinians sought a new alliance of the sciences united by, amongst other things, a common naturalism. The agenda of "Darwin's bulldog," Thomas Huxley (1825–1895) was to excise theology from natural history and thus to render it more like physics or astronomy. His was a conception of the unity of the sciences in which the cost of admission for the study of nature was the exorcising of its theological demons. His colleague John Tyndall (1820–1893) expressed the same desire to "wrest from theology the entire domain of cosmological theory" and at the same time, presumably, to wrest natural history from the hands of the clergy, who for so long had been among its major exponents.[113] Thus sacred history and natural theology, having introduced dynamic and unifying elements into natural history, were unceremoniously bade farewell. An important element of this transformation was to do with the identity of practitioners of natural history. For much of the nineteenth century, natural history was conducted, as it had been in the preceding two centuries, by nonprofessionals—among them a disproportionate number of members of the clergy. Huxley once wrote of his concern for a proposed research fund that it be "a scientific fund and not a mere naturalists' fund," explaining that "the word 'naturalist' unfortunately includes a far lower order of men than chemist, physicist, or mathematician . . . every fool who can make bad species and worse genera is a 'Naturalist.'"[114] It was Huxley's intention to make the study of nature "scientific" by establishing its independence from clerical dominance, and legitimising a new set of nonecclesiastical authorities.[115] "Science" was argued to be an activity that excluded theology, and a postfacto justification for the expulsion of theology from natural history was achieved through the creation of the myth of a longstanding rational and secular "science" that had been engaged in a perennial warfare with theology. So successful was this strategy that even now it is difficult to envisage a natural history that from its inception in the sixteenth century to its decline in the nineteenth, was often governed by theological principles and pursued by those with theological motivations. As for the new alliance of disciplines clustered under the umbrella "science," there remain tensions, but these are largely obscured by the standard usages of "science," "scientist," "laws of nature" and "scientific method." It cannot be assumed, despite the rhetoric of advocates of a scientific biology, that the "laws" that govern biological processes are close analogs of those that operate in the realm of physics, as Chambers and Huxley seem to have held.[116] As for the wide applicability of the term "scientist," the Nobel Prize–winning physicist Ernest Rutherford (1871–1937) was said to have remarked once that "scientists were divided into two categories—physicists and stamp collectors."[117] The place of biologists in this dichotomy is clear.

Apart from the professionalization of natural history, the nineteenth century witnessed other important developments. Even before the appearance of *On the Origin of Species*, laboratory-based biological sciences had gradually been assuming the role of the "scientific" study of nature. Something of this process can be seen in the insistence of the editors of the *Zeitschrift für wissenschaftliche Zoologie* (*Journal for Scientific Zoology*), founded in 1848, who announced that their concern was to give their new journal "the most scientific character possible."[118] The editors were quite specific in their desire to exclude descriptions of new species, and "simple notes and natural history news," their primary interests being morphology and physiology. Traditional natural historians, "mere naturalists" as Huxley would have it, were advised to avoid disappointment and submit their unscientific dross elsewhere. Related to this was the intensification of scientific specialization, which helped ring in the changes for natural history by undermining its unity. In the seventh edition of the *Encyclopaedia Britannica* (1830–42), readers of the entry "natural history" were for the first time informed that this field was now divided into five distinct disciplines: meteorology, hydrography, mineralogy, botany, and zoology.[119] By now, experimental biology was flourishing, and new research laboratories were built in the major centers and universities. The venue of natural history, having been successively the humanist's library and the vicar's country parish, became the scientist's laboratory. Such developments are suggestive of a transformation of natural history into an experimentally based "science." More accurately, perhaps, natural history was simply eclipsed by the new "scientific biology."[120] Abandoned by the newly defined "scientists," and evacuated of its theological meanings, by the beginning of the twentieth century natural history had been alienated from the two enterprises that could lend it legitimation.

CONCLUSION

Natural history no longer plays an integral role in the study of nature. Indeed, it no longer exists as a serious academic discipline. The Adamic dream of a complete catalog of living creatures is further away than ever before, with a huge proportion of species still awaiting description and classification. Neither is it seriously claimed that the description and naming of these innumerable undescribed creatures will add much of scientific substance to the sum of human knowledge. The religious and moral justifications for the pursuit of natural history have held little appeal for some time. The logical contention of David Hume (1711–1776) that values cannot be derived from facts, along with the development of modern ethical theory by such figures as Immanuel Kant (1724–1804) and John Stuart Mill (1806–1873), have made claims about the moral authority of natural history doubtful. The contention that God's wisdom and goodness is evident in the specific adaptations of creatures, or in the general benevolence of the system of nature, has also become difficult to sustain in the wake of the general acceptance of Darwin' theory of natural selection—witness J. B. S. Haldane's (1892–1964) celebrated observation that the one thing that could be discerned about the Deity from the study of nature was his "inordinate fondness for beetles."[121]

Yet the legacy of natural history lives on. "Scientific creationism" bears more than a passing resemblance to the natural histories of Earth produced in the late seventeenth and early eighteenth centuries. On the other side are the recent attempts of evolutionary psychologists to derive moral principles from the putative facts of evolutionary history in a manner akin to early-modern moralizers of insect behaviors—as if Hume had labored in vain to distinguish facts and values. The closest descendent of nineteenth-century natural history amongst the academic disciplines is perhaps ecology, although more mathematically inclined ecologists might not be happy with this genealogy. "Ecology," according to one of its early practitioners, "simply means a scientific natural history."[122] At a popular level—and natural history has always enjoyed popular appeal—sales figures for the essays of Stephen Jay Gould, ratings for the television productions of David Attenborough, the burgeoning membership of societies dedicated to conservation, and the crowds of children who pack the dinosaur halls in museums of natural history are testament to the fact that traditional natural history flourishes in some form, even if as Francis Bacon had feared in the seventeenth century, it is more often than not as entertainment for the curious.

NOTES

1. Wolfgang Franz, *The History of Brutes* (London, 1672), 1. English translation of *Animalium Historia sacra* (Amstelodami, 1653).

2. David C. Lindberg, *The Beginnings of Western Science* (Chicago: University of Chicago Press, 1992), 348.

3. Edward Topsell, *The Historie of Foure-footed Beastes* (London, 1607), title page.

4. Ibid., 456–87.

5. Ibid., title page.

6. Paula Findlen, "Courting Nature," in *Cultures of Natural History*, ed. Nicholas Jardine, James Secord, and Emma Spary (Cambridge: Cambridge University Press, 1996), 57.

7. Topsell, *The Historie of Foure-footed Beastes*, sig. 3v.

8. On natural history in antiquity see G. E. R. Lloyd, *Science, Folklore, and Ideology: Studies in the Life Sciences of Ancient Greece* (Cambridge: Cambridge University Press, 1983); Roger French, *Ancient Natural History* (London: Routledge, 1994); Sorcha Carey, *Pliny's Catalogue of Culture: Art and Empire in the "Natural History"* (Oxford: Oxford University Press, 2003); Trevor Murphy, *Pliny the Elder's "Natural History": The Empire in the Encyclopedia* (Oxford: Oxford University Press, 2004).

9. C. W. Fornara, *The Nature of History in Ancient Greece and Rome* (Berkeley: University of California Press, 1983), 15.

10. Vincent of Beauvais, *Speculum naturale* (1220–44); Thomas of Cantimpré, *De natura rerum* (ca. 1228–44); Albertus Magnus, *De vegetabilibus et plantis, De animalibus* (1258–62).

11. For the claim that natural history was actually the "invention" of the Renaissance period, see Brian Ogilvie, "The Humanist Invention of Natural History," in *The Science of Describing: Natural History in Renaissance Europe* (Chicago: University of Chicago Press, 2006), 87–138.

12. *Physiologus*, trans. by Michael Curley (Austin: University of Texas Press, 1979); Lindberg, *The Beginnings of Western Science*, 348–53.

13. For examples, see Brian Copenhaver, "A Tale of Two Fishes: Magical Objects in Natural History from Antiquity through the Scientific Revolution," *Journal of the History of Ideas* 52 (1991): 389; Jerry Stannard, "Medieval Herbals and their Development," *Clio Medica* 9 (1974): 27.

14. Allan Debus, *Man and Nature in the Renaissance* (Cambridge: Cambridge University Press, 1978), 35.

15. William Turner, preface to *A New Herbal* (London, 1568); F. D. and J. F. M.

Hoeniger, *The Development of Natural History in Tudor England* (Charlottesville: University Press of Virginia for the Folger Shakespeare Library, 1969), 20–36.

16. Konrad Gesner, *Catalogus plantarum Latine, Graece, Germanice, et Gallice* (Tiguri, 1542); Leonard Fuchs, *De historia stirpium commentarii insignes* (Basilae, 1542); Hieronymous Bock, *Hieronymi Tragi, De stirpium, maxime earum quae in Germania nostra* (Argentorati, 1552).

17. Quote in Anthony Grafton, *New Worlds, Ancient Texts* (Cambridge, MA: Harvard University Press, 1992), 84.

18. Nicholas Monardes, *Joyfvll Nevves out of the Newe Founde Worlde*, trans. by John Frampton (London, 1577), title page and fol. 34v; also see Miguel de Asúa and Roger French, *A New World of Animals: Early Modern Europeans on the Creatures of Iberian America* (Aldershot, UK: Ashgate, 2005).

19. André Cailleux, "Progresson du nombre d'espèces de plantes décrites de 1500 à nos jours," *Revue d'histoire des sciences* 6 (1953): 44. Cf. Karen Reeds, "Renaissance Humanism and Botany," *Annals of Science* 33 (1976): 540.

20. Henry Power, preface to *Experimental Philosophy* (London, 1664).

21. Martin Luther, "To the Christian Nobility," in *Three Treatises* (Philadelphia: Fortress Press, 1970), 92.

22. W. E. Peuckert (ed.), *Paracelsus: Die Geheimnisse: Ein Lesebuch aus seinen Schriften* (Leipzig: Dieterich, 1941), 172–78.

23. Francis Bacon, *Novum Organum*, Pt 1, §22, in *The Works of Francis Bacon*, ed. James Spedding, Robert Ellis, and Douglas Heath, (London: Longman, 1857–74), 4:108.

24. Francis Bacon, *The Advancement of Learning* and *New Atlantis*, ed. Arthur Johnston (Oxford: Clarendon, 1974), *Advancement*, Bk I, iv, 10, (30).

25. Nehemiah Grew, preface to *Musaeum Regalis Societatis* (London, 1681).

26. John Ray and Francis Willughby, preface to *The Ornithology of Francis Willughby* (London, 1678).

27. Bacon, *Great Instauration*, 4:30f.

28. Ogilvie, *Science of Describing*; G. Pomata and N. G. Siraisi, eds., *Historia: Empiricism and Erudition in Early Modern Europe* (Cambridge, MA: MIT Press, 2005).

29. Steven Shapin, *A Social History of Truth: Civility and Science in Seventeenth-Century England* (Chicago: University of Chicago Press, 1994); Simon Schaffer, "Godly Men and Natural Philosophers," *Science in Context* 1 (1987): 55–85.

30. Barabara Shapiro, "Testimony in Seventeenth-Century English Natural Philosophy: Legal Origins and Early Development," *Studies in History and Philosophy of Science* 33 (2002): 243–63; R. M. Sargeant, "Scientific Experiment and Legal Expertise: The Way of Experience in Seventeenth-Century England," *Studies in History and Philosophy of Science* 20 (1989): 19–45; Gianna Pomata, "*Praxis Historialis:* The Uses of *Historia* in Early Modern Medicine," in Pomata and Siraisi, *Historia*, 105–46.

31. Claude Perrault, *Memoir's [sic] for a Natural history of Animals* (London, 1688), a translation of *Mémoires pour servir à l'histoire naturelle des Animaux* (Paris, 1671–76), sig. A2v.

32. Sprat, *History of the Royal Society* (London, 1667), 91.

33. Bacon, *De Augmentis*, in *The Works of Francis Bacon*, 4: 292–95. Cf. *Works* 3:361; see also Robert Hooke, preface to *Micrographia* (London, 1665).

34. Bacon, *Novum Organum*, 4:169.

35. Bacon, *De Augmentis*, 4:341.

36. Ibid., 340.

37. William Ashworth, "Natural History and the Emblematic World View," in *Reappraisals of the Scientific Revolution*, ed. David Lindberg and Robert Westman (Cambridge: Cambridge University Press, 1990).

38. Grew, preface to *Musaeum Regalis Societatis*. Cf. Ray and Willughby, preface to *The Ornithology of Francis Willughby*; William Wotton, *Reflections upon Ancient and Modern Learning*, 2nd ed. (London, 1697), 258.

39. See Peter Harrison, *The Bible, Protestantism, and the Rise of Natural Science* (Cambridge: Cambridge University Press, 1998), 1–4, 161–84 passim.

40. John Gerard, *The Herball or Generall Historie of Plantes*, rev. and enlarged ed. (London, 1633), dedication. Cf. Levinus Lemnius, *An Herbal for the Bible* (London, 1587), 1; Wotton, *Reflections upon Ancient and Modern Learning*, 257–58.

41. Anthony Grafton, "The Availability of Ancient Texts," in *The Cambridge History of Renaissance Philosophy*, ed. Charles Schmitt and Quentin Skinner (Cambridge: Cambridge University Press, 1988), 767–91.

42. St. Bredwell, "To the Reader," in Gerard, *The Herball or Generall Historie of Plantes*.

43. John Tradescant, *Musaeum Tradescantianum* (London, 1656).

44. Topsell, *The Historie of Foure-footed Beastes*, epistle dedicatory.

45. Bacon, *Advancement of Learning*, Bk I, vi, 6 (p. 38).

46. Sprat, *History of the Royal Society*, 349–50. Cf. John Parkinson, *Paradisi in Sole* (London, 1629), epistle to the reader.

47. Bacon, *Valerius Terminus*, in *The Works of Francis Bacon*, 3:222. Cf. Grew, preface to *Musaeum Regalis Societatis*.

48. Lisbet Koerner, "Carl Linnaeus in His Time and Place," in Jardine, Secord, and Spary, *Cultures of Natural History*, 157.

49. Cunningham, "The Culture of Gardens," in Jardine, Secord, and Spary, *Cultures of Natural History*, 49.

50. John Brooke, *Science and Religion: Some Historical Perpectives* (Cambridge: Cambridge University Press, 1991), 232; Koerner, "Carl Linnaeus in His Time and Place," 156.

51. John Prest, *The Garden of Eden: The Botanic Garden and the Re-creation of Paradise* (New Haven, CT: Yale University Press, 1981), 23.

52. Walter Charleton, *Exercitationes de Differentiis & Nominibus Animalium* (Oxoniae, 1677), sig. d2v; John Rowland, epistle dedicatory to Edward Topsell, *The History of Four-footed Beasts and Serpents* (London, 1658).

53. Jim Bennett and Scott Mandelbrote, *The Garden, the Ark, the Tower, the Temple* (Oxford: Museum of the History of Science, 1998); Carla Yanni, *Nature's Museums: Victorian Science and the Architecture of Display* (London: Athlone, 1999), 15–17.

54. In Topsell, *The Historie of Foure-footed Beastes*, sigs. 2r–3r.

55. [Christopher Merrett?], *The Character of a Compleat Physician, or Naturalist* (London, 1680), 2–3; quoted in Harold J. Cook, "Physicians and Natural History," in Jardine, Secord, and Spary, *Cultures of Natural History*, 92.

56. Franz, *The History of Brutes*, 1–3, cf. 53f. This was the view of church father St. Augustine, *On Christian Doctrine*, II.xxxix.59.

57. Ralph Austen, *A Treatise of Fruit-Trees, together with The Spiritual vse of an Orchard* (Oxford, 1653), 14.

58. In Topsell, *The Historie of Foure-footed Beastes*, sig. 2r.

59. Robert Boyle, *A Discourse touching Occasional Meditations*, in *The Works of the Honorable Robert Boyle*, 6 vols., ed. Thomas Birch (Hildesheim: Georg Olms, 1966), 2:340.

60. Peter Harrison, "The Virtues of Animals in Seventeenth-Century Thought," *Journal of the History of Ideas* 59 (1998): 463–85.

61. See, e.g., George Fairholme, *A General View of the Geology of Scripture* (London, 1833), 28; John Brooke and Geoffrey Cantor, *Reconstructing Nature: The Engagement of Science and Religion* (Edinburgh: T & T Clark, 1998), 59.

62. Topsell, *History of Foure-Footed Beasts and Serpents* (1607 ed.), epistle dedicatory.

63. Cf. John Johnston, *An History of the Wonderful Things of Nature* (London, 1657), sig. a3v.

64. Ray and Willughby, preface to *The Ornithology of Francis Willughby.*

65. J. Topham, "Beyond the 'common context': the production and reading of the Bridgewater Treatises," *Isis* 89 (1998): 233–62; John M. Robson, "The Fiat and Finger of God," in *Victorian Faith in Crisis*, ed. Richard Helmstadter and Bernard Lightman (London, Macmillan, 1990), 71–125.

66. Fontenelle, "Eloges," *Oeuvres diverses* (La Haye, 1728–9), 3:7–8.

67. Abbé Dominique Réverend, *La Physique des Anciens* (Paris, 1701); Abbé Noël Antoine Pluche, *Spectacle de la Nature* 5th ed. (London, 1750).

68. Bernard Nieuwentijt, *L'existence de Dieu, démontrée par les merveilles de la nature* (Paris, 1725).

69. *Théologie des Insectes, ou Demonstration des Perfections de Dieu dans tout ce qui concerne les Insectes*, trans. by P. Lyonnet (La Haye, 1742).

70. Carl Linnaeus, *Reflections on the study of Nature* (1754), trans by. J. E. Smith (London, 1785).

71. Johnston, *An History of the Wonderful Things of Nature*, epistle dedicatory.

72. William Wallace, "Traditional Natural Philosophy," in *The Cambridge History of Renaissance Philosophy*, 213.

73. Ann Blair, *The Theatre of Nature: Jean Bodin and Renaissance Science* (Princeton, NJ: Princeton University Press, 1997), 35.

74. Gerard, *The Herball or Generall Historie of Plantes*, dedication.

75. Ray and Willughby, preface to *The Ornithology of Francis Willughby.*

76. Johnston, *An History of the Wonderful Things of Nature*, epistle dedicatory. For actual figures see John Gascoigne, "The Eighteenth-Century Scientific Community: A Prosopographical Study," *Social Studies of Science* 25 (1995): 575–81.

77. On *historia* and its relation to *scientia* during this period, see Pomata and Siraisi, *Historia*, esp. the introduction.

78. Hobbes, *Leviathan* I, 9, in *The English Works of Thomas Hobbes of Malmesbury*, ed. Sir William Molesworth, (London, 1839–1845), 3:71–72.

79. Aristotle *Parts of Animals* 639a–640a; *Posterior Analytics* 87b; Lindberg, *The Beginnings of Western Science*, 63.

80. Bacon, *Great Instauration*, 4:28.

81. Bacon, *De Augmentis*, 4:295.

82. Ibid., 299.

83. Bacon, *Great Instauration*, 4:28–29.

84. Bacon, *De Augmentis*, 4:362.

85. See Richard Yeo, "Classifying the Sciences," in *The Cambridge History of Science*, vol 4. *The Eighteenth Century*, ed. Roy Porter (Cambridge: Cambridge University Press, 2001), 241–66.

86. John Sergeant, preface to *The Method in Science* (London, 1696).

87. Perrault, *Memoir's [sic] for a Natural history of Animals*, sig. ar.

88. Ibid., sig. br.

89. Ibid.

90. John Gascoigne, "The Royal Society and the Emergence of Science as an Instrument of State Policy," *British Journal for the History of Science* 32 (1999): 172.

91. John Woodward, preface to *An Essay toward a Natural History of the Earth* (London, 1695).

92. Thomas Burnet, *Sacred Theory of the Earth*, 2nd ed. (London, 1691), title page.

93. John Lyon and Phillip Sloan, *From Natural History to the History of Nature: Readings from Buffon and his Critics* (Notre Dame, IN: University of Notre Dame Press, 1981), 121; see also John Gascoigne, "The Study of Nature," in *The Cambridge History of Eighteenth-Century Philosophy*, 2 vols., ed. Knud Haakonssen (Cambridge: Cambridge University Press, 2006), 2: 854–72.

94. George-Louis Leclerc, Comte de Buffon, *Sur les Oiseaux*, in *Oeuvres complètes de Buffon*, nouvelle ed. (Paris, 1817–19), 9:3.

95. Pierre-Louis Moreau de Maupertuis, preface to *Vénus Physique* (The Hague, 1746); Corneille de Pauw, *Recherches philosophiques sur les Américains*, vol 1, *Discours Préliminaire* (London, 1774). Cf. Buffon, *Animaux communs aux deux continents*, in *Oeuvres*, 6:515–21.

96. Buffon, *Les Animaux Domestiques*, in *Oeuvres*, 6:61.

97. Ibid., 62.

98. *Les Époques de la nature* (1779), ed. Jacques Roger (Paris: Editions du Mus, 1962), 43.

99. William Whiston, *A New Theory of the Earth* (London, 1696), 361–81.

100. Leibniz, *Monadology*, §87, and Leibniz to Arnauld, October 6, 1687; both in *Discourse on Metaphysics, Correspondence with Arnauld and Monadology*, trans. G. Montgomery (Ithaca: Cornell University Press, 2009), 268, 234.

101. Anne Conway, *The Principles of the Most Ancient and Modern Philosophy* (London, 1692), 59, 74, 96.

102. Leibniz to Arnauld, March 1690, in Montgomery, *Discourse on Metaphysics*, 244, 229–30. Cf. *Monadology*, §88, in Montgomery, *Discourse on Metaphysics*. On these general themes see Peter Harrison, "Animal Souls, Metempsychosis, and Theodicy in Seventeenth-Century English Thought," *Journal of the History of Philosophy* 31 (1993): 519–44.

103. Charles Bonnet, *Palingenesie*, VI.5, in *Oeuvres d'Histoire naturelle et de philosophie*, 18 vols., (Neuchatel, 1779–83), xv–xvi.

104. Ibid, I.1.

105. *Palingenesie*, III.3, in *Oeuvres*, XV, 218–20.

106. Louis-Jean-Marie Daubenton, "Histoire Naturelle," in *Encyclopédie, ou Diction-*

naire Raisonné des Sciences, ed. Denis Diderot and Jean d'Alembert (Neufchatel, 1765), 8:228; quoted in Gascoigne, "The Study of Nature."

107. Peter Roget, *Animal and Vegetable Physiology Considered with Reference to Natural Theology* (Philadelphia: Carey, Lea & Blanchard, 1839) 1: viii, ix. On natural theology as the unifying factor in the early nineteenth century sciences, see Bernard Lightman, "The Story of Nature: Victorian Popularizers and Scientific Narrative," *Victorian Review* 25 (2000): 1–29.

108. David Hull, *Darwin and his Critics: The Reception of Darwin's Theory of Evolution by the Scientific Community* (Cambridge, MA: Harvard University Press, 1973), 63.

109. Robert Chambers, *Vestiges of the Natural History of Creation* (London, 1844), 184–85, 231, 360.

110. Quoted in James Moore, *The Post-Darwinian Controversies* (Cambridge: Cambridge University Press, 1979), 194.

111. Hull, *Darwin and his Critics*, 16–36.

112. Peter Bowler, "The Decline of Theistic Evolution," in *The Eclipse of Darwinism: Anti-Darwinian Evolution Theories in the decades around 1900* (Baltimore, MD: The Johns Hopkins University Press, 1983), 44–57.

113. John Tyndall, quoted in David N. Livingstone, "Science, Region, and Religion," in *Disseminating Darwinism: The Role of Place, Race, Religion, and Gender*, ed. Ronald L. Numbers and John Stenhouse (Cambridge: Cambridge University Press, 1999), 7–38 (quote, 13).

114. Leonard Huxley, *The Life and Letters of Thomas Henry Huxley* (New York: D. Appleton, 1900), 1:177.

115. Frank Turner, "The Victorian Conflict between Science and Religion: A Professional Dimension," *Isis* 49 (1978): 356–76.

116. George Levine, "Scientific Discourse as an Alternative to Faith," in Helmstadter and Lightman, *Victorian Faith in Crisis*, 236f.

117. Quoted in Ross, "*Scientist:* The Story of a Word," *Annals of Science* (1962) 18: 65–85 (82).

118. Lynn Nyhart, "Natural History and the New Biology," in Jardine, Secord, and Spary, *Cultures of Natural History*, 427–28.

119. Macvey Napier, ed., *Encyclopaedia Britannica*, 7th ed. (Edinburgh, 1830–48), s.v. "natural history" (15:738).

120. On the significance of the term "biology" and its equivalents see Pietro Corsi, "Biologie," in *Lamarck, philosophe de la nature*, ed. Pietro Corsi, Jean Gayon, Gabriel Gohau and Stéphane Tirard (Paris: Presses Universitaires de France, 2006), 37–64.

121. Quoted in George Hutchinson, "Homage to Santa Rosalia, or why are there so many kinds of animals?" *American Naturalist* 93 (1959): 146.

122. Charles Elton, *Animal Ecology* (London: Methuen, 1927), 1; quoted in Jean-Marc Drouin and Bernadette Bensaude-Vincent, "Nature for the People," in Jardine, Secord, and Spary, *Cultures of Natural History*, 423.

CHAPTER 6

Mixed Mathematics

Peter Dear

THE QUADRIVIUM AND "MIXED MATHEMATICS"

The category "mixed mathematics" is nowadays perhaps best known by historians from its use by d'Alembert in the "Discours préliminaire" to the great eighteenth-century *Encyclopédie*. Its origin is sometimes attributed to Francis Bacon, d'Alembert's great hero, at the start of the seventeenth century.[1] However, the concept and its associations were well established long before the eighteenth century and were not rooted in the works of Bacon.

From the sixteenth century well into the nineteenth, "mixed mathematics" designated an approach to making natural knowledge that distinguished it from qualitative natural philosophy. The latter had been associated with causal explanations for phenomena, which before the seventeenth century had frequently been denied to the mathematical sciences; mathematics, on this view, had been useful for describing and coordinating the quantitative characteristics of things, including geometrically describable features, but not for addressing questions concerning the natures of things themselves. Mathematics might tell you how something behaved, but not why: geometrical optics, for example, could describe the reflection of images from mirrors, but could not tell you what light was and why it behaved in the ways it did.

Mixed mathematics was so called because it combined "pure" mathematics with specific subject-matters usually considered by physics. Pure mathematics, according to Aristotle and everyone following him in the medieval and early-modern university, comprised arithmetic and geometry, the first concerned with numbers, or "discrete quantity," the second with spatial extension, or "continuous quantity." Pure mathematics was therefore said to be about "general magnitude," quantity in itself rather than quantities *of* something. But this perspective also recognized that mathematics was not restricted to the "pure" variety, and that many areas of mathematics were indeed concerned with quantities of specific sorts of things. Astronomy had been recognized as a mathematical discipline since antiquity, because it applied geometrical reasoning to the motions of the heavens. Music, on the other hand, was a mathematical science to the extent that consonances could be understood in terms of number ratios, typically as applied to different string lengths of equal tension (3:2 for the interval of the fifth, for example; or 2:1 for an octave). As prototypes of mathematical sciences that used, in the first case, geometry, and in the second case, arithmetic, therefore, astronomy and music became established in the liberal arts education of classical antiquity as two of the four mathematical arts: geometry, arithmetic, astronomy, and music. However, these four by no means exhausted the list of mathematical subjects. Others, especially optics (using ray diagrams to represent the relationships between seen objects, vision, and images) and mechanics (meaning statics and referring to simple machines like the lever),[2] were also incorporated under these ideal headings; by the seventeenth and eighteenth centuries the list had become very long indeed, encompassing any subject of study that made central use of "pure" mathematical reasoning regarding measurable or quantifiable subject matter.

In the early Middle Ages in the Latin west, the four prototypical mathematical subjects became known, thanks to the late-fifth-century writer Boëthius, as the *quadrivium*, and together with the *trivium* (grammar, rhetoric, and logic) constituted the "seven liberal arts."[3] By the time of the Renaissance, the quadrivium, seldom much emphasized in the arts curriculum of the medieval university, began to be reemphasized in some quarters as part of the humanist revival of classical culture. The most important name here is that of Johannes Müller, known as Regiomontanus, who printed a number of texts in the second half of the fifteenth century that praised the project of restoring the mathematical project of the ancients. Regiomontanus placed especial (although by no means exclusive) stress on the mathematical science of astronomy, and a work that he wrote with Georg Peurbach, the *Epitome of the Almagest* (published posthumously in 1496) served as Copernicus's chief access to Ptolemy's work when he devised the first version of his new heliocentric system of the universe.[4]

The terms being used in the sixteenth century to designate those parts of the quadrivium that went beyond arithmetic and geometry were especially owing to Aristotle's discussions of them in the *Posterior Analytics* and the *Metaphysics*. Aristotle had argued that a science ought properly to be built up deductively from its fundamental principles, and that the conclusions of that science should only employ terms homogeneous with those of its principles. This doctrine, called *metabasis* in Greek, thus envisaged multiple sciences, each logically self-contained. In the case of the mathematical sciences of nature, however, Aristotle saw that this argument could not apply: a conclusion in geometrical optics about light rays would be rooted in the principles of geometry, not statements about light rays themselves, and hence the requisite homogeneity of terms would not apply. Nonetheless, Aristotle recognized such arguments as legitimately scientific, and hence established for them the category of "subordinate sciences" (in later Latin usage, *scientiae subalternatae*). Thus the proofs of a mathematical science of nature that employed geometrical demonstrations were regarded as subordinate to those of geometry itself, that is, dependent upon them, and analogously for mathematical sciences making use of arithmetical demonstrations.[5] In the thirteenth century, Thomas Aquinas had employed the term *scientiae mediae* ("middle sciences") to designate the same category, since such a science was located between its superior discipline and its inferior subject matter, between pure mathematics and physics. In the sixteenth century, Niccolò Tartaglia, in his Italian *Euclid* (1569), referred to these "impure" mathematical disciplines as "mixed" (*miste*).[6] The establishment of the term "mixed mathematics" by the beginning of the seventeenth century simply used the preexisting category in its specific reference to mathematical *scientiae mediae* or *subalternatae* (Aquinas recognized a few nonmathematical instances of the latter), and applied to them the new term, in Latin, *mathematicae mixtae*, to designate the entire field.[7] Thus the important Jesuit mathematician and pedagogue Christoph Clavius described astronomy and music in the quadrivium as representing "mixed" mathematics in the widely read "Prolegomena" to his *Opera mathematica* of 1612; a work on logic of 1608 by the Lutheran scholar Gregor Horst also discusses what he calls *disciplinae Mathematicae mixtae*—each, and no doubt many others, doing so evidently independently of Bacon's use of the English term "mixed mathematics" in *The Advancement of Learning* (1605).[8] The term rapidly became standard in the works of Jesuit as well as non-Jesuit mathematical writers, displacing the previous terms but not itself representing any conceptual reformation.[9]

Around the middle of the sixteenth century, philosophical arguments had been put forward that challenged the traditional Aristotelian assumption that the mathematical disciplines, whether "pure" or "subordinate," were truly sciences. Among others, the humanist churchman Alessandro Piccolomini and the Jesuit theologian and philosopher Benito Pereira were prominent proponents of the position that mathematical demonstrations could not be scientific in Aristotle's sense because they failed to specify the *causes* responsible for the truth of their conclusions—causal necessary demonstration being the strict Aristotelian definition of a science.[10] Unsurprisingly, this claim proved unpopular among mathematicians, including some of Pereira's fellow Jesuits. Clavius defended what he clearly saw as attacks on the intellectual standing of mathematics in policy documents concerning the establishment of the curriculum of the new Jesuit colleges, writing in the 1580s of the importance of dissuading philosophy teachers from downplaying the status of the mathematical disciplines by denying that they were truly sciences at all. Clavius's protestations aside, later Jesuit mathematicians also provided counterarguments to undermine the claims of mathematical nonscientificity, those of Giuseppe Biancani from 1615 being particularly well known and read throughout the rest of the seventeenth century. Among other points, Biancani noted that in the case of mixed mathematics, Aristotle had been at pains to explain, through his idea of subordinate sciences, how they too counted as true sciences; he thereby used Aristotle's authority as part of his defense. Biancani also argued that mathematical arguments, both in mixed and in pure mathematics, did in fact employ causes of the usual Aristotelian kinds (formal, final, material, and efficient) in proving their conclusions. After all, what kinds of demonstrations were as necessary and conclusive as those of mathematics? Certainly not those of natural philosophy.[11]

Seen in this light, attempts during the second half of the sixteenth century to extend the reach of the mathematical sciences into realms previously reserved for natural philosophy (*physica*, or physics in its Aristotelian sense) appear as a kind of cultural muscle-flexing on the part of mathematicians. Most notably, the so-called Archimedean revival in Italy in that period involved exploitation of the works of Archimedes and Pappus as ancient exemplars of a new, theoretically grounded mechanics that had ambitions to develop an understanding of moving bodies as well as of statical situations. The figures of Tartaglia, Commandino, Baldi, Benedetti, Guidobaldo dal Monte, and finally Galileo not only regarded mathematics as a respectable way of generating real understanding of the natural world, but also saw it as a means of accomplishing things of which the natural philosophers were incapable—as shown by Aristotle's mathematically demonstrable errors regarding falling bodies.[12] Both Benedetti and Galileo, perhaps independently,[13] had made "thought-experiment" arguments to show that Aristotle's apparent position that the speed of fall of heavy bodies was proportional to their absolute weights was logically incoherent; they preferred analyses drawing on the mathematical mechanics of Archimedes.

The same sort of move was also made in the second half of the sixteenth century in astronomy, starting with the signal attempt by Copernicus to recast physical understanding on mathematical-astronomical grounds, and pushed to perhaps its furthest extent by Johannes Kepler in the early seventeenth century. Kepler's Lutheran, but also mathematical, understanding of natural philosophy was in its specifics unusual for mathematicians in this period. His astronomical and associated optical work took mathematical entities and concepts as substantive constituents of an astronomy that he approached as a part of physics, or natural philosophy, as well as an enterprise of mathematical modeling. His astronomy was a theological endeavor that sought God in the construction of the universe and incorporated both qualitative causal explanations for celestial motions and formal mathematical accounts of celestial operations that put mathematics at the center of physical knowledge. To that end, he wrote that the astronomer "directs all his opinions, both by geometrical and by physical arguments, so that truly he places before the eyes an authentic form and disposition or furnishing of the whole universe."[14] Kepler saw himself, in effect, as a philosophical astronomer.

Galileo, who was not very interested in mathematical astronomy of Kepler's kind, began to call himself a "philosophical astronomer" when he embarked on his telescopically based campaign against Ptolemy and Aristotle; again, the idea was to show that the astronomer could tell the natural philosopher what was what, rather than being subordinated to him. It is telling that, when Galileo negotiated with the secretary to the Grand Duke of Tuscany for a court position in 1610, he famously made it clear that he was not interested in being just another court mathematician (a status held by Kepler). Instead, he would be the "Philosopher and Mathematician" to the Grand Duke. In the correspondence, Galileo was careful to point out that he had "studied a greater number of years in philosophy than months in pure mathematics."[15] His concern to stress his "philosophical" credentials suggests that, while annoyed at the higher status of his philosophical colleagues in the universities, Galileo had also internalized many of the values that justified his inferior position as a mathematician: he apparently regarded being a "philosopher" as more important, as well as more prestigious, than being a "mathematician." At the same time, in asserting his expertise in philosophy, he had contrasted philosophy with *pure* mathematics, in effect explicitly leaving out mixed mathematics. He seems to have seen his philosophical credentials as following from his expertise in mixed mathematics, taken as a legitimate part of a truly philosophical science of nature. Pure mathematics could not count as natural philosophy under any circumstances, of course, because it did not talk about the changeable natural things that defined physics for Aristotle; it only dealt with the unchanging ideal entities of pure quantity.[16]

The ambiguity surrounding the philosophical and cultural status of mathematical sciences in this period was therefore one that mathematically inclined scholars, whether academic or not (and, increasingly, they were not), attempted to resolve as much by practical and rhetorical accomplishments as by philosophical innovations concerning such issues as causal explanation. The Aristotelian conceptual categories in the terms of which mathematics in the study of nature were usually debated remained nonetheless common resources for all sides in the argument. Galileo's famous work on free fall, parabolic trajectories, and related topics, published in his *Discorsi* of 1638, drew on the legitimating resource of Archimedean mechanics but was widely regarded among his mathematical peers as retaining methodological imperfections and uncertainties. Thus, Galileo's interlocutor and critic Giovanni Battista Baliani wrote in his *De motu naturali* (On natural motion) of 1646 about the importance of discovering the appropriate principles on which a true science could be based, and the difficult role of experience in establishing or confirming them. "I have begun to search," Baliani said of such principles; "others will decide whether I have discovered anything."[17] After a lengthy discussion of the behavior of heavy bodies, Baliani then reflects upon the status and character of the knowledge that he has proposed.

> Hitherto, it seems to me that I have said as much as I can concerning the science of the natural motion of solid weights, as from certain properties familiar to the senses many things unknown have been deduced and disclosed. For according to Aristotle in this way alone is every science treated, as is seen in practice with Euclid, and others, who treat true and simple sciences: from whom [we see that] the geometer does not deal with the nature of quantity, nor the musician with the nature of sound, nor the optician with the nature of light, nor the mechanic with the nature of weight. But truly my mind

> is not entirely satisfied if it does not grasp, or at least investigate, the prior causes from which these effects finally result.[18]

Baliani's subsequent causal account of natural acceleration, because it concerned "not effects, but the natures of things," could not be affirmed as certain, being a matter of physics, not demonstrative mathematics.[19]

PHYSICO-MATHEMATICS

Just a few decades after the term "mixed mathematics" appeared, somewhat mysteriously, around the beginning of the seventeenth century, another term of similar signification came into general use: "physico-mathematics," together with its cognates. But while "mixed mathematics" had carried little greater specificity (or extension) than its predecessors such as "middle sciences," "physico-mathematics" tended sometimes to trade on a new cognitive claim. While "mixed mathematics" continued to be very widely used, and with essentially the same meaning related to the old quadrivium as well as to Aristotle's definition of subordinate sciences, "physico-mathematics" could also serve to make stronger methodological and cognitive claims for the mathematical sciences. Just as Bacon's use of "mixed mathematics" heralded a widespread, though evidently independent, use of that term, so the Dutch schoolmaster and philosopher Isaac Beeckman's term "physico-mathematics" in a private journal entry in 1618 was thereafter widely used to mean very different things by writers many of whom apparently owed nothing to Beeckman's private notation. The matter is obscure, however, since the term found its way into the work of René Descartes and Marin Mersenne, each of whom knew and corresponded with Beeckman, and whose own work and correspondence was in turn widely distributed among the learned in the 1620s, '30s, and '40s. In fact, Beeckman's first known use of "physico-mathematics" occurs in a reference to the young Descartes, whom he had recently met in the Netherlands:

> This Descartes has been educated with many Jesuits and other studious people and learned men. He says however that he has never come across anyone anywhere, apart from me, who uses accurately this way of studying that I advocate, with mathematics connected to physics. Neither, furthermore, have I told anyone apart from him of this kind of study.[20]

The novel terminology appears in the form of Beeckman's marginal note to this paragraph: "Very few physico-mathematicians."[21]

"Physico-mathematics" meant to Beeckman a curious amalgam of conjectural micro-mechanical explanations (of the sort that Descartes himself subsequently made famous) and, at least in principle, mathematical formalisms to discipline the explanatory models so developed.[22] As the term was adopted by others shortly thereafter, however, its application reverted to the more widely understood categories associated with the middle sciences of mixed mathematics. The attraction of this new term seems to have derived, for many, from its provocative attachment of the prefix "physico" to the mathematical sciences that addressed aspects of the natural world. Where Galileo had referred to himself as a "philosophical astronomer," so others began to refer to their own work as "physico-mathematics" or "physico-mathematical," and to convey much the same idea.[23]

By the 1640s the term had become widespread among mathematical and other philosophical writers, including several Jesuits, such as Athanasius Kircher. One of the more influential loci for its use was the circle around Marin Mersenne. Mersenne published two major works that employed it in their titles: *Cogitata physico mathematica* (1644) and its sequel, usually known as *Novarum observationum . . . tomus III* (1647), but the chief new part of which is called *Reflexiones physico-mathematicae*.[24] Both works deal with a multiplicity of mixed mathematical sciences, including the new mathematical science of falling bodies and projectiles. One of Mersenne's regular correspondents, Jean Le Tenneur, shortly thereafter published *De motu naturaliter accelerato tractatus physico-mathematicus* (1649), which uses the term in very much the same way, here specifically in relation to fall and acceleration.[25] Others, however, had begun to emphasize the prefix as a way of designating a "mathematical" approach to physics itself: Kircher, in his 1631 *Ars magnesia*, a "physico-mathematical" disquisition on the nature and effects of magnets, used the structural form of so-called theorems and problems to present largely qualitative assertions based on accounts of experimental procedures.[26] In this usage, "physico-mathematics" was concerned much more with a kind of experimentalism than it was with mixed mathematics. In general, however, despite the evident rhetorical flexibility of the term, it usually designated the well-understood terrain of mixed mathematics. Isaac Newton's mentor and predecessor as Lucasian Professor at Cambridge, Isaac Barrow, in a published version of his mathematical lectures originally given in the 1660s, explained the division of mathematics into "pure" and "mixed" and noted that studies falling under the latter category were sometimes dubbed, in Latin, *physico-mathematicas*.[27]

The two terms, "mixed mathematics" and "physico-mathematics," acquired overlapping and flexible, although constrained, senses in the seventeenth century. However, the "physical" claims made by the latter necessarily also served to weaken, or alter, the strict causal "Aristotelian" meaning of physics itself. And here another ambiguity arises: where the older grounds for separating physics from mathematics had been that mathematics did not participate in the causal explanatory mission of physics, so by the late seventeenth century a reason for combining them was now, not that mathematics really was causal after all (the argument of early seventeenth-century Jesuit mathematicians, among others), but that physics itself no longer needed to be strictly causal. In effect, this was Newton's mature position, as laid out in the *Philosophiae naturalis principia mathematica* (*Principia*) of 1687 and its subsequent defenses, as well as in his optical work.

The "mathematical principles of natural philosophy" rode roughshod over Aristotelian *metabasis*; natural philosophy, physics, should not have been able to rest on "mathematical" principles—this had been the very predicament that had led Aristotle to invent subordinate sciences. But Newton's natural philosophy was not presented as any kind of subordinate science. Indeed, the natural philosophy, whose mathematical principles were presented in the *Principia,* was by no means the entirety of Newton's philosophy of nature. He considered much more widely qualitative and matter-theoretical aspects of nature that did not figure (or figure significantly) in the *Principia* and yet that he took as seriously as the mathematical aspects. Some of these elements of his "speculative" philosophy appeared later in the "Queries" appended to successive editions of his *Opticks* from 1704 onwards, as well as in the General Scholium added to the *Principia's* second edition of 1713. But Newton represented the mode of philosophizing found in the *Principia* itself, as well in the main text of the *Opticks*, as the results of "experimental philosophy."[28] Experimental philosophy, an expression familiar in work by other early fellows of the Royal Society besides Newton, especially Robert Boyle, implied in Newton's hands also an intimate integration with mathematics. Thus, for example, Newton calibrated and measured gravitational force in relation to mechanical forces such as centrifugal force by conducting measuring experiments with conical pendulums. He insisted that he could demonstrate phenomena of nature, typically associated with forces of various kinds, through conclusive experimental and observational means, but that he could not—as yet—determine the causes of those phenomena. Most famously, he explained that he was unable to demonstrate the cause of gravitational attraction; he also proclaimed nescience in his optical studies concerning the nature of light, similarly restricted to demonstrations of properties and measurable phenomena. This did not mean that he had no ideas on such questions, but simply that he could not prove them conclusively using mathematical-experimental means. He was careful, too, not to rule out the possibility of such causal discoveries in the future, of a sort broadly resembling Beeckman's original corpuscular "physico-mathematics"; the methodological caution of Newton's "experimental philosophy" was not rooted in skeptical principles.

In his introductory "Ode" to the *Principia*, Edmond Halley described the work as a "mathematico-physical treatise."[29] An earlier expression that one of the founders of the Royal Society, John Wilkins, once used to describe the charge of the nascent Society was "Physico-Mathematicall-Experimentall Learning."[30] Wilkins's combination of such apparently disparate categories in that elaborate label captures well the relationship between "physico-mathematics" and "experimental philosophy" subsequently sought by Isaac Newton.

MATHEMATICS, MECHANICAL CAUSES, AND NATURAL PHILOSOPHY

As John Heilbron observes in his contribution to this volume, the label of physico-mathematics, like that of mixed mathematics, continued to be used throughout the eighteenth century. This continuation seems to have merged with the specifically Newtonian sense of the term as the century progressed, but never displaced, or even truly rivaled, "mixed mathematics" as the more commonly used expression. D'Alembert's use of "mixed mathematics" in his taxonomy of the sciences in the *Encyclopédie* certainly chimed with his admiration for Francis Bacon and the latter's own inventory in the *Advancement of Learning*, but, as we have seen, it had long since become usual and uncontroversial. As d'Alembert himself remarked, "Some divisions, such as those of mathematics into pure and mixed mathematics, that we have in common with Bacon, are found everywhere, and consequently belong to everyone."[31] The relationship between mixed mathematics and natural philosophy that the term "physico-mathematics" could be used to assert remained ambiguous nonetheless; it all turned on the crucial traditional claims of "physics" to identify *causes*. Were physical causes as understood in the later seventeenth or eighteenth centuries in any sense continuous with the older, Aristotelian sorts of causes that had underpinned the old demarcation of mathematics from physics? Or did "causal" explanation now mean something fundamentally different?

Newton's own pronouncements on this matter are important not just because they relate to his own scientific work but also because of the influential role they played thereafter. Besides eighteenth-century self-styled followers of Newton, the nineteenth century too took notice of the great man's metaphysical and methodological assertions. In that later period, for example, British physicists such as Faraday and Maxwell used Newton's words to lend additional legitimacy to their own arguments, besides the generally unquestioned admiration for Newton's theoretical achievements among nineteenth-century physical scientists. Newton famously received severe criticisms of his work in the *Principia* from the great mathematical philosophers Christiaan Huygens and Gottfried Wilhelm Leibniz, as well as from lesser lights, and their objections were certainly not due in those cases to a dislike of the mathematical sciences as philosophical tools.[32] The difference between Newton and these critics was simply that Newton explicitly eschewed in the *Principia* any claims to explain, in his mathematical demonstrations, the cause, or true nature, of gravitational attraction. He could demonstrate mathematically and experimentally the properties and characteristics of gravitational behavior, but not, he argued, its causes; on those he could only speculate. Here is an important point to note: Newton never claimed that natural philosophy was nothing more than its mathematically demonstrable aspects. His book was entitled *Mathematical Principles of Natural Philosophy* to signal that it dealt *only* with the mathematical principles, not to suggest that natural philosophy's principles were exclusively mathematical. In his preface to the first edition, Newton remarked that "since the moderns . . . have undertaken to reduce the phenomena of nature to mathematical laws, it has seemed best in this treatise to concentrate on *mathematics* as it relates to natural philosophy."[33] Demonstrability, always the trump card of the mathematical sciences, was central to Newton's accomplishments in his own eyes.

Among the responses directed at Newton's critics after the appearance of the *Principia* was a particularly pointed passage in Roger Cotes's editorial introduction to the second edition in 1713. The passage responds to complaints by Huygens and others that, while Newton had demonstrated that motions in the solar system could be accounted for on the supposition that they were brought about by an inverse-square force law quantitatively identifiable with gravitational attraction, he had not provided an explanation of the underlying cause of that force. Because he had not addressed the causes involved, Newton's work was not, in Huygens's view, natural philosophy. Cotes replied:

> But will gravity be called an occult cause and be cast out of natural philosophy on the grounds that the cause of gravity itself is occult and not yet found? Let those who so believe take care lest they believe in an absurdity

> that, in the end, may overthrow the foundations of all philosophy. For causes generally proceed in a continuous chain from compound to more simple; when you reach the simplest cause, you will not be able to proceed any further. Therefore no mechanical explanation can be given for the simplest cause; for if it could, the cause would not yet be the simplest. Will you accordingly call the simplest causes occult, and banish them? But at the same time the causes most immediately depending on them, and the causes that in turn depend on these causes, will also be banished, until philosophy is emptied and thoroughly purged of all causes.[34]

In effect, Newton's means of dealing with the distinction between causal physics and descriptive mathematics was to analyze the concept of cause that supposedly stood between them. His argument, as refracted here through his spokesman, Cotes, was that a proper understanding of physical causation shows that any and all causal analyses will ultimately terminate in primitive items that cannot be further reduced—if not gravitational attraction, then something else prior on the causal chain, but there will always be something that cannot be further explained in terms of something simpler. So if his critics held that he was not dealing in real causes by stopping at gravitational attraction, his reply was that any such analysis will end that way; they themselves would therefore be open to analogous criticisms of any of their own explanations. The point was simply a variation on the well-known account in Aristotle's *Posterior Analytics* of deductive demonstrations: all must begin from starting principles that cannot themselves be proven, or an infinite regress would result.[35]

Newton could make the shift from logical to causal analysis because his model of causation was no longer that of the Aristotelians, whose four causes—formal, final, efficient, and material—usually possessed a self-sufficiency that rendered them immune to regress arguments. For example, a formal-cause argument such as "Socrates is mortal because he is a man" took its validity from the supposed self-evidence of Socrates' humanity, as well as the implicit definitional property of mortality possessed by men. By contrast, Newton's type of causation was that of those he called "the moderns," by which term he referred to ontological mechanists in the Cartesian tradition, including most notably Huygens himself. If their kind of mechanical causation was the only physical causation permissible in natural philosophy, then Newton's response to their criticisms of him could undercut the claimed irregularity of his own work.[36] Mechanistic physics, in other words, had already taken the crucial steps in redefining causation in a way that now permitted Newton's experimental philosophy, with its mixed mathematical procedures, to be represented as a central part of natural philosophy. Causes were not what they used to be. Newton's exploitation of what amounted to a failure of strict mechanism in physics enabled the elevation of the phenomenological demonstrations of mixed mathematics to a primary role in natural philosophy.[37]

This was not always Newton's position; his eschewal of causes in defending his mathematical work was tactical as much as it was principled. In his early optical work (1672), later presented in the *Opticks* as a similar sort of mathematical-experimental demonstration to that of the *Principia*, he tried to make inferences from conclusions about colors and white light to the determination of the ontological nature of colors themselves: "These things being so, it can be no longer disputed, whether there be colours in the dark, nor whether they be the qualities of the objects we see, no nor perhaps, whether Light be a Body. For since Colours are the *qualities* of Light, having its Rays for their intire and immediate subject, how can we think those Rays *qualities* also, unless one quality may be the subject of and sustain another; which in effect is to call it *Substance*."[38]

Newton thus attempted to extend into the realm of natural philosophy the inferences that could be drawn from work in optics, a branch of mixed mathematics. His remarks were immediately and appropriately understood to support a view of light as material, in contrast to the wave theories of his optical critics Huygens and Robert Hooke. "I suppose," he wrote in a reply to Hooke, "*the Science of Colours* will be granted *Mathematicall* & as certain as any part of *Optiques*."[39] The argument about the substantiality of light was a significant claim, therefore, in the context of mechanical natural philosophy—even though Newton's argument about "qualities" and "substance" was couched in scholastic Aristotelian terminology, and would perhaps not have been possible in a strictly mechanical idiom.

Another, and much more consequential, of Newton's attempts to intrude the procedures of mixed mathematics into natural philosophy concerns his assault on Cartesian physics in Book II of the *Principia*. Again, a mechanical idiom set the ground rules. Descartes had tried to explain celestial motions in terms of vortices of subtle matter or ether that would sweep planets and other objects around in their orbits. Like all matter for Descartes, that comprising the vortices admitted of no void space: spatial extension and matter were identical. Newton made it his business to demonstrate the nonexistence of this ether by showing that it exerted no resistance to bodies supposedly moving through it, and thereby to demonstrate the existence of the vacuum. He did this through mathematical arguments about resistance to motion through a medium that were in large part predicated on precise experimental work with pendulums.[40]

Newton thought that he had established a substantive natural philosophical point by means of the experimental techniques of mixed mathematics. He could not have done so if natural philosophy were not already widely regarded, in the Cartesian tradition, as mechanically (and hence mathematically) intelligible.

IN THE EIGHTEENTH CENTURY

The post-Newtonian philosophical dispensation, as adopted in the opening decades of the eighteenth century in England, integrated mathematics with physics through the medium of "experimental philosophy," and the result was "mixed mathematics" or, sometimes, "physico-mathematics." The Newtonian lecturer Jean Desaguliers remarked that Newton himself "look'd upon Geometry as no farther useful than as it directs us how to make Experiments and Observations, and draw Consequences from them when made; so that the Improvement of Philosophy must be the result of mix'd Mathematics, that is, of Mechanics and Geometry."[41] Perhaps even more tellingly, John Harris, in his largely scientific and technical dictionary, the *Lexicon technicum*, described "*Mixt Mathematicks*, which is interwoven every where with Physical Considerations."[42] By contrast, the first edition of Chambers's *Cyclopedia* (1728) still represented the mixed mathematics as consisting of such traditional areas as optics, music, mechanics, astronomy, and geography, as well as hydrostatics and pneumatics, but, curiously, listing statics itself (including the "Doctrine of Motion") as a branch of mathematics along with arithmetic and geometry, quite distinct from the "*Artificial* and *Technical*" parts of knowledge into which mixed mathematics fell.[43] Chambers's taxonomy of knowledge continued to make a sharp demarcation between contemplative and practical knowledge, therefore: "Staticks is wholly scientifical, as it takes up with the mere Contemplation of Motion: Mechanicks, on the contrary, is an *Art*, as it reduces the Doctrines of Staticks into Practice."[44] His formal definition of mixed mathematics echoes that of Harris but tellingly excises the reference to physics. "Mix'd *Mathematics* consider Quantity as subsisting in material Beings, and as continually interwove."[45]

Indeed, mixed mathematics had never traditionally been understood as dealing with physical causes, and Chambers's definition easily and uncontroversially reverted to the primary meaning prior to the advent of physico-mathematics. That understanding is reflected in the avant-garde *Encyclopédie*, and not only in d'Alembert's quasi-Baconian discussions of mixed mathematics, as John Heilbron's chapter in this volume notes. Under "*Philosophie*," we read:

> It is time to pass to the second point of this article, which is concerned with establishing the meaning of the term "philosophy" and providing it with a good definition. To philosophize, is to give the reason of things, or at least to search for it, for to the extent that one restricts oneself to seeing and reporting what one sees, one is only an historian. When one calculates and measures the proportions of things, their sizes, their values, one is a mathematician; but he who stays to discover the reason for things, and for their being thus rather than some other way, that is the philosopher properly so-called.[46]

Mathematics simply could not constitute part of natural philosophy; quantity was not physical. D'Alembert to some degree maintained this view himself, in contrast to the assumptions of the English Newtonians, by distinguishing categorically between mixed mathematics and physico-mathematics. Speaking of quantity as the "object of mathematics," he recognized three divisions of its study: independent of real and abstract things (pure mathematics); considered in "real and abstract beings" (mixed mathematics, which includes statics); or "considered in their [the beings'] effects investigated according to real or supposed causes"—this last, causal variety being labeled as "physico-mathematics."[47]

D'Alembert's scrupulous taxonomy therefore strove to maintain a distinction between physical, causal knowledge and other kinds associated with noncausal mathematics; physico-mathematics still left him with a place to locate his own work in Newtonian-style rational mechanics as a kind of natural philosophy. Buffon, in the 1740s, had already expressed similar reservations, lumping together natural history and mathematics, much like the *Encyclopédie*'s discussion of philosophy, as exemplars of *non*-philosophy. And Buffon was not afraid to make it clear that an absence of true philosophy was a serious shortcoming in the study of nature.

> In this century itself, where the Sciences seem to be carefully cultivated, I believe that it is easy to perceive that Philosophy is neglected, and perhaps more so than in any other century. The arts that people are pleased to call scientific have taken its place; the methods of calculus and geometry, those of botany and natural history, in a word formulas and dictionaries preoccupy almost everyone. People imagine that they know more because of having increased the number of symbolic expressions and learned phrases, and pay

no attention to the fact that all these arts are nothing but scaffolding for achieving science, and not science itself.[48]

The "scientific arts," according to Buffon, were those bodies of technique that were good for calculating and classifying but that did not provide insight into the proper concerns of natural philosophy—namely, causes and the natures of things.

The continued existence during the middle part of the century of, especially, French complaints regarding mathematics and its philosophical shortcomings appears to have subsided by century's end, as Heilbron argues in his contribution to this volume. What had overcome the traditional complaints, and their associated conceptualizations, was a revision of what should count as natural philosophy, and this was driven by the Newtonian philosophical ideology already discussed. The stereotype of merely descriptive mathematics versus causal or explanatory natural philosophy had become much less clear-cut, in part because of the role of (loosely) mathematical instruments in the work of experimental philosophy.[49] The experimental cabinets of public as well as college lecturers in natural philosophy in the eighteenth century—Desaguliers, Dufay, Nollet, and the rest—were filled with gadgets that did things, rather than explained things in themselves; and they sometimes did things that could even be measured.[50]

This is the sense in which Christian Licoppe has argued that Benjamin Franklin's work on electricity can be seen as fundamentally mathematical and Newtonian even though that work was to a great extent qualitative and apparently unmathematized.[51] The ambiguous category of "physico-mathematics" became ever more identified with operationalism, that is, learning how to make experimental setups perform in predictable and controllable ways. Such a natural philosophy did not require explicit association with mathematical formulations in order to import the kind of causal nescience that had long been perfectly respectable for traditional mixed mathematical sciences.

THE AFTERLIFE OF MIXED MATHEMATICS AND THE RESHAPING OF NATURAL PHILOSOPHY

This chapter has purposely focused upon the term itself "mixed mathematics," on the understanding that such labels are not applied meaninglessly, even when they may be applied thoughtlessly. The use of the term in the seventeenth and eighteenth centuries bespeaks the sense that it served some communicative function, if only by virtue of its connotations and the disciplinary alliances that it suggested. In time, its slipperiness (which had evidently proved functional), like that of "physico-mathematics," resulted in its general abandonment among the professionalizing scientists of the nineteenth century, with the notable exception of the British. The remaining bastion of mixed mathematics throughout the nineteenth century was the University of Cambridge, where it had persisted in the mathematical tripos through the reforms of William Whewell around midcentury, taking on new life at the heart of a British physics that came to be associated with Maxwell's Cavendish Laboratory in the 1870s and thereafter.[52]

If, as John Heilbron has suggested, mixed mathematics in its association with scientific instruments contributed centrally to the quantification of the physical sciences (often, "experimental natural philosophy") in the decades around 1800,[53] and pointed to an ever-growing absorption into physics of concepts and techniques that had once counted strictly as "mathematics," one might conveniently see mixed mathematics as having always been modern physics in disguise. However, on that view, modern physics would have to be seen in its common late-nineteenth-century guise as a positivistic discipline concerned with appearances (what Heilbron himself has dubbed "descriptionism") and eschewing deep understanding of "things in themselves."[54]

In his chapter in this volume, Heilbron sees the mid-eighteenth century as representing the narrow point of a meta-temporal hourglass from which a much-restricted natural philosophy began to expand again as it absorbed Lavoisier's new quantified chemistry and adopted the mixed-mathematical traditions into itself. What "natural philosophy" understood in this way had lost as a consequence of this process was precisely its claim to provide causal understanding. In practice, the "mathematization" of the physical sciences in that period meant the growth of a natural philosophy eviscerated of its causal pretentions. Heilbron's chapter sums up this view as its story proceeds down through the nineteenth century: "An instrumentalist attitude toward theory often accompanies quantification. Mixed mathematics may be precise, but, as Aristotle knew, there is no truth in it."[55]

The old Aristotelian view of mixed mathematics as a body of disciplines that did not speak of the causes of the things with which they dealt had never itself been unchallenged, as Kepler bears powerful witness; a Platonizing tendency associated with the mathematical sciences had always had the potential to tinge the philosophical understanding of mathematical aspects of the physical sciences. One of the issues that emerges when one looks over the centuries from the sixteenth to the twentieth is the unsettled character of notions of "causes" and "essences" themselves. For those physicists in the later nineteenth and twentieth centuries who steered well clear of German idealism, for example, the concept of "things in themselves" as their ultimate quarry (whether actually capturable or not) need not have made any sense at all; for them, a positivistic focus on instruments and measurement was, and could not but be, enough. The dispute between Bohr and Einstein may be taken as exemplary of a version of this issue.[56] And one person's deep causes might be another's instrumental conveniences. The philosophical distinctions that had characterized the early-modern demarcation between mathematics and physics—even for Newton—had never been absolute, but were always open to renegotiation and disagreement. So, similarly, it has never been entirely clear exactly what the concept of "natural philosophy" means.

Thus (to return to the fading of the category "mixed mathematics" in the nineteenth century), the German physics community, unlike its British rival, spoke throughout that period not of "mixed mathematics" as a crucial adjunct of physics, but of "applied mathematics" (*angewandte Mathematik*), evidently to refer to much the same array of techniques and concepts.[57] German applied mathematics made use of mathematical models and conceptualizations to attain its ends, just as did the mixed mathematics of Cambridge-trained physicists, but it tended to be regarded more as a set of tools to be used in physics rather than a research area with its own boundaries and coherence.[58] One of the arguments for the continuing preeminence of mixed mathematics in Cambridge even by the end of the century was that, being intimately associated with mechanics, it appealed to the foundational conceptual elements of physical understanding.[59]

One final observation about this curious history: to the extent that early-modern natural philosophy had been in part characterized, and codified academically, in the terms of its distinction from practical knowledge, it had needed be kept clear of more than just mathematics. But in the seventeenth century, instrumental practicality had begun to find its way into newer, rather heretical reconceptualizations of what natural philosophy ought to be—Francis Bacon being the most obvious spokesman. By the later seventeenth century, the idea that natural philosophy had some special relationship to practical know-how had become widely accepted in the circles represented by the scientific societies and academies of the period. Meanwhile, another new category of natural knowledge had appeared by midcentury: "experimental philosophy." Experimentalism taken as a set of knowledge-generating tools was an important aspect of how natural philosophy began to be implicated with practical know-how. The accompanying association of mixed mathematics with experimental procedures rendered natural philosophy itself an increasingly contested territory. The nineteenth-century German label "applied mathematics," as an equivalent to "mixed mathematics," itself captured an important part of these developments, therefore: physics could not but be in part directed to instrumental utility ("application") in the nineteenth century, because it had taken mixed mathematics into its heart. There it has remained. But the compromised category of natural philosophy has not survived the upheaval.

NOTES

1. Gary I. Brown, "The Evolution of the Term 'Mixed Mathematics,'" *Journal of the History of Ideas* 52 (1991): 81–102; Richard R. Yeo, *Encyclopaedic Visions: Scientific Dictionaries and Enlightenment Culture* (Cambridge: Cambridge University Press, 2001), 64; on d'Alembert's usage, Lorraine J. Daston, *Classical Probability in the Enlightenment* (Princeton, NJ: Princeton University Press, 1988), 53–56.

2. W. R. Laird, "The Scope of Renaissance Mechanics," *Osiris* 2 (1986): 43–68; Laird, *The Unfinished Mechanics of Giuseppe Moletti: An Edition and English Translation of His Dialogue on Mechanics, 1576* (Toronto: University of Toronto Press, 2000). These sources discuss the promotion in sixteenth-century Italy of mechanics as a mathematical science rather than an art.

3. David L. Wagner, ed., *The Seven Liberal Arts in the Middle Ages* (Bloomington: Indiana University Press, 1983).

4. Paul Lawrence Rose, *The Italian Renaissance of Mathematics: Studies on Humanists and Mathematicians from Petrarch to Galileo* (Geneva: Droz, 1975); Michael H. Shank, "Regiomontanus on Ptolemy, Physical Orbs, and Astronomical Fictionalism: Goldsteinian Themes in the 'Defense of Theon against George of Trebizond,'" *Perspectives on Science* 10 (2002): 179–207.

5. Richard D. McKirahan Jr., "Aristotle's Subordinate Sciences," *British Journal for the History of Science* 11 (1978): 197–220; cf. Moletti, in Laird, *Unfinished Mechanics*, 189–191.

6. Tartaglia, discussing Pierre d'Ailly's commentary on Sacrobosco, in his *Euclide* (1569), f01.6^{r}, quoted in Ann E. Moyer, *Musica scientia: Musical Scholarship in the Italian Renaissance* (Ithaca, NY: Cornell University Press, 1992), 132–33.

7. James A. Weisheipl, "The Interpretation of Aristotle's *Physics* and the Science of Motion," in *The Cambridge History of Later Medieval Philosophy*, ed. Norman Kretzman, Anthony Kenny and Jan Pinborg (Cambridge: Cambridge University Press, 1982), 521–36, on 525; see also W. R. Laird, "The *Scientiae mediae* in Medieval Commentaries on Aristotle's *Posterior Analytics*" (PhD diss., University of Toronto, 1983).

8. Christophorus Clavius, "In disciplinas mathematicas prolegomena," *Opera Math-*

ematica (Mainz, 1612), 1:3; Francis Bacon, *The Advancement of Learning; New Atlantis*, ed. Arthur Johnston (Oxford: Clarendon Press, 1974), *Advancement*, Bk II, viii, 2 (pp. 96–97); see also Sachiko Kusukawa, "Bacon's Classification of Knowledge," in *The Cambridge Companion to Bacon*, ed. Markku Peltonen (Cambridge: Cambridge University Press, 1996), 60; Gregor Horst, *Institutionum logicarum libri duo* (Wittenberg, 1608), 59.

9. Ioannis de Guevara, *In Aristotelis Mechanicas commentarii* (Rome, 1627), esp. 24, on the question of the "mixed" character of mechanics (between physics and mathematics); cf. Moletti, in Laird, *Unfinished Mechanics*, 79; other references in Peter Dear, *Discipline and Experience: The Mathematical Way in the Scientific Revolution* (Chicago: University of Chicago Press, 1995), 40. See also, on Guevara's geometrical procedures, William A. Wallace, "The Problem of Apodictic Proof in Early Seventeenth-Century Mechanics: Galileo, Guevara, and the Jesuits," *Science in Context* 3 (1989): 67–87.

10. Daniele Cozzoli, "Alessandro Piccolomini and the Certitude of Mathematics," *History and Philosophy of Logic* 28 (2007): 151–71; Paolo Mancosu, *Philosophy of Mathematics and Mathematical Practice in the Seventeenth Century* (New York: Oxford University Press, 1996); Nicholas Jardine, "Epistemology of the Sciences," in *The Cambridge History of Renaissance Philosophy*, ed. Quentin Skinner Charles Schmitt, Eckhard Kessler, and Jill Kraye (Cambridge: Cambridge University Press, 1988), 685–711; Rivka Feldhay, "The Use and Abuse of Mathematical Entities: Galileo and the Jesuits Revisited," in *The Cambridge Companion to Galileo*, ed. Peter Machamer (Cambridge: Cambridge University Press, 1998), 80–145, esp. 83–100.

11. Josephus Blancanus, *De mathematicarum natura dissertatio* (Bologna, 1615). See Dear, *Discipline and Experience*, esp. 32–44; Mancosu, *Philosophy of Mathematics*, includes a translation of the *dissertatio*.

12. See selections in Stillman Drake and I. E. Drabkin, eds., *Mechanics in Sixteenth-Century Italy: Selections from Tartaglia, Benedetti, Guido Ubaldo, and Galileo* (Madison: University of Wisconsin Press, 1969); see also Rose, *Italian Renaissance of Mathematics*; Domenico Bertoloni Meli, *Thinking with Objects: The Transformation of Mechanics in the Seventeenth Century* (Baltimore, MD: The Johns Hopkins University Press, 2006), esp. ch. 1–2; W. R. Laird, "Archimedes among the Humanists," *Isis* 82 (1991): 629–38.

13. Bertoloni Meli, *Thinking with Objects*, 63–65.

14. Johannes Kepler, *Epitome astronomiae Copernicanae*, in Nicholas Jardine, *The Birth of History and Philosophy of Science: Kepler's A Defence of Tycho Against Ursus with Essays on Its Provenance and Significance* (Cambridge: Cambridge University Press 1984), 250. See, among others, J. V. Field, *Kepler's Geometrical Cosmology* (London: Athlone, 1988); James R. Voelkel, *The Composition of Kepler's* Astronomia Nova (Princeton, NJ: Princeton University Press, 2001). On the mathematical and physical characters of astronomy in this period, see also the classic article by Robert S. Westman, "The Astronomer's Role in the Sixteenth Century: A Preliminary Study," *History of Science* 18 (1980): 105–47.

15. Albert Van Helden, introduction to *Sidereus Nuncius, or, the Sidereal Messenger*, by Galileo Galilei, (Chicago: University of Chicago Press, 1989).

16. Peter Machamer, "New Perspectives on Galileo," in *New Perspectives on Galileo*, ed. Robert E. Butts and Joseph C. Pitt (Dordrecht: D. Reidel, 1978), 161–80; James G. Lennox, "Aristotle, Galileo, and 'Mixed Sciences,'" in *Reinterpreting Galileo*, ed. William A. Wallace (Washington DC: University of America Press, 1986), 29–51; W. Roy Laird,

"Galileo and the Mixed Sciences," in *Method and Order in Renaissance Philosophy of Nature: The Aristotle Commentary Tradition*, ed. Daniel A. Di Liscia, Eckhard Kessler, and Charlotte Methuen (Aldershot, UK: Ashgate, 1997), 253–70, who stresses the differences between Galileo's concern with demonstrative truth and the concern by Jesuits and others with the hypothetical dimensions of a "mixed" science like astronomy.

17. J. B. Baliani, *De motu naturali* (Genoa, 1646), 8: "Rimari caepi; an deprehenderim aliorum erit iudicium" (unless otherwise noted, all translations in this chapter are mine).

18. Ibid., 97–98: "Hactenus mihi videor de scientia motus naturalis gravium solidorum satis pro viribus dixisse, dum ex quibusdam proprietatibus sensui notis, plures ignotae deductae, & patefactae sunt: in hoc enim solummodo ex Aristotele omnis scientia versatur; ut in praxi apud Euclidem, & alios, qui veras, & simplices scientias tractant, videre est: unde nec agit Geometra de natura quantitatis, nec Musicus de natura soni, nec perspectivus de natura luminis, nec mechanicus de natura ponderis. At vero meus intellectus non omnino acquiescit, ni causas priores, à quibus his effectus demum proveniunt, si non assequatur, saltem investiget . . ."

19. Ibid., 98: "non effectus, sed rerum naturae"; 98–102. See, for a discussion of Baliani's physical account of fall, Serge Moscovici, *L'expérience du mouvement: Jean-Baptiste Baliani disciple et critique de Galilée* (Paris: Hermann, 1967), 56–72. Baliani's account resembles Fabri's; see Dear, *Discipline and Experience*, 138–44 on Honoré Fabri, Baliani, and their related criticisms of Galileo's assertions concerning falling bodies; also Bertoloni Meli, *Thinking with Objects*, 120–26; Carla Rita Palmerino, "Two Jesuit Responses to Galileo's Science of Motion: Honore' Fabri and Pierre Le Cazre," in *The New Science and Jesuit Science: Seventeenth-Century Perspectives*, ed. Mordechai Feingold (Dordrecht: Kluwer, 2003), 187–227.

20. Beeckman, *Journal tenu par Isaac Beeckman de 1604 à* 1634, ed. Cornelis de Waard (The Hague: Martinus Nijhoff, 1939–1953), 4 vols., 1:244. "Hic Picto cum multis Jesuitis alijsque studiosis virisque doctis versatus est. Dicit tamen se nunquam neminem reperisse, praeter me, qui hoc modo, quo ego gaudeo, studendi utatur accuratèque cum Mathematicâ Physicam jungat. Neque etiam ego, praeter illum, nemini locutus sum hujusmodi studij." "Picto" is Descartes: see *Journal*, 1:237.

21. Ibid., 1:244: "Physico-mathematici paucissimi." Beeckman generally preferred the formulation "mathematico-physica"; see, e.g., *Journal*, 4:196, 200, a letter of 1630 to Descartes where Beeckman is concerned about his priority. See esp. Desmond M. Clarke, *Descartes: A Biography* (Cambridge: Cambridge University Press, 2006), 46–52, and Klaas van Berkel, "Descartes' Debt to Beeckman: Inspiration, Cooperation, Conflict," in *Descartes' Natural Philosophy*, ed. Stephen Gaukroger, John Schuster, and John Sutton (London: Routledge, 2000), 46–59.

22. Stephen Gaukroger, *Descartes: An Intellectual Biography* (Oxford: Clarendon Press, 1995), ch. 3; Stephen Gaukroger and John A. Schuster, "The Hydrostatic Paradox and the Origins of Cartesian Dynamics," *Studies in History and Philosophy of Science* 33 (2002): 535–72, esp. 550–58; William R. Shea, *The Magic of Numbers and Motion: The Scientific Career of René Descartes* (New York: Science History Publications, 1991), 77–86.

23. Dear, *Discipline and Experience*, 168–79; note the disagreement with Gaukroger and Schuster, "Hydrostatic Paradox," 537n2, who underestimate the new term's flexibility.

24. See bibliographical discussion in Robert Lenoble, *Mersenne ou la naissance du mécanisme* (Paris: J. Vrin, 1943), xxix–xxx.

25. Carla Rita Palmerino, "Infinite Degrees of Speed: Marin Mersenne and the Debate over Galileo's Law of Free Fall," *Early Science and Medicine* 4 (1999): 269–328.

26. Dear, *Discipline and Experience*, 172–73.

27. Ibid., 178.

28. See Peter R. Anstey, "Experimental Versus Speculative Natural Philosophy," in *The Science of Nature in the Seventeenth Century: Patterns of Change in Early Modern Natural Philosophy*, ed. Peter R. Anstey and John A. Schuster (Dordrecht: Springer, 2005), 215–42, esp. 232–36; Alan E. Shapiro, "Newton's 'Experimental Philosophy'," *Early Science and Medicine* 9 (2004): 185–217. Cf. Rob Iliffe, "Abstract Considerations: Disciplines and the Incoherence of Newton's Natural Philosophy," *Studies in History and Philosophy of Science* 35 (2004): 427–54.

29. Isaac Newton, *The Principia: Mathematical Principles of Natural Philosophy: A New Translation and Guide*, trans. by I. Bernard Cohen and Anne Whitman (Berkeley: University of California Press, 1999), 379; Halley's Latin is "opus . . . mathematico-physicum": Isaac Newton, *Isaac Newton's Philosophiae naturalis principia mathematica*, ed. Alexandre Koyré and I. Bernard Cohen, based on the 1726 ed. (Cambridge, MA: Harvard University Press, 1972), 1:12.

30. Barbara J. Shapiro, *John Wilkins, 1614–1672: An Intellectual Biography* (Berkeley: University of California Press, 1969), 192.

31. Jean Le Rond d'Alembert, "Observations on Bacon's Division of the Sciences" (1751), in *Preliminary Discourse to the Encyclopedia of Diderot*, trans. by Richard N. Schwab (Indianapolis: Bobbs-Merrill, 1963), 159.

32. Alexandre Koyré, "Huygens and Leibniz on Universal Attraction," in *Newtonian Studies* (Chicago: University of Chicago Press, 1965), 115–38; Roberto De A. Martins, "Huygens's Reaction to Newton's Gravitational Theory," in *Renaissance and Revolution: Humanists, Scholars, Craftsmen and Natural Philosophers in Early Modern Europe*, ed. J. V. Field and Frank A. J. L. James (Cambridge: Cambridge University Press, 1993), 203–13.

33. Newton, *Principia*, 381. Cf. Niccolò Guicciardini, "Newton: Between Tradition and Innovation," in *Reading the Principia: The Debate on Newton's Mathematical Methods for Natural Philosophy* (Cambridge: Cambridge University Press, 1999), 99–117; Mordechai Feingold, "Mathematicians and Naturalists: Sir Isaac Newton and the Royal Society," in *Isaac Newton's Natural Philosophy*, ed. Jed Z. Buchwald and I. Bernard Cohen (Cambridge, MA: MIT Press, 2001), 77–102.

34. Newton, *Principia*, 392.

35. See, e.g., Aristotle *Posterior Analytics* I.3.

36. D'Alembert, "Observations on Bacon's Division of the Sciences."

37. Alan Gabbey, "The Mechanical Philosophy and Its Problems: Mechanical Explanations, Impenetrability, and Perpetual Motion," in *Change and Progress in Modern Science*, ed. Joseph C. Pitt (Dordrecht: D. Reidel, 1985), 9–84; Gabbey, "Newton, Active Powers, and the Mechanical Philosophy," in *The Cambridge Companion to Newton*, I. Bernard Cohen and George E. Smith (Cambridge: Cambridge University Press, 2002), 329–57.

38. *Philosophical Transactions* #80 (1672), 3085; A. Rupert Hall, "Beyond the Fringe: Diffraction as Seen by Grimaldi, Fabri, Hooke and Newton," *Notes and Records of the*

Royal Society 44 (1990):17–19, suggests that this passage may have been prompted by Newton's reading a review of Francesco Grimaldi's *Physico-mathesis de lumine* (1665); Grimaldi had also discussed the substantiality of light (Dear, *Discipline and Experience*, 235–36).

39. Isaac Newton, *The Correspondence of Isaac Newton*, ed. H. W. Turnbull (Cambridge: Cambridge University Press, 1959–1977), 1:187.

40. See Bertoloni Meli, "A New World System," *Thinking with Objects*, 255–87, 349–52.

41. Jean Desaguliers, *A Course of Experimental Philosophy* (London, 1734–44), 1:69; see Larry Stewart, *The Rise of Public Science: Rhetoric, Technology, and Natural Philosophy in Newtonian Britain, 1660–1750* (Cambridge: Cambridge University Press, 1992).

42. John Harris, *Lexicon technicum*, 2nd ed. (London, 1708), s.v. "Mathematicks."

43. Ephraim Chambers, *Cyclopedia* (London, 1728), 1: ii, iii ("Doctrine of Motion"). See in general Yeo, *Encyclopaedic Visions*.

44. Chambers, *Cyclopedia*, 1:144.

45. Ibid., 2:509.

46. *Encyclopédie, ou, Dictionaire raisonné des sciences, arts et des métiers* (Paris, 1751–1765), 12: 512, s.v. "Philosophie": "Il est tems de passer au second point de cet article, où il s'agit de fixer le sens du nom de la Philosophie, & d'en donner une bonne definition. Philosopher, c'est donner la raison des choses, ou du moins la chercher, car tant qu'on se borne à voir & à rapporter ce qu'on voit, on n'est qu'historien. Quand on calcule & mesure les proportions des choses, leurs grandeurs, leurs valeurs, on est mathématicien; mais celui qui s'arrête à découvrir la raison qui fait que les choses sont, & qu'elles sont plutôt ainsi que d'une autre maniere, c'est le philosophe proprement dit."

47. D'Alembert, "Detailed Explanation of the System of Human Knowledge" (*Encyclopédie*, 1751), in *Preliminary Discourse*, 152; see also Daston, *Classical Probability in the Enlightenment*, 54.

48. Georges-Louis Leclerc, comte de Buffon, *Histoire naturelle, générale et particulière*, (Paris, 1749), 1: 52: "Dans ce siècle même où les Sciences paroissent être cultivées avec soin, je crois qu'il est aisé de s'apercevoir que la Philosophie est négligée, & peut-être plus que dans aucun autre siècle; les arts qu'on veut appeller scientifiques, ont pris sa place; les méthodes de Calcul & de Géométrie, celles de Botanique & d'Histoire Naturelle, les formules, en un mot, & les dictionnaires occupent presque tout le monde; on s'imagine sçavoir davantage, parce qu'on a augmenté le nombre des expressions symboliques & des phrases sçavantes, & on ne fait point attention que tous ces arts ne sont que des échafaudages pour arriver à la science, & non pas la science elle-même . . ."

49. Jim Bennett, "Practical Geometry and Operative Knowledge," *Configurations* 6 (1998): 195–222.

50. On natural philosophy lecturers, see John L. Heilbron, *Electricity in the Seventeenth and Eighteenth Centuries: A Study in Early Modern Physics* (Berkeley: University of California Press, 1979). Even Jesuit colleges often boasted experimental cabinets in the eighteenth century: see Marcus Hellyer, "The Spread of Experiment," *Catholic Physics: Jesuit Natural Philosophy in Early Modern Germany* (Notre Dame, IN: University of Notre Dame Press, 2005), 181–201, 286–89.

51. Christian Licoppe, *La formation de la pratique scientifique: Le discours de*

l'expérience en France et en Angleterre, 1630–1820 (Paris: Éditions La Découverte, 1996), esp. 169–74; also Jessica Riskin, "Poor Richard's Leyden Jar: Electricity and Economy in Franklinist France," *Historical Studies in the Physical and Biological Sciences* 28 (1998): 301–36.

52. John Gascoigne, *Cambridge in the Age of the Enlightenment: Science, Religion and Politics from the Restoration to the French Revolution* (Cambridge: Cambridge University Press, 1989); Gascoigne, "Mathematics and Meritocracy: The Emergence of the Cambridge Mathematical Tripos," *Social Studies of Science* 14 (1984): 547–84; Harvey W. Becher, "William Whewell and Cambridge Mathematics," *Historical Studies in the Physical Sciences* 11 (1980): 1–48; Alex D. D. Craik, *Mr. Hopkins' Men: Cambridge Reform and British Mathematics in the 19th Century* (London: Springer, 2007); Andrew Warwick, "The Reform Coach: Teaching Mixed Mathematics in Georgian and Victorian Cambridge," in *Masters of Theory: Cambridge and the Rise of Mathematical Physics* (Chicago: University of Chicago Press, 2003), 49–113.

53. John L. Heilbron, "A Mathematicians' Mutiny, with Morals," in *World Changes: Thomas Kuhn and the Nature of Science*, ed. Paul Horwich (Cambridge, MA: MIT Press, 1993), 106–7. Cf. John V. Pickstone, "Working Knowledges before and after Circa 1800: Practices and Disciplines in the History of Science, Technology, and Medicine," *Isis* 98 (2007): 489–516, esp. 508–9.

54. John L. Heilbron, "Fin-de-siècle Physics," in *Science, Technology, and Society in the Time of Alfred Nobel*, ed. Carl-Gustav Bernhard, Elizabeth Crawford, and Per Sèrböm (Oxford: Pergamon Press, 1982), 51–71; see also Theodore M. Porter, "Ether Squirts and the Inaccessibility of Nature," in *Karl Pearson: The Scientific Life in a Statistical Age* (Princeton, NJ: Princeton University Press, 2004), 178–214.

55. See "Natural Philosophy" by John Heilbron (in this volume, 173–199).

56. There is too much literature to summarize here. See, for a light entrée, Peter Dear, "How to Understand Nature? Einstein, Bohr, and the Quantum Universe," in *The Intelligibility of Nature: How Science Makes Sense of the World* (Chicago: University of Chicago Press, 2006), 141–72, 203–4; Jan Faye, "The Bohr-Høffding Relationship Reconsidered," *Studies in History and Philosophy of Science* 19 (1988): 321–46, discusses an important Danish philosophical connection for Bohr.

57. See, e.g., Heilbron, "A Mathematicians' Mutiny," 101–2, on the early part of the period; Iwan Rhys Morus, "A Revolutionary Science," in *When Physics Became King* (Chicago: University of Chicago Press, 2005), ch. 2, 22–53, is a good overview. The *Journal für die reine und angewandte Mathematik*, known informally as "Crelle's journal," was founded in 1826: Christa Jungnickel and Russell McCormmach, *Intellectual Mastery of Nature: Theoretical Physics from Ohm to Einstein* (Chicago: University of Chicago Press, 1986), 1:37. Cf. H. M. Mulder, "Pure, Mixed and Applied Mathematics: The Changing Perception of Mathematics through History," *Nieuw archief voor wiskunde* 8 (1990): 27–41.

58. Consult Jungnickel and McCormmach, *Intellectual Mastery of Nature*, esp. vol. 2.

59. See, e.g., Warwick, *Masters of Theory*, 220–21, on the mathematics coach Edward Routh; also William Thomson and Peter Guthrie Tait, *Treatise on Natural Philosophy*, vol. 1 (Oxford: Clarendon Press, 1867).

CHAPTER 7

Natural Philosophy

John L. Heilbron

The subject of this chapter is the changing character of natural philosophy and the attitudes of its cultivators toward it from the restabilization of Europe around 1660 to the end of the eighteenth century. What follows is merely an indication of a possible way to arrange the material. It begins with a definition and periodization of natural philosophy. These suggest a development in the shape of an hourglass, the long axis representing time increasing downward and the cross section the corresponding breadth of natural philosophy. The widest part at the top stands for the situation during the last decades of the seventeenth century; the constriction in the middle, the situation in the decades around 1750; and the progressive widening at the bottom, the years 1770–1815. The analogy limps in many ways, perhaps most grievously in suggesting that natural philosophy returned in 1800 to the same responsibilities it had a hundred years earlier. The two states differed greatly in coverage as well as in content. The image of the hourglass has the merit, however, of emphasizing that the later expansion of natural philosophy followed an earlier constriction, during which it lost its traditional connections within the body of knowledge.

The beginning of the expansion dates from around 1770. At its end around 1815, natural philosophy had taken on many of the traits of classical physics. The reasons for and nature of the inflection around 1770 thus have a strong claim on the historian of science. The forces at play in forming modern science—the interests of nation-states, individual entrepreneurs, and the military—also figured in the transformation of natural philosophy into classical physics. The lower flare of the hourglass is framed by the Seven Years' War (1756–1763) and the revolutionary and Napoleonic wars, which made greater demands on applied natural philosophy and mathematics than earlier conflicts had done. This period, and especially the transition around 1770, is the focus of the hints and indications in this paper. It coincided with the so-called age of the democratic revolution, with which it had ties not merely coincidental.[1]

Reformers of the age, whether of government or natural philosophy, emphasized control, efficiency, and utility, and rejected traditional categories, customs, and privileges that did not meet their test of rationalization. A confluence of the two sorts of reforms and a symbol for their relationship was the metric system of weights and measures, designed by the leading academicians of France, defined by the most exact instruments and experimental techniques of the day, and imposed by the Revolution on a people slow to celebrate the marriage of democracy and decimalization. Here we have rationalization in both a political and mathematical sense, and a resistance to its imposition by people who preferred familiar practices and attitudes. As in political theory and government, so in natural philosophy and its applications, the old persisted as the new gained strength. The cross sections of the hourglass were not homogeneous. For example, John Locke's *Elements*, composed in 1706 as "an abstract or summary of whatever is most material in natural philosophy," was reprinted in both England and France as an introductory text long after its conception of the subject was outmoded.[2] The historian takes account of these survivals by presenting counterexamples to his or her thesis, and by arguing that they represent a diminishing trend.

If the figure of the hourglass be allowed for the development of natural philosophy, that of a string gradually closing into a circle may do for the relationships among natural philosophers. In the middle of the string are the leading members of the leading academies; on their right, proceeding clockwise, professors, then other teachers, public lecturers, booksellers, and instrument makers; on the left of the high academicians, proceeding counterclockwise, come lesser academicians, habitués of intellectual salons, cameralwissenschaftler, agriculturalists, manufacturers, and inventors with claims or connections to natural philosophy. According to this superstring theory, as the eighteenth century proceeded, gaps along it filled and the ends joined. The filling intensified after 1770. The joining of the ends may be expressed symbolically by the investigation and development of the steam engine with separable condenser by the former instrument makers James Watt and Matthew Boulton.

The second section below cuts up the string of natural philosophers into some of its better-known parts. The first section breaks open the hourglass.

THE HOURGLASS OF PHYSICS

1: Natural Philosophy

"Natural philosophy" is not-very-plain English for Latin *physica* and French *physique*. It concerns natural knowledge. It does not carry any implication of natural theology, although many natural philosophers believed that their studies of God's handiwork brought them close to their Creator. "Cull the sweets of religion, as you roam through the flowery paths of natural philosophy," Margaret Bryan admonished the girls to whom she taught physics, "and guard your religious and moral principles against all innovations." The correct term for "culling the sweets," or (to quote Samuel Johnson) for contemplating "divinity enforced or illustrated by natural philosophy" was "physico-theology."[3]

To know that natural philosophy was physics and not natural theology is not to know much. The *Encyclopédie* of Diderot and d'Alembert takes us only a little further. It defined *physique* (alias *philosophie naturelle*) in the manner of Aristotle and the scholastic philosophers as the "science of the properties of [all] natural bodies, their phenomena and their effects," and so was marginally more restrictive than Locke's natural philosophy, which included the study of angels. When the *Encyclopédie* proffered its definition in 1765, however, physics books no longer covered everything from animalcula to the stars. Charles Hutton repeated it sixty years later in his *Philosophical and mathematical dictionary* of 1815, a year before Jean Baptiste Biot published his voluminous *Traité de physique expérimentale et mathématique*, the first full textbook of physics restricted to domains and methods recognizably modern.[4] The effective meaning of physics or natural philosophy during the eighteenth century cannot be determined by examining dictionaries.

The title of Biot's text suggests a periodization of natural philosophy as well as its content in his time. First came experiment, then the combination of experiment with mathematics. That is roughly what happened from the middle of the seventeenth century to the end of the eighteenth in textbooks and also in real life. Jacques Rohault's *Traité de physique*, a work of Cartesian inspiration first published in 1671, was exemplary for the decades around 1700. It reached its sixth French edition in 1692 and its third English translation, with prophylactic notes by Newton's paladin

Samuel Clarke, in 1735. It described and explained demonstrations illustrative of "all the effects that Nature produces." Nonetheless, its encompassing subject—the "speculative knowledge of all natural bodies," as John Harris defined "physicks" in 1704—remained largely bookish and qualitative. Experiment had become central to the investigation, but not yet to the transmission, of natural knowledge. During the last third of the seventeenth century, the Accademia del Cimento of Florence, the Royal Society of London, the Académie des sciences of Paris, the Accademia fisicomatematica romana, and the Collegium Experimentale of Altdorf had made the pinching and probing of nature with instruments a respectable and even a virtuous way to obtain natural knowledge.[5]

The members of these early academies usually performed experiments together or before one another as eyewitnesses so that the outcomes could be certified as contributions to knowledge. The most prized experiments, because the most clear and striking to several spectators at once, involved the air pump and other sturdy instruments able to create reliable and striking effects. A narrowing of the scope of natural philosophy resulted: experimental philosophy, which emphasized mechanical and pneumatic experiments, became an important and then the leading sector of natural philosophy.[6]

Newton's discoveries in optics fit easily into this program, as did demonstrations of the sorts of forces he supposed at work in the universe. The first textbook of natural philosophy in this narrow sense was W. J. 'sGravesande's *Physica elementa mathematica experimentis confirmata, sive introductio ad philosophiam neutonianam* (1720–1721). It and updated texts on similar lines by Pieter van Musschenbroek, 'sGravesande's successor as professor of mathematics and astronomy at the University of Leiden, had a large circulation in several languages. Despite their titles, they contain very little mathematics. They also omit everything that now falls in the domains of biology and geology, and almost everything we would call chemistry and meteorology, all of which figure in Rohault's *Physique*. The same process can be followed in texts independent of the market-leading Dutch Newtonians. The tradition descending from the Collegium Curiosorum, which derived from the Accademia del Cimento, continued in Christian von Wolff's *Allerhand nützliche Versuche*, published almost simultaneously with 'sGravesande's *Physica*. Wolff admitted mechanics, hydrostatics, pneumatics, meteorology, fire, light, color, sound and magnetism, animals as subjects for experiment in vacuo, and sense organs as examples of optical and mechanical principles—a more generous selection than 'sGravesande's but still impoverished as an account of natural phenomena in general.[7]

The assimilation of experimental philosophy with natural philosophy was accomplished by the distinctive science of the eighteenth century, electricity, and by the creation of new audiences for the demonstration of natural knowledge. As the practice of witnessing experiments diminished at the major learned societies in favor of reports printed in their publications, a new audience grew up consisting of well-off people curious enough about the latest findings in natural philosophy to spare an evening for it from cards. Their expectations put an even higher premium on experiments easily seen and appreciated by groups. Schools provided another considerable consumer of demonstrations. Instrument makers and booksellers furnished the necessaries. The fine points of anatomical or microscopic investigations had little to offer philosophical showmen. They stuck to mechanics, hydrostatics, and pneumatics, and their applications as exhibited in models of machines—pile drivers, driving bells, fire engines, steam engines, and so on. At midcentury, demonstrations of electricity joined and then soon led the repertoire. Lecture courses of the time typically bore the title "Natural and experimental philosophy," meaning experiments and their interpretations.[8]

The electrical fire producible by electrostatic generators and the shocks procurable from condensers after the invention of the Leyden jar in 1745 made wonderful games and displays. The electrical kiss, the simulation of lightning, the electrocution of birds, and shock therapy became the signatures of natural philosophy. According to Daniel Jenisch's *Geist und Character des achtzehnten Jahrhunderts*, which portrayed the dying century from the perspective of 1800, "Electricity and its wonderful effects are without doubt the most remarkable and useful discoveries with which the eighteenth century has enriched [natural philosophy]."[9] (Jenisch was a prolific polymath priest and professor, some of whose works won prizes from the Berlin Academy of Sciences.[10]) Together with other branches of *physica particularis* (pneumatics, heat, light, magnetism, and often also meteorology) and *physica generalis* (statics and dynamics of solids and liquids, and perhaps qualitative Newtonian astronomy), electricity defined the subject of natural philosophy around 1750.[11]

During the last third of the eighteenth century the narrowly focused natural and experimental philosophy of midcentury underwent a major expansion in two directions. For one, it again took in parts of chemistry, particularly the theory and practice of the gases discovered in the 1760s and 1770s and, later, the principles of Lavoisier's system and nomenclature. These provinces of chemistry and the sorts of experiments that supported them were closer to the concerns and methods of experimental philosophy than the older chemical recipes and reasonings. Lavoisier himself claimed and was accorded the title of *physicien*, an apt usage for the time that caused distress two hundred years later among historians of chemistry who fretted that a "physicist" might have reconstructed their science. Lavoisier meant that he was a natural philosopher.[12]

The second direction of expansion of physics after 1770 took place across its boundary with mathematics. Mixed mathematics, or, as the *Encyclopédie* and its echo Hutton had it, "physico-mathematics," included quantified portions of astronomy, optics, acoustics, and mechanics.[13] As it pushed quantification in the subsciences of heat, electricity, and magnetism, natural philosophy began to acquire parts of mixed mathematics and some of the arrogant rhetoric of the mathematician. "Everyone now agrees that a physics lacking all connections with mathematics . . . would only be an historical amusement, fitter for entertaining the idle than for occupying the mind of a philosopher." Thus Franz Karl Achard, a prominent member of the Berlin Academy of Sciences who pretended to read his electroscope to six places, addressed his colleagues in 1773. "Having thus learnt to distinguish knowledge from what has only the appearance of it," added J. A. Deluc, a man as exact and exacting as his barometer, "we shall be led to seek for exactness in everything."[14] More and better instruments made possible the measurement of newly devised quantities, like electrical force and tension, and increased the precision of the determination of familiar ones, like temperature and pressure. The result was exact natural philosophy, to use an historian's rather than an actor's term. The neologism is meant to emphasize that most of the measurements done around 1800 did not serve as checks on theory, indeed could not have been predicted by any theory then available. "Mathematical physics," that is, quantitative theories predictive of the phenomena, was more the goal than the content of Biot's *Traité*. "The law of a phenomenon can only be established experimentally . . . To make it absolutely certain and rigorous, it must be related to the general laws of mechanics . . . When such a reduction can be effected completely, it necessarily makes clear the FORCES that produce the phenomena, which is as far as human science can go."[15]

Natural experimental philosophy was thus richer around 1800 than it had been during the dominance of experimental philosophy in the middle decades of the eighteenth century. It had acquired some lost ground by sharing progress in chemistry and it had begun to fill out its mechanics, optics, acoustics, and so on by capture from mixed mathematics. In another aspect also it experienced a partial return to its umbrella status of 1700. Rohault's *Physique* had the unity of Cartesian qualitative corpuscularism. The natural philosophers around 1800 had two unifying schemes.

The more widely accepted, the system of imponderables, associated each of the main divisions of *physica particularis*—light, heat, magnetism, electricity—with a special fluid or fluids composed of weightless or unweighable particles that repelled and/or attracted one another and the particles of ponderable matter. Although the fluid(s) associated with different sets of phenomena did not interact, they had this much in common besides imponderability: the forces acting between their particles were central distance forces of the type that Newton had invoked in his gravitational theory. The system rested precariously on a few measurements like Coulomb's on electric and magnetic forces and more securely on a formidable array of analogies among electricity, heat, radiant heat, and light.[16]

The other integrating system bruited around 1800 was the *Naturphilosophie* of F. W. J. Schelling, which, insofar as it took in all of nature and marginalized experiment and mathematics, was a throwback to 1700 and even earlier times. But, like the old global natural philosophy, it soon moved to exploratory experimentation and realized some fine discoveries. It achieved its greatest successes with the battery that Alessandro Volta—one of the great acrobats on the network of imponderables—made public in 1800, just after Schelling had set the basis of his speculative physics. Like Cartesianism, *Naturphilosophie* rested on a priori premises and was intended to be true of nature. In this last respect it again differed from the system of imponderables, whose advocates admitted freely that they did not know even in the exemplary case of electricity whether its effective cause was two fluids, one fluid, or none.[17] An instrumentalist attitude toward theory often accompanies quantification. Mixed mathematics may be precise, but, as Aristotle knew, there is no truth in it.

2. Natural Philosophers

The character of natural philosopher is more difficult to set forth than the nature and periods of natural philosophy. A common escape, which makes natural philosophers "practitioners" of natural philosophy, will not do. "Practitioner" has too much the air of "professional," as in "general practitioner." Johnson gives clergymen and idolaters ("papistical practitioners") as examples. He also allows the use of the term for an "exercised, thorough-paced practitioner of . . . vices," or anyone else who does something habitually.[18] Although some eighteenth-century people practiced natural philosophy as an art or profession and others used it to practice upon the gullible, most who claimed the title natural philosopher did not consider natural philosophy a vocation. In 1718 the Royal Society asked its members to identify their interests, "to show what all are good and bad for." Almost 60 percent could not manage to name a subject. Half of those who did, some forty Fellows, specified "mathematics and natural philosophy," not because they "practiced" them but because Newton had. Except for a very few, they were at most dilettantish cultivators of their favorite sciences.[19] Even those who devoted much of their time to natural philosophy avoided the stain of practice. For the same reason, "men of science," as they liked to call themselves in the nineteenth century, rejected William Whewell's neologisms "scientist" and "physicist" as smacking of "dentist," a paid practitioner of an unpleasant profession.[20]

On the continent, *physicien* and its counterparts *Physiker*, *Naturkundiger*, and *fisico* may have had a shade more of the professional about them than natural philosopher. The salaried members of the leading academies, specifically those of Paris, Berlin, and St. Petersburg, were required to contribute original papers to their academies' publications. Most of the members of the provincial academies in France were professional men in the eighteenth-century sense—lawyers, doctors, clergy, military men—and improving agriculturalists. It was much the same in the economic and patriotic societies in the Germanies, with, perhaps, a greater proportion of small landholders.[21] No more in Europe than in England was a term with the professional overtones of "scientist" required. *Scientifique* used as a noun and *Naturwissenschaftler*, the equivalents of "scientist," were introduced in the later nineteenth century.

Few among those who called themselves natural philosophers or *physiciens* followed the *Encyclopédie*'s dictate about philosophizing: "A true philosopher does not see through others' eyes." The Royal Society's arms bear a similar slogan, *Nullius in verba*, which the learned recognized as Horace's injunction not to swear allegiance to any school or master. "The philosopher," continues the *Encyclopédie*, "is to give the reasons for things, or, at least, to seek them. Those who stop at reporting what they see are only historians. Those who calculate and measure the proportions of things, their sizes and values, are only mathematicians. But he who arrives at discovering the reasons why things are as they are and not otherwise, is a true philosopher." What applied to the genus philosopher applied also to the species natural philosopher, since physics, according to the *Encyclopédie*, made up a third of philosophy.[22]

The natural habitat of the true natural philosopher was the academy of science. There are indications that the proportion of true natural philosophers in them increased after 1770. In 1776, the Royal Society called a moratorium on the election of foreign members who, on average, were less truly philosophical than even the English; and when it lifted the ban, it required evidence of productive engagement with natural philosophy (or some other science) for admission. Another example is the Società Italiana delle scienze, founded in 1782 to strengthen ties among the natural philosophers of the peninsula. Despite its mission of inclusiveness, the society applied the tightest tests of the time, to the approval of those who passed it. The polymath peripatetic Jesuit physico-mathematician Roger Boscovich wrote the society's founder: "I'm glad that your Società Italiana is progressing and I entirely agree that it will be very good to restrict the number [of its members] so as not to make it a Noah's arc." Similarly, the Hollandsche Maatschappij der Wetenschappen, which had begun in 1758 with a membership representative of all the literate animals in nature, had restricted itself by the end of the century to "professionals who are professors, or have acquired their reputations by [their] work."[23] The rise in the proportion of professionals and professors in the academies coincided with an explicit acknowledgment of the arrival of exact natural philosophy. In 1785 the Paris Academy added a new class of "experimental physics" to its group of mathematical sciences. Likewise, in 1807 the Hollandsche Maatschappij established a new division for "experimental and mathematical sciences."

The Hollandsche Maatschappij's specification of ideal academic types included two novelties: professionals and professors. By professionals they appear to have meant natural philosophers like Martinus van Marum, a physician by training, director and curator of instruments for the Maatschappij and also for Tylers Tweede Genootschap, a small general-purpose learned academy with a substantial museum and collection of apparatus. Van Marum demonstrated, lectured, experimented, and measured (though not often) in return for several salaries. Professionals also included men like Van Marum's associate Adriaan Paets van Troostwijk, a merchant by trade and a serious chemist by avocation. His professionalism in natural philosophy lay in the new field of pneumatics, especially the ventilation of mines, prisons, and factories, and the action of electricity on the gas types, about which he manufactured much exact misleading data.[24] As for professors, their appearance among desirable academics indicated an important development, even a revolution, in the Republic of Letters. The great academic movement of the eighteenth century filled a niche not occupied by the universities. Professors taught; academicians worked to advance knowledge and to apply it usefully. As the *Encyclopédie* laid it down, "An academy is by no means intended to teach or profess any art, no matter what it is, but to perfect it." That implied a risk. "An academician must invent and improve," so said one of the founders of the

Göttingen Societät der Wissenschaften, "or inevitably betray his nakedness." Ordinary teachers, "those who aspire only to be professors," faced no such danger.[25]

Universities tended to oppose the establishment of academies in their neighborhoods for fear of challenges to their authority. The Hollandsche Maatschappij knew that well. The University of Leiden had opposed its founding. The professors claimed that the Royal Society had undercut the prestige of Oxford and Cambridge and that the University of Paris had hardly been heard from since the foundation of the Paris Academy of Sciences. Relations were much better in Bologna, where professors could be seconded to paid posts in the Accademia delle scienze, and Göttingen, where leading professors also belonged to the Societät der Wissenschaften. Some professors elsewhere added their bits to knowledge with the help of administrators eager to demonstrate their enlightenment. The English-Hanoverian administration practiced such a policy toward the University of Göttingen almost from its foundation in 1737 as did the Austrian government of Lombardy toward the University of Pavia after 1770. The most fruitful contributions to natural philosophy made in Italy in the later eighteenth century came from Bologna, where Galvani worked as a professor and academician, and Pavia, where Volta enjoyed something like the life of a modern professor, with research grants, a good budget for instruments, and expenses for foreign travel.[26] A quantitative indicator of the increasing identification of physics professors with their subject and its advancement can be given for Protestant Germany. During the first quarter of the eighteenth century, three-fourths of those who held chairs of physics left them for other careers, usually in medicine. During the last quarter, three-fourths made their entire careers teaching, and sometimes also adding to, physics.[27]

In addition to salaried academicians and professors who did some research, full-time natural philosophers might be found in observatories, botanical gardens, and chemical laboratories. The number engaged in research increased after 1750 as the pace of foundation of the institutions most closely associated with the creation of new knowledge in natural philosophy—the academy and the observatory—stepped up. The number of fellow travelers on the ark of science also grew: the reporters of marvels, witnesses of demonstrations, supporters of academies, observers of weather, authors of systems, agricultural improvers, natural theologians, collectors, inventors, hobbyists, and, not least, sharpers and charlatans. This throng included many who called themselves natural philosophers. And who could deny them the title? No formal criteria for enrollment existed, not even election into a learned society, and, in any case, rejection by one usually could be repaired by application to another. Few who desired an affiliation with natural philosophy shared the fate of the unrequited philosopher whose tombstone, now in a museum in Dijon, reads "Ci-gît un qui ne fut rien, pas même académicien."

THE STRING OF PHILOSOPHERS

1. Physics for Display

In the spring of 1798, Deluc wrote his great friend and colleague Georg Christoph Lichtenberg, professor of mathematics and physics at the University of Göttingen:

> I told you, mon cher Monsieur, that there is no one I prefer to talk physics with than you; because I've always seen that like me (if I dare say so) you love physics for itself; which is true of very few people. Some do physics to talk about it, or to make a reputation, or to get a job; but the feeling and love for the truth are rare.[28]

Together with one other consideration, Deluc's rough typology of motivations—doing physics for display, reputation, career, and love—can serve as a classification of natural philosophers. That other consideration, which allows a subclassification of the careerists, was the Enlightenment's programs of rationalizing arts, sciences, and social institutions, and of sweeping away customs and concepts that had no deeper foundation than habit and tradition. Rationalization coincided with the interests of bureaucratizing governments eager to curtail wasteful inefficiencies of the ancien régime. The tougher, leaner, more mathematical experimental philosophy of the later eighteenth century both served and profited from this coincidence. Particularly after the devastation that the Seven Years' War inflicted on Austria and Prussia and the humiliation it visited on France, the larger states recognized the need for a better accounting and more efficient use of their resources. At the same time, more powerful steam engines came into service in mills and mines, machines and weapons became more precise, agricultural practice waxed scientific, and communications improved. Engineers, agriculturalists, and technocrats of all kinds were in demand. The entrance of the expert, the first step toward realizing the rationalization demanded by Enlightenment, and the beginning of the effective mathematizing of natural philosophy, were coeval. The mixed mathematicians and natural philosophers engaged included (to choose a few stereotypes) French engineers, British agriculturalists, clerics in and out of Italy, and German bureaucrats.

The natural knowledge to which the Paris Academy and the Royal Society devoted themselves aroused great interest in the wider society when administered in homeopathic doses. It gave people something to talk about. A large literature, usually arranged in dialogues, gave instruction in natural-philosophical conversation. The earliest and most successful of these manuals, Fontenelle's *Entretiens sur la pluralité des mondes*, eavesdrops on a beautiful marquise and her handsome young tutor, who stroll through her gardens on summer evenings discussing cosmology. Fontenelle was an *homme de lettres*—a poet and playwright—before becoming a full-time natural philosopher. That he did in 1699, on the reformation of the Paris Academy of Sciences, which chose him as its permanent secretary. His duties were to publish the papers of the academicians, preface their annual *Mémoires* with a précis of their most important results, and celebrate each in turn, when his turn came, in an eloquent obituary notice. Although Fontenelle ventured a work or two on mathematics, his role as natural philosopher was organizational and rhetorical. He owed his appointment to the clarity and eloquence of his writings—which shows sufficiently his distance from most academics of our time.

Fontenelle's imitators included Francesco Algarotti and the abbé Noël Antoine Pluche, whose *Spectacle de la nature* (*Nature Displayed*) has a count and countess as instructors, a young nobleman as their tutee, and no mathematics or method. Pluche had learned that "mathematical Discourses, or a chain of Dissertations . . . frequently satiate and disgust"; therefore he chose interlocutors for their ability to "furnish conversation."[29] Like Pluche, Algarotti ruled out mathematics as giving a work "too scientific an Air." Also like Pluche, he purveyed natural philosophy to help "polish and adorn Society." And like Fontenelle, to whom he dedicated his *Newtonianismo per le dame*, he took his dialogue from conversations between a beautiful and witty noblewoman and a stand-in for himself. They talked and basked under the summer sun at a villa overlooking Lake Garda.[30]

In sunless England the parlor became as full of talk about natural philosophy as about the weather. When Sir Richard Steele saw the first orrery in the workshop of its maker in 1716, he immediately perceived its importance as a centerpiece of conversation. Even the dullest of his countrymen, he said, could learn the elements of astronomy from it in an hour and have natural-philosophical conversation for a lifetime. "This one consideration should encite any numerous Family of Distinction to have an Orrery as necessarily as they would have a clock."[31] Many English introductions to natural philosophy are arranged as conversations between a young lady and her tutor, as in the dialogue between the fictional fair Euphrosyne and her brother in Benjamin Martin's *The young gentleman and lady's philosophy* (1772). These characters had real-life counterparts. Polly Stevenson, the daughter of Benjamin Franklin's London landlady, preferred talk to gossip. She asked Franklin for instruction. She addressed him exactly as if she were Euphrosyne, although her letters preceded Martin's text by a decade. Franklin responded almost in the words of Euphrosyne's brother. Was it a case of life anticipating art, or art copying life? Joseph Addison recorded a visit to a realistic-fictional household where the ladies made jam while reading aloud from Fontenelle. "It was very entertaining for me to see them dividing their speculations between jellies and stars, and making a sudden transition from the sun to an apricot, or from the Copernican system to the figure of a cheese-cake."[32]

2. Physics for Place and Profit

Reputations made in the leading salons in Paris often determined the outcome of elections to the Paris Academy of Sciences and influenced the award of royal pensions. Fontenelle rose via the salon as he gained a reputation for his dialogues and showed that he could talk well himself. D'Alembert, mathematician, natural philosopher, and raconteur, became the darling of the salons and the most influential academician in France. No doubt Fontenelle and d'Alembert had an abiding interest in mathematics and natural philosophy; but they also were ambitious, cultivated their reputations, and parleyed themselves into governors of the Republic of Letters.[33] Many more modest citizens of the same republic likewise entered the patronage system with the help of reputations for knowing more about nature than other people did.

One of the most insistent of nature lovers on the make was the abbé Jacques Casanova. Well educated in a seminary in Venice, he soon improved on Fontenelle's method of seducing women with philosophy and a telescope and taught himself enough chemistry to enter aristocratic and even royal circles where natural philosophers played with alchemy. He traveled with a case full of philosophical instruments and projects ranging from mixed mathematics (calendar reform for Catherine of Russia, a lottery for Frederick of Prussia, surveying for lesser landholders) to applied natural philosophy (mining, hydraulics, waterways, printing, dyeing, silk making). He tried to develop a reputation as a mathematician by solving the ancient problem of the doubling of the cube, which is no more impossible than squaring the circle. He tried to couple natural philosophy to classical literature by detailing a hydraulic system by which Hercules could have cleaned the Augean stables. He wrote a journey to the center of Earth in which, in only five volumes, his adventurers explained the natural philosophy of the Enlightenment to intelligent kobolds.[34] Like Casanova, however, they talked more than they performed.

A good living could be made by preparing people to talk well about nature. The highest ranking of these philosophical elocutionists by virtue of the status of his clients was the abbé Jean-Antoine Nollet. These included duchesses by the carriageful, the royal children, and the royal artillery. Nollet was no mere entertainer. He added to his income by making instruments, writing books, and collecting his salary as a member of the Paris Academy. His paler counterpart in England, Jean Théophile Desaguliers, the son of Huguenot refugees, began as curator of experiments at the Royal Society, of which he was a Fellow, demonstrated the Newtonian philosophy to visiting dignitaries, spread it among his fellow masons, and lectured to London audiences who could afford the guinea per course. Several other members of the royal societies of Paris and London made a living lecturing in the metropolises and also, in England, on the road. Benjamin Martin, who aspired to but did not achieve membership in the Royal Society, was one of these entrepreneurs. His description of a tour in the country suggests adventures of almost Casanovan quality as well as the philosophical attainments of the English countryside. "There are many places I've been so barbarously ignorant, that they have taken me for a Magician; yea, some have threaten'd my life, for raising Storms and Hurricanes: Nor would I show my face in some Towns, but in company with the Clergy and Gentry, who were of the Course." One intrepid lecturer made it all the way to Philadelphia, where his show awakened Franklin's interest in electricity.[35]

Further down the scale of purveyors of natural philosophy were the buskers and jugglers who made "*agréable*" and "*curieuse*" synonyms for "*physique*." The *Almanach dauphin* for 1777 lists four "*physiciens*." One was and another was to become a member of the Paris Academy; a third operated a cabinet of curiosities, and the fourth, "known for his extreme sleight of hand," gave magical shows. After the Seven Years' War, which depressed the business of the natural philosophical lecturer, some demonstrators rose to the summit previously reached only by Nollet, and, in the unhappy case of Jean François Pilâtre de Rozier, even higher. Proprietor of a *musée* in Paris with seven hundred subscribers, Pilâtre de Rozier saw the promise of the new *montgolfières* whose launchings brought crowds to Paris in the early 1780s. He attempted to cross the English Channel in a basket hanging from a balloon suspended from a large bag of hydrogen.

Theory outran practice: the hydrogen caught fire and the latter-day Icarus crashed before leaving France. Back in Paris demonstrators of nature faced another sort of danger: one another. In the year of Pilâtre de Rozier's death, 1785, at least sixteen public courses of good-quality lectures on natural philosophy were offered in the City of Lights. In contrast, lectures did not resume strongly in London after the war but found increasing audiences in the industrializing midlands.[36] As the strong interest in the midlands and in Paris suggests, the audiences of the better lecturers wanted instruction useful in advancing themselves as well as diversion for their guinea or louis d'or.

French military engineering under Jean-Baptiste Vaquette de Gribeauval offered one avenue of advance. Alone among French artillerists, Gribeauval proved almost an equal to the Prussians in gunnery during the Seven Years' War. He was put in charge of French artillery during the peace. Under his direction, the curriculum in the central artillery schools switched from elementary mathematics to more advanced forms and their applications. A new multi-volume textbook, Etienne Bézout's *Cours de mathématiques à l'usage du corps royal de l'artillerie,* was adopted in 1770. It ran from arithmetic and algebra through Newtonian mixed mathematics. Instruction in chemistry and mineralogy, subjects newly made useful by Lavoisier, strengthened in the 1770s and 1780s.[37] The state engineering school at Mézières also exposed its students to mathematics that tested their strength, accustomed them to thinking, and reduced their effectiveness in the work place. There Charles Bossut did for the curriculum for geographical engineers what Bézout accomplished at the same time for the artillery schools.[38] Biot later singled out two graduates of Mézières as the instigators of the quantification of physics in France. These were Coulomb and Charles Borda, a naval officer whose contributions to natural philosophy included a delicate surveying instrument (the repeating circle) that could make measurements as accurately as the large English theodolites. Both Coulomb and Borda became members of the Paris Academy. The policy of nurturing engineers on mathematics continued at the École Polytechnique, set up in 1796 as a preparatory school for the higher engineering faculties. The amount of time devoted to mathematics and its applications in the school's curriculum increased from 25 percent in 1799 to 50 percent in 1812.[39]

Pierre Simon de Laplace was largely responsible for the mathematizing of the curriculum at the École Polytechnique. He thus reinforced, with the greater power that his elevated position under Napoleon gave him, the direction he had pursued in the 1780s at Gribeauval's artillery schools, where he served as an examiner. Whatever value this reinforcement had for engineering, it gave an irresistible push to the quantification of natural philosophy. Most of the notable French physico-mathematicians active between 1800 and 1830 taught or studied or did both at the École Polytechnique. A short list would include François Arago, Biot, Sadi Carnot, J. B. J. Fourier, L.J. Gay-Lussac, Etienne Malus, and S. D. Poisson. All except Sadi Carnot, who died before he could be elected, became members of the academy in its resurrection after the Terror as the First Class of the Institut de France. There they could meet with the most successful graduate of the artillery schools of the ancien régime, the emperor Napoleon, the patron of Laplace and Volta, who supposed himself a natural philosopher, and had a Casanovan research program to prove it. "Spending the night between a beautiful woman and a clear sky, and the day reconciling observations and calculations, seems to me heaven on earth."[40]

The work of the higher technical schools and their graduates stimulated a market for textbooks in applied mathematics and other products of high technology.[41] The most obvious of these other products were exact instruments, like Borda's circle, which P. F. A. Méchain and J. B. J. Delambre employed in their triangulation of France that defined the standard meter. Meanwhile, Gribeauval perfected his favorite instrument. In the late 1780s, he and his artisans could make the ingredients of a firing mechanism to such exactitude that they could disassemble one and put it together again using parts of others selected at random. Earlier methods of manufacture had required a lengthy adjustment of every part to its neighbors. Gribeauval's system of interchangeable parts required precision tools and machinists who could work to tolerances of a hundredth of an inch or so. Like the standard meter, Gribeauval's interchangeable parts carried rationality in the service of society to a point beyond the wildest imaginings of the *philosophes*.

The English genius did not run to mathematics and centralization. To be sure, it had an important Military Academy at Woolwich near London, where, after Hutton became professor in 1773, soldiers could learn more mathematics than they needed. Like his French counterparts Bézout and Bossut, Hutton wrote textbooks on all branches of mathematics, pure and applied. Also like them, he undertook experiments in ballistics and ranked among the leading exponents of natural knowledge; they belonged to the Paris Academy, he to the Royal Society. Hutton's service to his science extended to the popularizations and entertainments that recommended it to the wider society. From 1773 to 1818 he edited the *Ladies' diary*, an almanac filled with puzzles and problems, mainly in geometry and algebra but also in the mathematical branches of natural philosophy, on which many would-be mathematicians, some women, honed their skills. One of them was the pious schoolteacher Margaret Bryan.[42]

Among the users of Hutton's books for beginners or other guides to arithmetic were agriculturalists intent upon increasing the yields of their farms. The enclosure of farmlands, which rapidly accelerated after the Seven Years' War, amalgamated scattered strips into more manageable holdings and made possible experiments in all aspects of agricultural practice. Improving farmers improved by rationalized cut-and-try: by varying planting techniques and fertilizer recipes, quantifying input and output, and exchanging information. Fundamental to it all was the so-called Norfolk system of mixed husbandry and crop rotation. New implements for planting, hoeing, plowing, and harvesting; methods of soil improvement; techniques of selective breeding; innovations in processing byproducts like hides; and shops to keep everything in repair completed the apparatus for scientific agriculture. In 1771, Arthur Young, soon to be a Fellow of the Royal Society, arrived in Norfolk on his travels to observe best farming practice. He praised in particular the experimental methods of Sir John Turner, whose son-in-law belonged to the family of Martin Folkes, onetime president of the Royal Society, who were long-time improving landowners in Norfolk.[43]

Young continued his travels, spreading the word through books and also in his *Annals of agriculture*, which offered farmers a forum for relating their latest improvements. Among these contributors was King George III. Whereas the base-born emperor of France fancied himself an astronomer and mathematician, the royal-born king of England did not disdain to improve the muck that made the wealth of his nation. In 1793, the further to promote the increase and dissemination of agricultural science, the landowners set up a Board of Agriculture with Young as its secretary. To help alleviate the condition of the rural poor, the men around the Board of Agriculture proposed to acquaint cottagers with efficient ways to use fuel, prepare food, and manage their allotments of land.[44] They had the help of the cosmopolitan Count Rumford, an American Tory in the service of the King of Bavaria. Rumford had a reputation for making nourishing soups and smokeless fireplaces. He also had the reputation of a natural philosopher. He visited England in 1798 in order to read to the Royal Society, of which he was a Fellow, his memoir on the production of unlimited heat from friction. He took the opportunity to draw up plans for an institution to transmit, through lectures, the latest science of soups and chimneys, and also of manuring, tanning, soap-making, and the like. The Royal Institution opened in 1799. It was to make capital contributions to natural philosophy if not to scientific farming or the relief of the poor. Among its first professors were Thomas Young, re-inventor of the wave theory of light, and Humphry Davy, who isolated sodium and potassium via electrolysis and invented the miner's safety lamp.[45]

The same sort of rational cut-and-try that made scientific agriculture underpinned the inventions of the group living around Birmingham who called themselves the Lunar Society because they met when the full moon illuminated their travels. The best known of the productions of the group was James Watt's steam engine. Watt invented its principle in 1765, experimented on steam according to the standards he had learned by talking with the professor of chemistry at Glasgow, Joseph Black, and developed a prototype in line with suggestions from members of the Lunar Society. Like Watt, the society's organizer, Matthew Boulton, had made philosophical instruments (thermometers and electrical apparatus) and designed steam engines. Unlike Watt, Boulton had a factory, which, in 1775, had reached the industrial equivalent of a Norfolk model farm, and where, once perfected, Watt's engines could be made. Manufacture began around 1780.

The lunar group included several other philosophical artists: the physician Erasmus Darwin and a landed gentleman, Richard Lovell Edgeworth, both of whom dabbled in steam engines, scientific instruments, agricultural implements and materials, and balloons—that is, all the high-tech sectors of the day; Josiah Wedgwood, whose knowledge of instruments, chemistry, and experimental technique made a pottery famous throughout Europe; and James Keir, a partner of Watt's, who translated Macquer's *Dictionary of chemistry* into English. Keir enriched the work with a treatise on the gas types, in which for the first time (so he claimed) "this branch of experimental philosophy" was set forth systematically; in contrast to chemistry proper, which he rated "little more than a collection of facts . . . not yet capable of synthetic or analytic modes of explanation." In 1775, the greatest of all gas detectors, Joseph Priestley, then already famous for his work and books on electricity, optics, and pneumatics, moved into the neighborhood to teach at a dissenting academy and joined the group. All these gentlemen—Boulton, Edgeworth, Keir, Priestley, Watt, and Wedgwood—continued to experiment with heat, electricity, and gases and to send reports of their findings to the Royal Society, of which all were members.[46]

Until the suppression of the Society of Jesus, its schools dominated secondary education in Italy, France, southern Germany, and Austria. The larger of these institutions offered instruction at the same level as universities. The pedagogues of the order included many prominent mathematicians during the seventeenth century and several—Roger Boscovich, Girolamo Saccheri, and Leonardo Ximenes, among others—in the eighteenth.[47] The Jesuits gave an important place to applied mathematics in their curriculum for this-worldly reasons. Mathematics attracted sons of elite families, who might need to know something about calculation, architecture, surveying, and fortification, as well as fencing and dancing (which were also taught in Jesuit schools), for their future roles in society. Secondly, as a means of teaching controversial and even condemned subjects, especially Copernican astronomy, mathematics was magic. The Jesuits turned Aristotle's doctrine that mathematics offers an exact description of phenomena, but no demonstration of the truth, into an argument for presenting astronomical systems as harmless but useful hypotheses. Their pedagogical mission reinforced their Jesuitical epistemology: especially in the eighteenth century, when they began to lose market share to other teaching orders, they had to keep their curriculum up-to-date. Thus the Jesuits tended to multiply mathematicians, many of whom, including members of the society, found niches in which to contribute to natural philosophy outside the order's institutions.

Two examples will indicate the possibilities. Boscovich began his extraordinarily varied career as a professor of mathematics at the Jesuits' main seminary, the Gregorian College in Rome. Then, acting as a civil engineer, he drew up plans for repairing St Peter's, canalizing the Tiber, and building harbors. On order from the Vatican, he made a trigonometric survey of the Papal States. He taught mathematics at the Jesuit college in Pavia and founded the Jesuit observatory in Milan. He became, briefly, director of optics for the French Navy. He was a member of the Paris Academy and the Royal Society. Ximenes learned enough mathematics at the Gregorian College to be sent by his general as tutor in the subject to the family of a Florentine duke. He also wrote some mathematics, published an almanac, and made a small name for himself. That was enough to bring him an ill-paying job as professor of mathematics at the University of Florence and many commissions from local governments to drain swamps, manage wetlands, canalize rivers, and repair buildings. For a year or so he turned the cathedral of Florence into a laboratory of experimental physics and a solar observatory in order to answer the then vexed question whether the inclination of Earth's axis changes slowly in time.[48]

The history of the Jesuits agrees with that of natural philosophy in suffering a sea change around 1770. Their expulsion from several European states in the 1760s and their suppression by the pope in 1773 freed governments from the meddling of the society at the cost of a rich source of teachers, informers, and disciplinarians. The well-organized Jesuits fell before greater powers newly bent on obtaining efficient control of their resources. Their order remained alive, however, in the most absolute states, Prussia and Russia, and was resurrected as part of the restoration of the ancien régime in 1814–1815. The official nonexistence of the Jesuits thus agreed perfectly in time with the development of exact natural philosophy and its deployment by the European states. The sort of technical work done by Boscovich and Ximenes was continued and expanded, but not by Jesuits.

On the other end of the scale of organization, ubiquitous independent abbés became important actors in natural philosophy. Some of them priests, others only deacons, they owed their access to the Republic of Letters to the education they received, and often also to the contacts they made, in the church. Four have entered this account so far: Nollet, the experimental philosopher, instrument maker, and showman of the old school, who died in the epochal year 1770; his successor, Pilâtre de Rozier, killed in a natural-philosophical stunt in 1785; Casanova, who retired from the world in 1785 to write his memoirs; and Bossut, professor of exact science to engineers, a former student of the Jesuits who lived just long enough to see the society reborn in 1814.

Among the many other abbés who helped advance natural philosophy was François Rozier. He began with scientific agriculture and found his vocation in publishing. His *Observations* (later *Journal*) *de physique*, which started in 1771, at first admitted most of physics in its old sense, including chemistry, meteorology, mineralogy, anatomy, physiology, zoology, agriculture, and the mechanical arts, usually in the form of experimental philosophy, not natural history. Rozier excluded mathematics, including its usual applications. It is therefore particularly instructive to follow the gradual restriction in subject matter, and the increasing frequency of measurements, in the *Journal*. During the 1790s, Deluc was one of its primary contributors. His mixture of accuracy in observation and precision in measurement with qualitative speculations worthy of Baron Munchausen represent much of exact natural philosophy almost exactly. After a decade editing the *Journal de physique*, which put out twice as many pages a year as the memoirs of the Paris Academy of Sciences, Rozier handed it over to his nephew, another abbé, who continued its policy of fast publication of experimental results and helped to bring it closer to the emerging consensus about the nature of natural philosophy.[49]

The welter of German states and principalities, which abounded in universities and bureaucracies, and, after the Seven Years' War, in problems of reconstruction, provided a theater for a special combination of economic rationalization and natural knowledge. This combination, known as Cameralwissenschaft, after the camera, or meeting room, in which matters of state were discussed, had a place in German universities and special higher schools from the middle of the eighteenth century and, in isolated cases, even earlier. This science collected information about a state's agriculture, commerce, industry, population, territory, history, laws, and customs to serve as a basis for economic policy.

The seat of Cameralwissenschaft was the University of Göttingen. Its leading representative there from 1769 to 1809, A. L. Schlözer, had responsibility for the newly instituted subject of universal history. "Universal" signified not only everywhere but also everything: in contrast to the usual preoccupations of historians, which centered on religious, legal, and political history, Schlözer emphasized economic, demographic, and cultural aspects, that is, the material of statistics. For him statistics constituted history; in a famous aphorism set down in 1793, he declared, "history is statistics in motion, statistics history at rest." The Cameralwissenschaftler described the condition of the world at a chosen instant, the historian explained the change of the world condition in time. Statistics not only made the core of the work of the historian, the prince, and the bureaucrat, but also guaranteed the liberty of people governed by rulers so informed. Schlözer was a man of the Enlightenment. He believed that where free inquiry reigned, despotism could not enter.[50]

Universal history had close ties to natural philosophy at Göttingen. Schlözer's colleague in the history of everything, J. C. Gatterer, made contributions to meteorology. Schlözer himself took natural history as his model when he elaborated his idea of universal history. Whenever possible, the statistics that made history were to be quantified. Consequently, the coverage of statistics tended to be restricted to quantifiable aspects of life and Cameralwissenschaft to fall among the quantitative sciences of the late eighteenth century. Like exact natural philosophy, it earned the disgust of critics inspired by the Romantic movement, whose sensitivities were outraged by the reduction of individuals of different moral worth to normalizing calculations.[51]

The most successful and demanding of the subjects within Cameralwissenschaft was forestry. The Seven Years' War damaged many a forest from which princelings and large landholders had derived a principal part of their income. Earlier methods of forest management, which scarcely went beyond an estimate of the number of standing trees, gave way to plans for sustained yields drawn up by men with knowledge of sylviculture, economics, surveying, and practical mathematics.[52] As in agriculture, so in forestry, the increase in sustainable yields of lands managed scientifically—that is, in accordance with rationalized experience and calculation—directed interest and support to the study of natural philosophy and mixed mathematics.

Another quantity that brought the natural philosopher close to the people was the problem of counting them. The Swedish Academy of Sciences organized the first census in Europe in 1749. During the last third of the eighteenth century, enlightened governments everywhere pursued the number of its people; for wealth, according to mercantalist economics, was proportional to the size of the productive population. The relation between census data and natural knowledge may best be examined in France, where the population refused to be counted lest revelation of its prosperity raise its taxes. Numbering the king's recalcitrant subjects engaged the ablest natural philosophers and mixed mathematicians. Their rhetoric is gratifying to the historian of exact science. "Experiment, research, calculation are the probe of the sciences! What problems could not be so treated in administration! What sublime questions could not be submitted to the law of calculation!" Men of such science had only to read the "thermometer of public prosperity," "the barometer [of the nation]," that is, the number of citizens, to know the health of the state. The Paris Academy opened its pages to vital statistics; but the peasantry refused to allow itself to be counted even in so prestigious a place. The calculators had finally to invent sampling methods. Laplace worked out the effect of the size of the sample on the probable error of the estimated population. Full data from cooperating parishes combined with partial information from the rest then gave a number that might or might not have been the population of France.[53]

3. Physics for Love

By doing physics for love Deluc did not mean without hope of reputation or emolument. The two examples he offers, himself and Lichtenberg, stood high in the Republic of Letters and supported themselves by their knowledge of natural philosophy. After leaving his native Geneva in 1773 when his business failed, Deluc came into the grand sinecure of reader in natural philosophy to Queen Charlotte of England, who housed him in Windsor Castle with an income that met his needs. Lichtenberg was a professor and privy councilor. Both were members of the Royal Society and other prestigious academies. Their love for physics was the philosopher's devotion to seeking and saying the truth. In 1798, when Deluc enrolled Lichtenberg along with himself among the few lovers left in the world of physics, he had in mind as counterexamples the wanton mindless natural philosophers of Germany who had taken up with French chemistry. This

Jacobin science, as he and Lichtenberg called it, was dangerous as well as wrong. The true natural philosopher, the lover of wisdom, had to oppose it for the good of chemistry and the salvation of Europe.[54] This may show sufficiently that self-professed truth lovers are as liable to self-deception as other mortals. Still, there may be something useful in the concept.

Isaac Newton once drew up a list of nine "candid promoters of truth"; in his limited experience, most other so-called philosophers were "contentious and noisy for the sake of fame or preferment."[55] It would make a good game for historians of science to guess the names of the nine: Tycho, Kepler, Hevelius, Guericke, Galileo, Torricelli, Huygens, Cassini, and Bianchini. Bianchini? We may well regard the obscure monsignor Francesco Bianchini as Newton's model truth seeker, for he was the only one of the nine living when Newton compiled his list and the only one with whom Newton had ever talked. Now, this Bianchini represented everything that Newton most detested in religion and politics. He was a papist, indeed, the right hand of Pope Clement XI; Newton considered the Roman Catholic Church a font of evil and the pope the Whore of Babylon. Again, Bianchini was a friend and sometime guardian of the Stuart pretender James III, whom Clement supported in his claims to the throne of England; Newton, a fierce Whig, favored the Hanoverian succession that ended the Stuart reign.[56] We are left with the hypothesis that Newton truly admired Bianchini.

Bianchini was at home everywhere in the Republic of Letters. Accomplished as an astronomer, chronologist, geodecist, archeologist, and historian, he built the beautiful solar observatory in Santa Maria degli Angeli in Rome, straightened out calendars and early Christian history, made the first trigonometric survey of the Papal States, supervised digs and deciphered inscriptions, and composed a "universal history" of human kind during its first thirty-two centuries of existence. His reports as consultant to the Roman censorship could well serve as models for advisors of university presses. Perhaps Bianchini's report to Newton that he and some colleagues in Rome had managed to reproduce some contentious experiments from the *Optice* may help explain Newton's admiration for him. But then everyone admired Bianchini: Leibniz, Mabillon, and the academicians of Paris, who elected him one of their eight foreign associates, and the English savants whom he met during his visit to England in 1713. Apparently he won them all with a rare blend of affability, responsibility, and modesty.[57]

Bianchini may stand for many humbler citizens of the Republic of Letters who took an interest in natural knowledge not because there was a career or money in it, but because they believed that any educated person should cultivate it if he could. It would be futile to try to list those whom contemporaries valued as lovers of science for its own sake. We may be confident, however, that Joseph Priestley would figure among them. Petty motives pained him. "Did it depend upon me, it should never be known to posterity, that there had ever been any such thing as envy, jealousy, or caviling among admirers of my favorite study."[58] These words appear in Priestley's *History and present state of electricity*, drawn up with every effort at fairness to provide the basis for further advance in the subject. We know that Bianchini too was a thoughtful historian. Lichtenberg, whom Deluc ranked among truth lovers, followed both Priestly and the Göttingen Geist in cultivating the history of his science and general history. Their example prompts the eccentric thought that a Diogenes looking for lovers of science for its own sake would have a good chance of finding some among natural philosophers attracted to the study of history.

NOTES

1. R. R. Palmer uses phrases now stereotypical of accounts of scientific revolutions: "a crisis of community . . . struggles of opposed ideas and interests . . . incompatible conceptions of what the community ought to be" (*The Age of the Democratic Revolution* [Princeton, NJ: Princeton University Press, 1959–64], 1:21–22).

2. Editor's introduction to *Elements of natural philosophy* by John Locke (Berwick upon Tweed: R. Taylor, 1764), iii.

3. "Address to my pupils" in *Lectures on natural philosophy . . . with an appendix containing a great number and variety of astronomical and geographical problems; also some useful tables* by Margaret Bryan (London: T. Davison, 1806); Samuel Johnson, *A dictionary of the English language*, ed. H.J. Todd, 4 vols. (London: Longman et al., 1818), s.v. "Physicotheology." Cf. Simon Schaffer, "Natural Philosophy," in *The Ferment of Knowledge: Studies in the Historiography of Eighteenth-Century Science*, ed. G.S. Rousseau and Roy Porter (Cambridge: Cambridge University Press, 1980), 68–91.

4. *L'Encyclopédie, ou Dictionnaire raisonné des sciences, des arts et des métiers* (Paris: Briasson et al., 1751–72), 12:539; Charles Hutton, *Philosophical and Mathematical Dictionary* (London: F.C. Rivington et al., 1815), 2:185; Susan Faye Cannon, "The Invention of Physics," in *Science in Culture: The Early Victorian Period*, ed. Susan Cannon (New York: Science History Publications, 1978), 116.

5. J. L. Heilbron, *Electricity in the 17th and 18th Centuries: A Study of Early Modern Physics* (New York: Dover, 1999), 10–11, 140–41.

6. J. L. Heilbron, *Physics at the Royal Society during Newton's Presidency* (Los Angeles: William Andrews Clark Memorial Library/UCLA, 1983), 7–15.

7. Heilbron, *Electricity in the 17th and 18th Centuries*, 14–15.

8. Cf. Johann Carl Fischer, *Geschichte der Physik* (Göttingen: J.F. Röwer, 1801–08),

4:4–5, referring to the texts of Musschenbroek, Nollet, and others, and emphasizing their exclusion of natural history and chemistry.

9. Daniel Jenisch, *Geist und Charakter des achtzehnten Jahrhunderts* (Berlin: Akademie der Wissenschaften, 1800–1), 1:480.

10. Heinrich Döring, *Die gelehrten Theologen Deutschlands im 18. und 19. Jahrhundert* (Neustadt a.d. Orla: J. Wagner, 1831–35), s.v. "Jenisch."

11. J. L. Heilbron, "Experimental Natural Philosophy," in *The Ferment of Knowledge*, 361–67.

12. Arthur Donovan, "Lavoisier and the Origins of Modern Chemistry," *Osiris* 4 (1988): 215, 221–22, 226–28; Carlton E. Perrin, "Research Traditions, Lavoisier, and the Chemical Revolution," *Osiris* 4 (1988): 53–81; Perrin, "Chemistry as Peer of Physics," *Isis* 81 (1990): 262, 265–67; Anders Lundgren, "The Changing Role of Numbers in Eighteenth Century Chemistry," in *The Quantifying Spirit in the 18th Century*, ed. Tore Frängsmyr, J.L. Heilbron, and Robin E. Rider (Berkeley: University of California Press, 1990), 245–46, 257–63; Armand Seguin, "Observations générales sur le calorique," *Annales de chimie* 3 (1789): 150n, 227n.

13. *L'Encyclopédie*, 12:536–7; Hutton, *Philosophical and Mathematical Dictionary*, 2:185. The term "fisicomatematica" was in use in Italy in the seventeenth century.

14. J. L. Heilbron, *Weighing Imponderables and Other Quantitative Science around 1800* (Berkeley: University of California Press, 1993), 29; F.K. Achard, "Mémoire sur la mesure de force de l'électricité," *Journal de physique* 21 (1782): 196.

15. Jean Baptiste Biot, *Traité de physique expérimentale et mathématique* (Paris: Deterville, 1816), 1:xvi. Biot gave the example of Newton's derivation of Kepler's laws.

16. Heilbron, *Weighing Imponderables*, 6–16.

17. Ibid., 17–23.

18. Johnson, *A dictionary of the English language*, s.v. "practitioner."

19. Heilbron, *Physics at the Royal Society*, 35.

20. Sydney Ross, "Scientist: The Story of a Word," *Annals of Science* 18 (1962): 65–85.

21. Heilbron, *Electricity in the 17th and 18th Centuries*, 122, 126; Henry E. Lowood, "Patriotism, Profit, and the Promotion of Science in the German Enlightenment: The Economic and Scientific Societies" (PhD thesis, University of California, Berkeley, 1987).

22. *L'Encyclodédie*, 12:512–13. The other parts are theology and psychology. *L'Encyclopédie* credits Wolff with this tripartite division.

23. Heilbron, *Electricity in the 17th and 18th Centuries*, 77–8, 128–29; Boscovich to Lorna, July 20, 1784, in *Ruggiero Boscovich. Lettere ad Anton Maria Lorgna, 1765–1785*, ed. Ugo Baldini and Pietro Nastasi (Rome: Accademia dei XL, 1988), 112.

24. H.A.M. Snelders, *Dictionary of Scientific Biography*, vol. 13, 468–69.

25. Heilbron, *Electricity in the 17th and 18th Centuries*, 126–31, 134, 137, 163; the last quote from an official of the Petersburg Academy concerning recruitment of personnel in 1724.

26. Ibid., xv–xxiv, 412–13, 422–24.

27. Unpublished data, Office for History of Science and Technology, University of California, Berkeley.

28. Deluc to Lichtenberg, March 27, 1798, in Georg Christoph Lichtenberg, *Briefwechsel*, ed. Ulrich Joost and Albrecht Schöne (Munich: Beck, 1992), 4:832–33.

29. Cf. Noël Antoine Pluche, *Spectacle de la nature: or Nature Display'd*, 3rd ed.

(London: J. Pemberton et al., 1736), 1:vii. This translation gives only the first few volumes of Pluche, which was completed in 1746 in seven volumes; the main subjects are natural history and industrial arts. Cf. Alice Walters, "Conversation Pieces: Science and Polite Society in Eighteenth-Century England," *History of Science* 35 (1997): 121–54.

30. Francesco Algarotti, *Sir Isaac Newton's Philosophy Explain'd for the Use of the Ladies* (London: E. Cave, 1739), 1: i–ii, vi–vii, 2–4; and Algarotti, *Il newtonianismo per le dame*, 2nd ed. (Naples: G. Pasquale, 1739), ff. b.3r, b.4r, 2–6. The discussion begins with polite literature and turns to natural science when the noblewoman mentions Algarotti's ode in praise of Laura Bassi.

31. Quoted in Henry C. King and John R. Millburn, *Geared to the Stars: The Evolution of Planetariums, Orreries, and Astronomical Clocks* (Toronto: University of Toronto Press, 1978), 154.

32. Article from *The Guardian* dated September 8, 1713, quoted in Gerald Dennis Meyer, *The Scientific Lady in England, 1650–1760* (Berkeley: University of California Press, 1995), 24.

33. Thomas L. Hankins, *Jean d'Alembert: Science and Enlightenment* (Oxford: Oxford University Press, 1970), 14–27; Geoffrey V. Sutton, *Science for a Polite Society: Gender, Culture, and the Demonstration of Enlightenment* (Boulder, CO: Westview Press, 1995), 138–40, 184.

34. Otto Krätz and Helga Merlin, *Cassanova: Liebhaber der Wissenschaften* (Munich: Callwey, 1995), 32–34, 67, 73, 91, 96–99, 103–5, 110–17, 121, 129, 131–43, 164.

35. Benjamin Martin, *A Supplement: Containing Remarks on a Rhapsody of Adventures of a Modern Knight-Errant in Philosophy* (Bath: privately printed, 1746), 28–29n.

36. Heilbron, *Electricity in the 17th and 18th Centuries*, 16–17, 162–5.

37. Ken Alder, *Engineering the Revolution: Arms and Enlightenment in Eighteenth-Century France* (Princeton, NJ: Princeton University Press, 1997), 37–39, 70–74; Roger Hahn, "L'enseignement scientifique aux écoles militaires et d'artillerie," in *Enseignement et diffusion des sciences en France au XVIIIe siècle*, ed. René Taton (Paris: Hermann, 1964), 533.

38. René Taton, "L'Ecole du génie de Mézières," in *Enseignement et diffusion des sciences*, 576, 584–85, 593–94.

39. Biot, *Traité de physique expérimentale et mathématique*, 1:iv–v; Alder, *Engineering the Revolution*, 309.

40. Heilbron, *Weighing Imponderables*, 36.

41. Hahn, "L'enseignement scientifique aux écoles militaires et d'artillerie," 527, 529.

42. Margaret Bryan, *Lectures on natural philosophy*, second dedication ("Ever must your effective judgment be the touchstone of my literary fame").

43. Heilbron, *Physics at the Royal Society*, 36–40, 56–60; Arthur Young, *General view of the County of Norfolk* (London: Board of Agriculture, 1804), 75, 86, 168, 435–39; Naomi Riches, *The Agricultural Revolution in Norfolk* (Chapel Hill: University of North Carolina Press, 1937), 3, 14–15, 27–32, 77, 93–95.

44. Morris Berman, *Social Change and Scientific Organization: The Royal Institution, 1799–1804* (Ithaca, NY: Cornell University Press, 1978), 3, 11, 32–33, 39–40, 45.

45. Ibid., 23 (quote), 27, 32, 52, 58–61, 66, 68.

46. Robert E. Schofield, *The Lunar Society of Birmingham* (Oxford: Oxford University Press, 1963), 27–28, 45–48, 60–68, 72, 79–81, 189–200; James Keir, *A Treatise on the*

Various Kinds of Permanently Elastic Fluids, or Gases (London: T. Cadell, 1777), appended to Pierre Joseph Macquer, *A Dictionary of Chemistry*, 2nd ed. (London: T. Cadell, 1777), 1:vi–vii.

47. Heilbron, *Electricity in the 17th and 18th Centuries*, 405–6.

48. Danilo Barsanti and Leonardo Rombai, *Leonardo Ximenes: Uno scienziato nella Toscana lorense del settecento* (Florence: Medicea, 1987), 27–40, 107–11, 201–25.

49. C. C. Gillispie, *Science and Polity in France at the End of the Old Regime* (Princeton, NJ: Princeton University Press, 1980), 188–90.

50. Luigi Marino, *I maestri della Germania* (Turin: Einaudi, 1975), 346–58; Josephine Hecht, "L'idée de dénombrement jusqu'à la Révolution," *Pour une histoire de statistique* (Paris: Imprimerie nationale, 1976), 1:21–81 (on 43–44); Peter H. Reill, "Science and the Science of History in the Late Aufklärung," in *Aufklärung und Geschichte*, ed. Hans Erich Bödeker et al. (Göttingen: Vandenhoek and Ruprecht, 1986), 430–51 (quote, 443–45).

51. Reill, "Science and the Science of History," 439–41; Marino, *I maestri della Germania*, 348, 356–57; P. F. Lazarsfeld, "Notes on the History of Quantification in Sociology: Trends, Sources, and Problems," *Isis* 52 (1961): 292–94.

52. Henry E. Lowood, "The Calculating Forester: Quantification, Cameral Science, and the Emergence of Scientific Forestry Management in Germany," in *The Quantifying spirit in the 18th Century*, 315–23.

53. Karin Johannison, "Society in Numbers: The Debate over Quantification in 18th-Century Political Economy," in *The Quantifying spirit in the 18th Century*, 343–61 (on 358); Hecht, "L'idée de dénombrement jusqu'à la Révolution," 52, 57.

54. Deluc to Lichtenberg, November 18, 1793, in Lichtenberg, *Briefwechsel*, 4:187–88.

55. Newton to J. B. Mancke, 1724, in Isaac Newton, *The correspondence* (Cambridge: Cambridge University Press, 1977), 7:254–55.

56. Salvatore Rotta, *Dizionario biografico italiano*, s.v. "Bianchini."

57. A biography of this remarkable man is underway.

58. Joseph Priestley, *The history and present state of electricity*, 3rd ed. (London: C. Bathurst et al., 1775), v1:xi.

CHAPTER 8

Science and Medicine

Ronald L. Numbers

For well over a century physicians have celebrated ceremonial occasions by recounting how medicine escaped superstition and empiricism to become scientific. The story almost invariably begins with Hippocrates and recounts the achievements of such scientific worthies as Andreas Vesalius, William Harvey, Rudolf Virchow, Claude Bernard, Joseph Lister, Robert Koch, and the nonphysician Louis Pasteur. "For many centuries the practice of medicine was far closer to the arts of the primitive magician than to the logic of the scientist," declared one clinician in a typical narrative. "Fortunately, genius broke through the barriers of superstition and tradition sporadically; Hippocrates denied the divine etiology of disease, Vesalius dared look beneath the skin, Harvey turned attention from morphology to dynamics, and Claude Bernard expounded the experimental method . . . Then, gradually, science began to invade the wards."[1] Although some speakers grumbled about the extent to which "science" had eclipsed the "art" of medicine, most welcomed scientifically informed practice.

Until recently historians of both science and medicine typically portrayed medicine prior to the advent of "scientific medicine" in the late nineteenth century as more of a craft than a science. Although the term frequently went unexplained, its use left the strong impression that medicine remained *un*scientific until Pasteur and Koch rescued it from its lowly state. And even then the status of medicine—and its historians—remained in dispute. In the mid-1930s George Sarton and Henry Sigerist, the godfathers of the history of science and the history of medicine, respectively, engaged in a public spat over the standing of their respective fields. Sarton, envious of the support that The Johns Hopkins University was giving to Sigerist's Institute for the History of Medicine and smarting from Harvard University's refusal to create a parallel center for him, went out of his way to denigrate medicine, questioning its status as a "science" and insisting that it would "always remain an 'art'": "The historian of medicine who imagines that he is *ipso facto* a historian of science," he added gratuitously, "is laboring under a gross delusion." A peeved Sigerist sniffed that medicine was neither an "applied science" nor "a branch of science"—and thus, by implication, was none of Sarton's business.[2]

Since that unfortunate exchange, historians of science have tended to ignore rather than disparage the history of medicine. For their part, more and more historians of medicine since the late 1970s have abjured trying to determine whether medicine was (or is) essentially scientific in favor of attempting to understand what contemporaries meant when they appealed to science and what they hoped to gain by doing so.[3] This chapter, which looks at the changing and often contested relationship of medicine to "natural philosophy" and "science," follows in this ever-widening historiographical path. Beginning with the ancient Greeks, who first concerned themselves with the connection between medicine and philosophy, it moves quickly to the late Middle Ages, when the status of medicine as *scientia naturalis* emerged as an acute issue for professionalizing physicians. It continues through the modern period, focusing increasingly on the American scene and on the emergence—and meanings and uses—of "scientific medicine." By closely examining the historical nexus of science and medicine, I hope to illuminate how healers wrestling with disease and death conceptualized and categorized their activities.

FROM THE BIRTH OF PHILOSOPHY TO THE BIRTH OF THE CLINIC

Since the beginning of human history the formal status of medicine has been a matter of concern to only a small handful of persons, mostly elite medical practitioners. The lay public rarely entered discussions, and even for the overwhelming majority of healers—doctors and priests, drug sellers and midwives—the question remained far removed from their practical interests. In ancient Greece philosophers and physicians, from Plato and Aristotle to the co-called Hippocratic writers, considered medicine to be an art (*techne*), the possession of which, along with such activities as cooking, not only distinguished civilized humans from savages but separated doctors from unskilled laypersons. A few ancient writers speculated about the relationship between medicine and philosophy. "Most of those who study nature," wrote Aristotle, "end by dealing with medicine, while those of the doctors who practise their art in a more philosophical manner take their medical principles from nature."[4] Because the art of medicine focused on natural things, the first-century Roman Pliny, perhaps the first to write about medicine in Latin, regarded it as a part of natural history.[5] Galen, a century later, likened the art of medicine to that of archery, which relied more on practice than on reasoning.[6] The eleventh-century Persian physician Ibn Sīnā (known in Europe as Avicenna) began his immense and immensely influential *Canon of Medicine* by defining medicine as "the science by which we learn the various states of the human body, when in health and when not in health, whereby health is conserved and whereby it is restored, after being lost."[7]

Until the twelfth and thirteenth centuries, when the new universities started bestowing MD degrees on their medical graduates, there was little reason to discuss the relationship between medicine and *scientia*. The distinction between the learned *physicus*, who had studied the liberal arts and natural philosophy in the university, and the skilled but unlettered *medicus* (healer) became commonplace. As the highly regarded twelfth-century *medicus* William of Malmesbury noted, his reputation derived from successful practice, not from "*scientia*."[8] Eager for academic status and intellectual respectability in the new world of the university, physicians such as Taddeo Alderotti, the influential thirteenth-century doctor at the *studium* of Bologna, took the lead in promoting medicine as a subdivision of natural philosophy, distinct from, though informed by, *scientia naturalis*. Influenced by Avicenna, whose *Canon* served as the centerpiece of the medical curriculum, Taddeo defined medicine as "the science by which the dispositions of the human body are known . . . so that health may be maintained or, if lost, recovered." He highlighted the connection between medicine and natural philosophy in part to distinguish the work of physicians from "the usual practice that old women carry on."[9] The fourteenth-century French physician and surgeon Guy de Chauliac made much the same point. "If the doctors have not learned geometry, astronomy, dialectics, nor any other good discipline," he wrote, "soon the leather workers, carpenters, and furriers will quit their own occupations and become doctors."[10]

Not all medieval scholars applauded the efforts to hitch medicine to *scientia*. The twelfth-century Paris theologian Hugh of St. Victor pointedly assigned medicine—along with the making of fabrics and armaments, navigation, commerce, agriculture, hunting, and theatrics—to the mechanical arts (*mechanica*), which required more manual skill than intelligence, rather than to *theorica*, which comprised *theologia*, *mathematica*, and *physica*, the study of nature. Partly to avoid such stigmatizing associations, physicians often made a point of distancing themselves from surgeons, who bloodied their hands cutting sick bodies.[11] Through the late Middle Ages and into the Renaissance, the word science continued to be used colloquially to describe virtually any body of knowledge or skill. In the fifteenth century, for example, England's Privy Council, in issuing regulations for barbers who practiced and taught surgery, referred to "the science of barbery."[12]

Throughout the Renaissance, scholars continued to dispute the intellectual status of medicine and related disciplines such as anatomy. A common solution allowed the precepts, but not the practice, of medicine to count as natural philosophy. As Franciscus Valleriola, the celebrated sixteenth-century Provençal physician explained, the "work of healing . . . being uncertain and having an uncertain outcome," could never rise above conjecture. Science, in contrast, required incontrovertible demonstration. Some university-trained physicians, drawing on their training in Aristotelian logic, argued that their diagnoses rose to this level. Despite appearances, insists the historian Ian Maclean, the continuing debate over status was "not a vain squabble over words, but ha[d] substantial effects in the real world, which [could] be associated with the rising value attributed to therapeutics, to clinical precepting, and to the design of hospitals at this time."[13]

By the early-modern period, university-trained practitioners of "physic" had successfully linked themselves to natural philosophy or science, as they sometimes called it.[14] Harvey's discovery of the circulation of the blood in 1628 reinforced the connection, as did various chemical discoveries. Medical schools offered the only formal training available in such disciplines as anatomy, physiology, and botany, and medical professors held the only paying jobs in those fields.[15] In 1695 the professor of medicine at the University of Jena, Friedrich Hoffmann, declared that "as far as medicine uses the principles of physics [i.e., natural philosophy] it can properly be called a science."[16] By the early eighteenth century numerous observers were praising medicine as a science, increasingly understood as knowledge of the natural world. The author of the essay "Medicine" in the second edition of Chambers's *Cyclopedia* captured the growing enthusiasm for medicine as science when he credited "the experiments and discoveries of chymists and anatomists," especially those of "the immortal Harvey," with laying "a new and certain basis of the science."[17] Although most medical practitioners did not see themselves primarily as men of science, many viewed their art as resting on scientific truths. Skeptics, however, dismissed all the talk about natural philosophy and mathematics as mere intellectual foppery, aimed at building up a bigger practice. In the early eighteenth century the Dutch-English doctor Bernard Mandeville dismissed some so-called Newtonian physicians as mere "Braggadocio's, who . . . only make use of the Name of Mathematicks to impose upon the World for Lucre."[18]

Throughout the eighteenth century the British colonies of North America shared little in the growing excitement over medical science, although Cotton Mather's stunning success in inoculating his Boston neighbors against smallpox gave some grounds for optimism. No more than about one in ten colonial "doctors" had attended medical school, and apprentice-trained practitioners typically averaged little more than a year learning their art. In 1786 the prominent Philadelphia physician Benjamin Rush explained to a London colleague why American practitioners seemed so little attracted to science. "Philosophy does not here, as in England, walk abroad in silver slippers," he wrote; "the physicians (who are the most general repositories of science) are chained down by the drudgery of their professions; so as to be precluded from exploring our woods and mountains."[19]

Still, the medical community often played a leading role in the cultivation of natural history and natural philosophy. Medical men composed nearly one-fifth of the early membership of Philadelphia's American Philosophical Society, which devoted one of its six sections to "medicine and anatomy."[20] In 1765 the trustees of the College of Philadelphia created the first medical professorship in British North America and appointed the Edinburgh-trained John Morgan to fill it. The whole purpose of the venture was to provide training in the medical sciences, not to offer clinical training, which was still acquired by apprenticeship. In sharing his vision of medical education, Morgan emphasized the importance of "medical science," comprising such disciplines as anatomy, *materia medica*, botany, and chemistry. He urged that "medical researches and careful experiments" be undertaken in the field of natural history, "as natural history is one of the most essential studies to prepare a person for prosecuting medicine with success, and one of the most distinguished ornaments of a physician and man of letters."[21]

By the turn of the nineteenth century it was becoming commonplace to associate medicine with science, a term rapidly supplanting natural philosophy to designate the systematic study of nature. In surveying the progress of medicine in his *Brief Retrospect of the Eighteenth Century* (1803), the American clergyman Samuel Miller praised physicians for revolutionizing their "immense field of science," while candidly admitting that his fellow Americans had contributed little to the "science of medicine" during the first half of the century. By applying Lord Bacon's "plan of pursuing knowledge by observation, experiment, analysis and induction," medical men had introduced "a more precise, rigid and logical mode of philosophising" in place of "the wild and visionary hypotheses which disgraced the science of the preceding centuries."[22]

MEDICAL SCIENCE COMES OF AGE

Despite the growing conflation of medicine and science in the eighteenth century, some historians have insisted that the "birth" of scientific medicine—in Paris—did not occur until the last years of the century. Michel Foucault argued that medicine did not make "its appearance as a clinical science" until the birth of the clinic in the 1790s. Laurence Brockliss and Colin Jones have asserted that "medical science came of age" at the turn of the nineteenth century when "the newly established *écoles de santé* turned the discipline [of medicine] into a fully fledged empirical science."[23] In the years after the Revolution, French physicians increasingly hobnobbed with men of science, and cutting-edge medicine shifted gradually from the domestic bedside to the hospital and the laboratory. Inspired by the investigations of the brilliant young physician Xavier Bichat, the study of pathology flourished in Paris laboratories, and the techniques developed there quickly became "the touchstone of a new medical science," not only in France but throughout Europe.[24] On the heels of Bichat's pathological initiative Pierre Louis developed a quantitative method to test the efficacy of such practices as bloodletting. With the triumph of the numerical method, he declared in 1835, "We shall hear no more of medical tact, of a kind of divining power of physicians. No treatise whatsoever will continue to be the sole development of an idea, or a romance; but an analysis of a more or less extensive series of exact, detailed facts; to the end that answers may be furnished to all possible questions: and then, and not till then, can therapeutics become a science." Speaking two years later as president of the Société Médicale d'Observation, Louis described medicine as being entirely "a science of observation."[25]

Not all of Louis's colleagues approved of his efforts to construct a science of medicine based on quantification. Among the naysayers was T. C. E. Auber, who argued that medicine had "a much greater analogy with the moral, metaphysical, religious, and political sciences" than it did with "the physical, chemical, and mathematical sciences." Largely because of the "prodigious variation that created difficulties in the physiological sciences," another physician-critic, François Double, professed to see "no similarity" between medicine and the physical sciences. The positivist philosopher Auguste Comte scorned the notion of basing clinical decisions on empirically derived numerical calculations.[26]

Despite such reservations at home, the French influence spread rapidly to the major medical centers of Europe and across the Atlantic to North America. Laboratories, institutes, societies, and journals dedicated to experimental medicine, especially physiology, appeared in its wake. In Germany in 1842 several physicians launched the *Archiv für physiologische Heilkunde* to help build a scientific basis for medicine, declaring that "now is the time to establish a positive science from the existing material of experience, [a science] that does not seek to ground itself in the authorities but rather in the empirical evidence which allows phenomena to be understood, and which also avoids the illusions of praxis and will lead to a deliberate, certain therapy."[27] By promoting the experimental method, such activities, in the words of one German, "first raised medicine to a science."[28]

British physicians, in contrast to their French and German brethren, initially displayed little affection for making medicine scientific. Even though the fledgling British Association for the Advancement of Science created a section on "medical science" in the 1830s, it failed to thrive, largely because leading medical men declined to participate.[29] As the historian Gerald L. Geison observed, it was not until the early 1870s, when the Royal College of Surgeons began requiring familiarity with laboratory physiology of all candidates for membership and the University of London began requiring it for medical students, that the tide began to turn.[30]

Among historians of medicine Geison took the lead in stressing "the separation between medical science and medical practice before about 1870" and, consequently, the practical vacuity of all the talk about science. In his view, experimental science left virtually no imprint on medical practice before Joseph Lister used Louis Pasteur's insights to develop antiseptic surgery in the late 1860s, and even then elite physicians promoted science more for professional than practical reasons.[31] Arleen Tuchman, however, has challenged this view. While not denying that science did little to improve therapy before the last third of the century, she points to the many important developments in diagnostic technique before then. Even prior to the appearance of the germ theory, nineteenth-century physicians could diagnose specific diseases such as Bright disease and Addison disease; demonstrate that the "fevers" of the eighteenth century were actually distinct diseases such as malaria, typhus, and typhoid; see and often remove internal tumors in their early stages; and identify in general early warning signs of oncoming illnesses. To a large extent, they had acquired this knowledge through the techniques of auscultation, percussion, microscopy, and chemical analysis, aided by the use of such instruments as the ophthalmoscope, kymograph, laryngoscope, urometer, thermometer, and pleximeter.[32]

Given these science-inspired aids to the practice of medicine, physicians possessed very practical—as well as professional—reasons for embracing science.

Although physiology and pathology have grabbed most of the attention of future-looking historians, during the first two-thirds of the nineteenth century medical geography, broadly conceived, reigned as the queen of the medical sciences. This was especially true outside of Europe, in regions such as India, North Africa, and North America.[33] When, in 1787, Benjamin Rush and his friends created the College of Physicians of Philadelphia to advance "the science of medicine," they determined to do so not by establishing laboratories but "by investigating the diseases and remedies which are peculiar to our country; by observing the effects of different seasons, climates and situations upon the human body; by recording the changes that are produced in diseases, by the progress of agriculture, arts, population, and manners; by searching for medicines in our woods, waters and the bowels of the earth; by enlarging our avenues to knowledge, from the discoveries and publications of foreign countries."[34] Nearly a century later, as germs began pushing miasmas to the periphery of medical science, Nathan Smith Davis repeated much the same message: "We cannot advance one step in the study of the causes of disease and the laws that govern their action, without a knowledge of the earth, the air, the water, the products of vegable [*sic*] and animal growth and decay; or in more technical language, without entering directly into the departments of general science, called geology, meteorology, topography, hydrology, and general chemistry."[35] America's mushrooming medical schools rarely, if ever, taught medical geography as a separate subject, but early medical societies and journals frequently urged its cultivation, as did the office of the U.S. Surgeon General. For an age that traced the origin of much disease to filth and other environmental factors, searching for correlations between topography and disease made eminent scientific sense.[36]

Despite America's reluctant embrace of medical science, by the 1830s and 1840s events were nudging the medical community further into the arms of science. John Harley Warner has estimated that between 1815 and 1860 one thousand American physicians sailed to Paris to witness the wonders of medical science for themselves. Few of these medical pilgrims returned to replicate what they had seen, but the experience abroad left an indelible impression. "Merely to have breathed a concentrated scientific atmosphere like that of Paris," gushed young Oliver Wendell Holmes during his 1833 sojourn, "must have an effect on anyone who has lived where stupidity is tolerated, where mediocrity is applauded, and where excellence is defied."[37] Unfortunately for the cause of science, the much-mentioned American obsession with practice and "getting ahead" deterred even scientifically inclined physicians from engaging in research. As one Yankee medical professor warned a student contemplating a scientific career, "You will lose a patient for every experiment you make in the laboratory."[38]

Chauvinistic Americans took understandable pride in the mite they did contribute to antebellum medical science. They lauded the military surgeon William Beaumont, whose extensive experiments on the accessible stomach of Alexis St. Martin led in 1833 to a landmark treatise on the physiology of digestion, for his "devotion to the improvement of medical science."[39] And they credited "modern science" with discovering "that mental maladies are as susceptible of cure as corporeal." The resulting treatment, called moral therapy, produced a wave of asylum-building during the second third of the century.[40] But most of all, Americans celebrated the discovery of surgical anesthesia in Boston in 1846. Although occasionally dismissed as quackery, it soon came to symbolize the beginning of a new era in surgery, made possible by the demonstration of "one of the most glorious truths of science."[41]

By midcentury, gleams of optimism were beginning to appear, although reform-minded American physicians continued to contrast the high state of medical science in Europe with its dismal condition in America. A special committee on medical education appointed in 1849 by the newly founded American Medical Association (AMA) had the following to say:

> Many intelligent men have doubted whether medicine was, or could be, a science. This skepticism still prevails extensively amongst the educated. It was not without foundation until a comparatively recent period. But within the last half century, no one acquainted with the progress of medicine can hesitate to recognize its rapid expansion into a science, and that it is rapidly entering the circle of the positive sciences. It is now far more certain in the judgments and opinions of well-educated medical men, than are the law, theology, or the moral sciences.

The American doctors expressed regret that they and their country had not participated more actively in "the movement of progress," but, like so many reformers, they placed the blame solidly on the shoulders of medical educators: "It is vain to expect that medicine, as a science, can be widely known and diffused, when it is not taught as a science in the schools."[42]

In a country plagued by a surplus of low-quality proprietary medical schools, which lacked a single laboratory until 1871, the outlook for medical science looked bleak indeed.[43] Writing home from Paris in 1869, where he was attending Claude Bernard's lectures, the aspiring physiologist Henry P. Bowditch confessed to wanting to devote himself "entirely to the science of the profession." He worried that America was "in danger of being left very far in the rear" if it did not encourage more young people to take up "medical science," adding that he did not see why Americans were "not as capable of doing good work in a scientific way as any other people."[44]

The same year that Bowditch, with the assistance of his family, established the first physiological laboratory in the United States—at the Harvard Medical School in 1871—a senior colleague, Henry J. Bigelow, was already complaining of too much science in the curriculum. "In an age of science, like the present, there is more danger that the average medical student will be drawn from what is practical, useful, and even essential, by the well-meant enthusiasm of the votaries of less applicable science, than that he will suffer from want of knowledge of these," declared the distinguished surgeon. In his opinion, no medical student should "while away his time in the labyrinths of Chemistry and Physiology, when he ought to be learning the difference between hernia and hydrocele."[45]

During the years between the end of the Civil War and the outbreak of World War I, an estimated fifteen thousand American physicians traveled across the Atlantic to study in German medical centers. When they first began returning from German laboratories in the 1870s, they experienced difficulty in finding employment as medical scientists. On coming home in 1878, the pathologist William H. Welch told his sister that he "was often asked in Germany how it is that no scientific work in medicine is done in this country, how it is that many good men who do well in Germany and show evident talent there are never heard of and never do any good work when they come back here." The explanation, he continued, was simple: unlike Germans, Americans gave no encouragement to scientific work—and "the condition of medical education here is simply horrible."[46] The only remedy, concluded a like-minded New York colleague, was to turn American medical schools into "workshops of scientific medicine."[47]

Doing so began in earnest in 1893 with the opening of The Johns Hopkins School of Medicine under the leadership of Welch. Blessed with a large endowment, Johns Hopkins became the first real center for medical science in the country. In addition to creating chairs in anatomy, physiology, pathology, and pharmacology, it provided their occupants—recruited nationally—with well-equipped laboratories and salaries sufficient to free them from the burdens of practice. Before long Johns Hopkins students were spreading across the land, similarly transforming other medical schools. By the turn of the century the medical schools at Harvard, Pennsylvania, Chicago, and Michigan had joined Johns Hopkins as important medical research centers, but the country still lacked an institution comparable to the Koch Institute in Berlin or the Pasteur Institute in Paris. In 1901, however, Congress provided funds for a national hygienic laboratory to investigate infectious and contagious diseases, and, more important, John D. Rockefeller, the oil magnate, donated the first of millions of dollars to create the finest medical research facility in the country.[48] Convinced by reading William Osler's *Principles and Practice of Medicine* (1892) that "medicine could hardly hope to become a science until medicine should be endowed and qualified men could give themselves to uninterrupted study and investigation, on ample salary, entirely independent of practice," Rockefeller's chief advisor, Frederick T. Gates, resolved to create such a place. The resulting Rockefeller Institute for Medical Research not only freed researchers from practicing medicine but from teaching as well, allowing them to devote their entire lives to medical science. Its success soon inspired the creation of other American institutes for medical research and provided a model for the Kaiser Wilhelm Gesellschaft, which opened in Berlin in 1911.[49]

By this time scientific expertise was paying off clinically and economically. Soon after the young Harvard-trained physician Richard Cabot had mastered the analysis of blood using microscopic and serological techniques, he volunteered to accompany sick soldiers returning home from the Spanish-American War. Despite his youth, none of his colleagues could best him in diagnosing malaria and typhoid fever. "No one else knew anything," he boasted. "I was king." Unsurprisingly, Cabot went on to a lucrative career as one of Boston's leading medical consultants.[50]

"The intrusion of science into clinical medicine," as an influential American practitioner once described it, picked up momentum after 1913, when the Rockefeller-funded General Education Board launched an initiative to create full-time salaried positions for clinicians. "No single event has had a more profound effect of medical education and medical practice than the movement to full-time positions in clinical departments," concluded A. McGehee Harvey of Johns Hopkins. "Out of this emerged the clinical scientists, versed in the bedside practice of medicine and capable of applying the knowledge and techniques of the basic sciences to the study of human disease." Under the control of such hybrid scientists, hospitals became "temples of science" where clinical and laboratory research was conducted and medical students were educated.[51]

SECTS AND SCIENCE

The growing enthusiasm for "scientific medicine" did not stem exclusively from a desire to improve medical care. It also played an important rhetorical role in the battle for cultural and legal authority waged against sectarian or alternative practitioners. During the first half of the nineteenth century regular medicine in the United States came to be closely associated with such "heroic" practices as bleeding, blistering, puking, and purging, the last induced by large doses of calomel (*mercurous chloride*), a mineral drug that sometimes produced nasty side effects. Partially in response to this situation, there arose a host of sectarian movements that challenged the dominance of the regulars, from Thomsonianism and eclecticism to homeopathy and hydropathy. All foreswore phlebotomy. The Thomsonians and eclectics administered only botanical remedies; homeopaths relied on infinitesimal doses of drugs; hydropaths put their trust—and bodies—in water. The sectarian assault on medical privilege in the first half of the nineteenth century erased virtually all legal restrictions to the practice of medicine, and medical licensing did not begin to return until the 1870s. Throughout the century regular physicians—or allopaths, as they were frequently called—retained at least 80 percent of the medical marketplace, but with a rapidly developing physician surplus, sectarian competition became the regulars' greatest fear. Science became their salvation.[52]

One of the AMA's first acts after its founding in 1847 was to set up a Committee on Medical Sciences, primarily to stimulate the production of medical literature. The committee viewed scientific publications as "one of the broadest lines of distinction between physicians and all pretenders to the name." Thus, it repeatedly called for the "systematic, and thorough investigation of subjects connected with medical science," reasoning that "if the practitioners of Homoeopathy [*sic*], Thompsonianism [*sic*], Eclecticism, Animal Magnetism, Needle Cure, Lifting Cure, or Water Cure [were] required, as is done in Prussia and Austria, to be thoroughly educated in scientific medicine . . . they would be comparatively harmless."[53]

Local and state medical societies thought the same way. In an address before the Illinois State Medical Society in 1853 the prominent medical educator and reformer Nathan Smith Davis argued that "legitimate" medicine's "indissoluble connection with the whole chain of natural sciences" distinguished it from "all the special *pathys* and *isms* of the day." "True medical science," he declared, "is simply a part of the great science of nature, while the art of practice of medicine is the application of the knowledge derived from the study of such science, to the prevention and cure of disease."[54]

Unfortunately for the regulars, the sectarians had no intention of abdicating their own claim to science, particularly in a society that was granting science ever more cultural authority. Some, especially the homeopaths, claimed to be *more* scientific than the allopaths, because their "provings" (that is, experiments to see how healthy patients responded to various doses of drugs), constituted experimental medicine. For at least one homeopathic practitioner, the "science of medicine" amounted to nothing more than "a knowledge of how to cure the sick."[55] Wooster Beach, founder of the eclectic sect, referred to his practice as "the scientific system of medicine."[56] Delegates at a convention in 1852 to organize a national society of botanic physicians reflected this view when they extended an invitation to "all who believe that medicine is strictly a natural science, and as *systematic* as are other natural sciences."[57] One of Thomson's disciples contrasted the "the horrible, unwarrantable, murderous quackery" of the regulars with the "truly scientific practice" of botanics, based on "the immutable laws of Creative wisdom."[58]

When new systems of healing—Christian Science, osteopathy, and chiropractic—appeared in the late nineteenth century, at a time when scientific medicine was enjoying unprecedented popularity, they, too, cloaked themselves in the mantle of science. Mary Baker Eddy may have rejected the reality of disease, death, and matter itself, but she insisted on calling her system of healing Christian Science and on naming her major text *Science and Health* (1875).[59] Andrew Taylor Still likewise based his "Science of Osteopathy" on "a scientific principle."[60] D. D. Palmer, who in the late 1880s was trying to educate the world about "the science of magnetic healing," went on a few years later to found chiropractic, the most successful sectarian movement of the twentieth century. Palmer defined science as "knowledge; ascertained facts; accumulated information of causes and principles systematized." In his *Text-Book of the Science, Art and Philosophy of Chiropractic* (1910) he laid out his claim to being a genuine scientist:

> I ascertained these truths, acquired instruction, heretofore unrecognized, regarding the performance of functions in health and disease. I systematized and correlated these principles, made them practical. By doing so I created, brought into existence, originated a science, which I named Chiropractic; therefore I am a scientist.[61]

As Steven C. Martin has shown, chiropractic's claims to science rang less hollow at the turn of the twentieth century than they do now. Chiropractors may have rejected the new laboratory-based experimental science—the faculty at the Palmer School of Chiropractic, for example, dismissed chemical and microscopic studies as "useless"—but they actively pursued an older style of observational science, associated with the tradition of natural history and based on observation in their clinics and museums, which housed sizeable osteological collections. By the early 1910s chiropractors were promoting their system not merely as science but as "the only truly scientific method of healing."[62]

The embarrassing proliferation of medical sects, each claiming scientific truth, led many physicians, sectarian as well as regular, to search for common ground sanctified by science. As early as the 1860s some progressive homeopathic physicians were calling for less dogmatism and more freedom to participate in "the Republic of Scientific Medicine."[63] By the last years of the century even regular leaders were calling for unity based on science. "The tendency in America is to give up all sects in medicine and allow physicians to choose their professional associates purely on the ground of integrity and personal preference," Louis Faugères Bishop informed the New York Academy of Medicine in 1898. "The modern school of scientific medicine relies for its recognition and support upon a public, educated in modern philosophy, a public dominated by the scientific spirit of a scientific age."[64] Three years later AMA president Charles A. L. Reed proclaimed "the existence of a new school of medicine" based on scientific medicine. "It is as distinct from the schools of fifty years ago as is the Christian dispensation from its Pagan antecedents. It is the product of convergent influences, of diverse antecedents."[65] The sometime schoolmaster Abraham Flexner, in his famous indictment of North American medical schools in 1910, credited scientific medicine with erasing sectarian distinctives because "science and dogma" could not coexist. "No man is asked in whose name he comes—whether that of Hahnemann, Rush, or of some more recent prophet," he wrote. "To plead in advance a principle couched in pseudo-scientific language or of extra-scientific character is to violate scientific quality."[66] The stubborn persistence of Christian Science, osteopathy, and chiropractic—as well as a swarm of even newer healing alternatives in the twentieth century—gave the lie to such fatuous beliefs, but the ongoing struggle over sects and science illustrates just how malleable the term science was.

"SCIENTIFIC MEDICINE"

Probably no practice in the historiography of medicine has distorted a proper understanding of the relationship between science and medicine more than the uncritical (and perhaps unconscious) use of the term "scientific medicine" to describe what became of medicine in the late nineteenth century. Instead of looking at the origin and use of the slogan, too many historians have employed it descriptively, suggesting by their use that medicine did not become truly scientific until the late nineteenth century, when the term enjoyed wide circulation. As one historian has expressed the common view, "In the last third of the nineteenth century, in response to bacteriological discoveries and technological innovations that ushered in dramatic advances, the practice of medicine became a science."[67] If so, then anyone who thought that medicine had become a science earlier than the last third of the nineteenth century must have been mistaken or deluded. Some historians have written anachronistically of "scientific medicine" in the eighteenth century (or earlier), adopting it as a synonym for the then-used terms "philosophical" or "rational" medicine.[68] To alleviate such confusion, John Harley Warner has taken to referring to "the new scientific medicine" when speaking of developments in the last decades of the nineteenth century.[69]

The precise origins of the term "scientific medicine" are lost in the haze of history, but by the 1840s, decades before the germ theory arrived, it had come into common usage. In 1844 medical progressives in Berlin founded the Gesellschaft für wissenschaftliche Medicin (Society for Scientific Medicine), which provided a forum two years later for the budding young cellular pathologist Rudolf Virchow to address the subject of scientific medicine. Noting that the very phrase needed clarification, he explained that "Scientific medicine is compounded of two integrated parts—pathology, which delivers, or is supposed to deliver, information about the altered conditions and altered physiological phenomena, and therapy, which seeks out the means of restoring or maintaining normal conditions."[70] In 1865 Claude Bernard, perhaps the leading French promoter of scientific medicine, published his *Introduction à l'étude de la médecine expérimentale*, in which he argued that "scientific medicine, like the other sciences, can be established only by experimental means, i.e. by direct and rigorous application of reasoning to the facts furnished us by observation and experiments."[71]

Most partisans of scientific medicine, like historians, rarely defined what they meant by the term. Instead, they simply enumerated its benefits and accomplishments. For some clinicians the introduction of the clinical thermometer in the 1860s epitomized "the scientific approach to medical research."[72] For others, such as Nathan Smith Davis, the list of ways in which the practice of medicine had profited from "the advancement of its science" included "anaesthetics for the relief of pain, the nervous and arterial sedatives for allaying nervous restlessness and controlling the circulation, the mechanical appliances for dressing wounds and injuries, preventing or relieving deformities, and a thousand other things that have been added directly to the daily practice of our art."[73] Lord Lister nominated the adoption of surgical anesthesia in the 1840s, the development of antiseptic surgery in the 1860s, and the discovery of x-rays in the 1890s.[74] In 1900 the British physician Philip H. Pye-Smith, who believed that medicine's claim to be "a true science" rested on its ability to make accurate predictions, cited "the detection and treatment of plumbism [chronic lead poisoning], the diagnosis and cure of scabies and ringworm, the treatment of poisons by chemical antidotes, and of specific disease by attenuated inoculations"—along with the recent discovery of the origin of malaria—as "instances of strictly scientific medicine."[75] A few years later the New York physician Alexander Lambert described the therapeutic fruits of the germ theory as "one of the most brilliant achievements of scientific medicine." In his opinion, "the most complete adaptation of pure scientific work to clinical medicine" came in the early 1890s, when Emil von Behring and others working at the Koch Institute in Berlin developed antitoxins for diphtheria and tetanus that had been "conceived and worked out completely in the laboratory before [being] applied to medicine."[76]

The American press, as Bert Hansen has recently shown, generated its own list of breakthroughs in scientific medicine, which, he argues, led to "an entirely new idea bec[oming] embedded in popular consciousness: that medical research could provide widespread benefits." The first medical event to grab the reading public's attention was Pasteur's development in 1885 of a vaccine against rabies. In covering the dramatic story of four Newark children who had been rushed across the Atlantic to the Pasteur Institute in Paris after being bitten by an apparently rabid dog, New York–area newspapers told of local physicians, dubbed "the Newark Scientists," who infected rabbits with nerve tissue taken from the offending dog to determine its virulence. The excitement over the rabies vaccine was followed in quick order by the announcement in 1889 of Charles-Édouard Brown-Séquard's use of testicular extracts from animals to rejuvenate the elderly, Robert Koch's finding in 1890 of a purported cure for tuberculosis (tuberculin), the introduction of diphtheria antitoxin in 1894, based on the work of Émile Roux and Emil von Behring, and Wilhelm Röntgen's discovery of x-rays in 1895. By the turn of the century microscopes and white "lab coats" were serving the trademarks of the medical profession.[77]

For many physicians and large segments of the public, scientific medicine, however understood, came to represent all that was good in medicine. "All else which may go under the name of medicine," declared a president of the AMA, "is sham and fraud."[78] A Canadian physician addressing students at the McGill University Medical School argued that if practitioners of medicine did not "approach their patients with the certain method and moral force of a scientist," they were no better than "tradesmen who trade in medicine; men not much better than quacks, who by superficial observation and improper criticism remain at a low mental and moral level all life."[79] And so it seemed.

Scientific medicine became so inextricably identified with the laboratory that many practicing physicians felt left out. Thus in 1900 the editors of the *Boston Medical and Surgical Journal* called for a redefinition of the term to make it inclusive of the activities of clinicians observing at the bedside as well as scientists working in the laboratory. While acknowledging that the growth of laboratory medicine represented "the most important advance of the past few years," the editors noted that

> the use of the word "scientific" in a narrow sense as applied to laboratory work had done much to arouse prejudice and establish false standards. By degrees there had grown up a distinction, heard on every hand, between what was termed "scientific medicine," and "practical medicine," naturally with the implication that practical medicine as personified in the practitioner was not scientific. In other words, "scientific" as popularly used in medical writing and discourse demanded a laboratory and the accessories a laboratory implies, whereas the man who examined his patients from day to day in most painstaking and accurate fashion should be looked upon perhaps as a faithful and conscientious practitioner but not as a scientific investigator.

The image of the laboratory had become so dominant "that he who often aimlessly looks through a microscope, or with narrow vision describes a lesion, is placed upon a pedestal as a man of 'scientific' tendencies, while his colleague who faithfully describes a symptom is denied any such distinction," continued the editorial. "In the popular mind all that is accurate and 'scientific' has come to be regarded as a product of laboratory methods."[80]

Well into the twentieth century and even the twenty-first, some physicians continued to grouse about too little science in medicine and too little appreciation of what there was, no matter how much patronage and acclaim came their way. They saw scientific medicine under siege by a host of antivivisectionists, right to lifers, and alternative healers. They witnessed medical schools turning from scientific workshops into profit centers run by unsympathetic managers and falling "prey to unprecedented levels of greed, commercialism, and intellectual dishonesty" vaguely reminiscent of the prescientific days of the nineteenth century.[81] Still other physicians lamented the lost art of medicine, stolen by science. And always there were the commencement speakers and anniversary lecturers regaling audiences with the inspiring story of medicine's rise to the pinnacle of science.

NOTES

I want to thank Spencer Fluhman and Paul Erickson for their assistance in tracking down many of the sources used in this essay.

1. Dana W. Atchley, "Science and Medical Education," *Journal of the American Medical Association* 164 (1957): 542.

2. George Sarton, "The History of Science versus the History of Medicine," *Isis* 23 (1935): 313–20; Henry E. Sigerist, "The History of Medicine *and* the History of Science: An Open Letter to George Sarton, Editor of *Isis*," *Bulletin of the Institute of the History of Medicine* 4 (1936): 1–13.

3. See, e.g., Gerald L. Geison, "Divided We Stand: Physiologists and Clinicians in the American Context," in *The Therapeutic Revolution: Essays in the Social History of American Medicine*, ed. Morris J. Vogel and Charles E. Rosenberg (Philadelphia: University of Pennsylvania Press, 1979), 67–90; Russell C. Maulitz, "'Physician versus Bacteriologist': The Ideology of Science in Clinical Medicine," in *The Therapeutic Revolution*, 91–107; S. E. D. Shortt, "Physicians, Science, and Status: Issues in the Professionalization of Anglo-American Medicine in the Nineteenth Century," *Medical History* 27 (1983): 51–68; Christopher Lawrence, "Incommunicable Knowledge: Science, Technology and the Clinical Art in Britain, 1850–1914," *Journal of Contemporary History* 20 (1985): 503–20; and especially the writings of John Harley Warner: "Science in Medicine," *Osiris* 2nd ser., 1 (1985): 37–58; *The Therapeutic Perspective: Medical Practice, Knowledge, and Identity in America, 1820–1885* (Cambridge, MA: Harvard University Press, 1986); "Science, Healing, and the Physician's Identity: A Problem of Professional Character in Nineteenth-Century America," *Clio Medica* 22 ([1989?]): 65–88; "Ideals of Science and Their Discontents in late Nineteenth-Century American Medicine," *Isis* 82 (1991): 454–78; "The Idea of Science in English Medicine: The 'Decline of Science' and the Rhetoric of Reform, 1815–45," in *British Medicine in an Age of Reform*, ed. Roger French and Andrew Wear (London: Routledge, 1991), 136–64; "The History of Science and the Sciences of Medicine," *Osiris* 2nd ser., 10 (1995): 164–93; and *Against the Spirit of System: The French Impulse in Nineteenth-Century American Medicine* (Princeton, NJ: Princeton University Press, 1998).

4. G. E. R. Lloyd, *Magic, Reason and Experience: Studies in the Origins and Development of Greek Science* (Cambridge: Cambridge University Press, 1979), 96–97. See also Roy Porter and W. F. Bynum, "The Art and Science of Medicine," in *Companion Encyclopedia of the History of Medicine*, ed. W. F. Bynum and Roy Porter (London: Routledge, 1993), 1:3–11; and Jacques Jouanna, "The Birth of Western Medical Art," in *Western Medical Thought from Antiquity to the Middle Ages*, trans. and ed. Mirko D. Grmek and Antony Shugaar (Cambridge, MA: Harvard University Press, 1998), 22–71.

5. Roger French, *Ancient Natural History* (London: Routledge, 1994), 223–24.

6. Charles H. Talbot, "Medicine," in *Science in the Middle Ages*, ed. David C. Lindberg (Chicago: University of Chicago Press, 1978), 392.

7. Nancy G. Siraisi, *Avicenna in Renaissance Italy: The Canon and Medical Teaching in Italian Universities after 1500* (Princeton, NJ: Princeton University Press, 1987), 23–24.

8. Jerome J. Bylebyl, "The Medical Meaning of *Physica*," *Osiris* 2nd ser., 6 (1990): 16, 30 (Malmesbury). Charles H. Talbot attributes the distinction between the *physicus* and the *medicus* to the Arabs; see Talbot, "Medicine," 402.

9. Nancy G. Siraisi, "Taddeo Alderotti and Bartolomeo da Varignana on the Nature of Medical Learning," *Isis* 68 (1977): 27–39. For a somewhat fuller account, see Siraisi, *Taddeo Alderotti and His Pupils: Two Generations of Italian Medical Learning* (Princeton, NJ: Princeton University Press, 1981), 118–46. On the place of medicine in medieval classificatory schemes, see James A. Weisheipl, "The Nature, Scope, and Classification of the Sciences," in *Science in the Middle Ages*, 461–82.

10. Vern L. Bullough, *The Development of Medicine as a Profession: The Contribution of the Medieval University to Modern Medicine* (Basel: S. Karger, 1966), 95.

11. Darrel W. Amundsen, "Medicine and Surgery as Art or Craft: The Role of Schematic Literature in the Separation of Medicine and Surgery in the Late Middle Ages," *Transactions and Studies of the College of Physicians of Philadelphia* 5th ser., 1 (1979): 52–55; Weisheipl, "Nature, Scope, and Classification," 473–74.

12. Robert S. Gottfried, *Doctors and Medicine in Medieval England, 1340–1530* (Princeton, NJ: Princeton University Press, 1986), 37.

13. Ian Maclean, *Logic, Signs and Nature in the Renaissance: The Case of Learned Medicine* (Cambridge: Cambridge University Press, 2002), 70–76 (quote, 73).

14. Harold J. Cook, "The New Philosophy and Medicine in Seventeenth-Century England," in *Reappraisals of the Scientific Revolution*, ed. David C. Lindberg and Robert S. Westman (Cambridge: Cambridge University Press, 1990), 397–436; see also Cook, "Physicians and Natural History," in *Cultures of Natural History*, ed. N. Jardine, J. A. Secord, and E. C. Spary (Cambridge: Cambridge University Press, 1996), 91–105. On the history of medicine and natural philosophy, see Roger French, *Medicine before Science: The Rational and Learned Doctor from the Middle Ages to the Enlightenment* (Cambridge: Cambridge University Press, 2003).

15. Roger French, *William Harvey's Natural Philosophy* (Cambridge: Cambridge University Press, 1994); Jacques Roger, "The Living World," in *The Ferment of Knowledge: Studies in the Historiography of Eighteenth-Century Science*, ed. G. S. Rousseau and Roy Porter (Cambridge: Cambridge University Press, 1980), 255–83; see also Jacques Roger, *The Life Sciences in Eighteenth-Century French Thought*, ed. Keith R. Benson, trans. Robert Ellrich (Palo Alto, CA: Stanford University Press, 1997).

16. Quoted in Thomas H. Broman, "The Medical Sciences," in *The Eighteenth Century*, ed. Roy Porter, vol. 4 of *The Cambridge History of Science*, ed. David C. Lindberg and Ronald L. Numbers (Cambridge: Cambridge University Press, 2001), 534–35.

17. "Medicine," Chambers's *Cyclopaedia; or, An Universal Dictionary of Arts and Sciences*, 2nd ed. (1738), quoted in Roy Porter, "Medical Science and Human Science in the Enlightenment," in *Inventing Human Science: Eighteenth-Century Domains*, ed. Christopher Fox, Roy Porter, and Robert Wokler (Berkeley: University of California Press, 1995), 56.

18. Thomas H. Broman, *The Transformation of German Academic Medicine, 1750–1820* (Cambridge: Cambridge University Press, 1996), 10; Anita Guerrini, "Newtonianism, Medicine and Religion," in *Religio Medici: Medicine and Religion in Seventeenth-Century England*, ed. Ole Peter Grell and Andrew Cunningham (Aldershot: Scolar Press, 1996), 293–312 (quoting Mandeville on p. 297).

19. Whitfield J. Bell Jr., *The College of Physicians of Philadelphia: A Bicentennial History* (Canton, MA: Science History Publications, 1987), 7.

20. Edward C. Carter II, *"One Grand Pursuit": A Brief History of the American Philosophical Society's First 250 Years, 1743–1993* (Philadelphia: American Philosophical Society, 1993), 6–7; James H. Cassedy, *Medicine in America: A Short History* (Baltimore, MD: The Johns Hopkins University Press, 1991), 16. For a brief survey of science and medicine in America, see John Duffy, "Science and Medicine," in *Science and Society in the United States*, ed. David D. Van Tassel and Michael G. Hall (Homewood, IL: Dorsey Press, 1966), 107–34.

21. Whitfield J. Bell Jr., *John Morgan: Continental Doctor* (Philadelphia: University of Pennsylvania Press, 1965), 116–28.

22. Samuel Miller, *A Brief Retrospect of the Eighteenth Century* (New York: T. and J. Swords, 1803), 1: 201–2, 316–17.

23. Michel Foucault, *The Birth of the Clinic: An Archaeology of Medical Perception*, trans. A. M. Sheridan Smith (New York: Pantheon, 1973), xv; Laurence Brockliss and Colin Jones, *The Medical World of Early Modern France* (Oxford: Clarendon Press, 1997), 826.

24. Brockliss and Jones, *Medical World of Early Modern France*, 826, 832–33. Generally, however, medical participation in scientific societies seems to have declined during the eighteenth century; see James E. McClellan III, *Science Reorganized: Scientific Societies in the Eighteenth Century* (New York: Columbia University Press, 1985), 37–38.

25. Quoted in J. Rosser Matthews, *Quantification and the Quest for Medical Certainty* (Princeton, NJ: Princeton University Press, 1995), 19, 62–63. For recent assessments of the so-called Paris school, see Caroline Hannaway and Ann La Berge, *Constructing Paris Medicine* (Amsterdam: Rodopi, 1998).

26. Matthews, *Quantification and the Quest for Medical Certainty*, 30 (Auber), 65 (Double), 71 (Comte).

27. Ibid., 46.

28. Arleen Marcia Tuchman, *Science, Medicine, and the State in Germany: The Case of Baden, 1815–1871* (New York: Oxford University Press, 1993), 10–11, 162; see also John E. Lesch, *Science and Medicine in France: The Emergence of Experimental Physiology, 1790–1855* (Cambridge, MA: Harvard University Press, 1984); and Michael Hagner, "Scientific Medicine," in *From Natural Philosophy to the Sciences: Writing the History of Nineteenth-Century Science*, ed. David Cahan (Chicago: University of Chicago Press, 2003), 49–87.

29. "Is Medicine Entitled to Rank as a Branch of Science?" *London Medical Gazette* 26 (1840): 918–19; see also Jack Morrell and Arnold Thackray, *Gentlemen of Science: Early Years of the British Association for the Advancement of Science* (Oxford: Clarendon Press, 1981), 453–54, 509.

30. Gerald L. Geison, *Michael Foster and the Cambridge School of Physiology: The Scientific Enterprise in Late Victorian Society* (Princeton, NJ: Princeton University Press, 1978), 329.

31. Ibid., 26; see also Geison, "Divided We Stand," 67–90.

32. Tuchman, *Science, Medicine, and the State in Germany*, 163–64. On the meanings of scientific medicine in Victorian Britain, see Terrie M. Romano, *Making Medicine Scientific: John Burdon Sanderson and the Culture of Victorian Science* (Baltimore, MD: The Johns Hopkins University Press, 2002).

33. See Michael A. Osborne, "The Geographical Imperative in Nineteenth-Century French Medicine" (ch. 2), Mark Harrison, "Differences of Degree: Representations of India in British Medical Topography, 1820–c. 1870" (ch. 3), and Ronald L. Numbers, "Medical Science before Scientific Medicine: Reflections on the History of Medical Geography" (ch. 13), all in *Medical Geography in Historical Perspective*, ed. Nicolaas A. Rupke (London: Wellcome Trust Centre for the History of Medicine, 2000).

34. Bell, *College of Physicians of Philadelphia*, 8–9.

35. N. S. Davis, "An Address on the Nature of the Science and Art of Medicine, and Their Relations to the Various Important Interests of the People," *Chicago Medical Journal & Examiner* 40 (1880): 450.

36. On medical geography in America, see Gregg Mitman and Ronald L. Numbers, "From Miasma to Asthma: The Changing Fortunes of Medical Geography in America," *History and Philosophy of the Life Sciences* 25 (2003): 391–412; and James H. Cassedy, "Medical Geography of a Growing Nation," in *Medicine and American Growth, 1800–1860* (Madison: University of Wisconsin Press, 1986), 33–59; see also Cassedy, "Meteorology and Medicine in Colonial America: Beginnings of the Experimental Approach," *Journal of the History of Medicine* 24 (1969): 193–204.

37. Warner, *Against the Spirit of System*, 3.

38. S. Weir Mitchell, "Memoir of John Call Dalton, 1825–1889," National Academy of Sciences, *Biographical Memoirs* 3 (1895): 181.

39. Ronald L. Numbers and William J. Orr Jr., "William Beaumont's Reception at Home and Abroad," *Isis* 72 (1981): 590–612, esp. 593.

40. "Report of the Commissioners Appointed by the Governor of New Jersey, to Ascertain the Number of Lunatics and Idiots in the State" (Newark: M. S. Harrison, 1840), 9, quoted in David Gollaher, *Voice for the Mad: A Life of Dorothea Dix* (New York: Free Press, 1995), 186.

41. John Collins Warren, quoted in Sherwin B. Nuland, *Doctors: The Biography of Medicine* (New York: Alfred A. Knopf, 1988), 289–90. On the various responses to surgical anesthesia, see Martin S. Pernick, *A Calculus of Suffering: Pain, Professionalism, and Anesthesia in Nineteenth-Century America* (New York: Columbia University Press, 1985).

42. Samuel Jackson et al., "Report of the Special Committee Appointed to Prepare: A Statement of the Facts and Arguments Which May Be Adduced in Favour of the Prolongation of the Courses of Medical Lectures to Six Months," American Medical Association, *Transactions* 2 (1849): 361.

43. For histories of the various biomedical disciplines, see Ronald L. Numbers, ed., *The Education of American Physicians: Historical Essays* (Berkeley: University of California Press, 1980).

44. Henry P. Bowditch to Henry I. Bowditch, January 26, 1869, quoted in John Harley Warner and Janet A. Tighe, eds., *Major Problems in the History of American Medicine and Public Health* (Boston: Houghton Mifflin, 2001), 198.

45. Henry J. Bigelow, *Medical Education in America* (Cambridge: Welch, Bigelow &

Co., 1971), quoted in W. Bruce Fye, *The Development of American Physiology: Scientific Medicine in the Nineteenth Century* (Baltimore, MD: The Johns Hopkins University Press, 1987), 107–8.

46. Quoted in Simon Flexner and John Thomas, Flexner, *William Henry Welch and the Heroic Age of American Medicine* (New York: Viking, 1941) 112–13; see also Thomas Neville Bonner, *American Doctors and German Universities: A Chapter in International Intellectual Relations, 1870–1914* (Lincoln: University of Nebraska Press, 1963).

47. Alfred L. Loomis, "Inaugural Address on the Present Needs of Scientific Medicine," *Medical Record* 35 (1889): 141–44.

48. See Ronald L. Numbers and John Harley Warner, "The Maturation of American Medical Science," in *Scientific Colonialism: A Cross-Cultural Comparison*, ed. Nathan Reingold and Mar Rothenberg (Washington DC: Smithsonian Institution Press, 1987), 191–214.

49. Frederick T. Gates, quoted in E. Richard Brown, *Rockefeller Medicine Men: Medicine and Capitalism in America* (Berkeley: University of California Press, 1979), 106; George W. Corner, *A History of the Rockefeller Institute,1901–1953: Origins and Growth* (New York: Rockefeller Institute Press, 1964).

50. Christopher Crenner, *Private Practice: In the Early Twentieth-Century Medical Office of Dr. Richard Cabot* (Baltimore, MD: The Johns Hopkins University Press, 2005), 12.

51. Charles E. Rosenberg, *The Care of Strangers: The Rise of America's Hospital System* (New York: Basic Books, 1987), 332–33.

52. See, e.g., Ronald L. Numbers, "Do-It-Yourself the Sectarian Way," in *Medicine without Doctors: Home Health Care in American History*, ed. Guenter B. Risse, Ronald L. Numbers, and Judith Walzer Leavitt (New York: Science History Publications, 1977), 49–72.

53. American Medical Association, *Digest of Official Actions, 1846–1958* (Chicago: American Medical Association, 1959), 609–10; E. Giddings et al., "Report of the Committee on Medical Education," American Medical Association, *Transactions* 22 (1871): 127, 148.

54. N. S. Davis, "The Intimate Relation of Medical Science to the Whole Field of Natural Sciences," *Transactions of the Illinois State Medical Society* 3 (1853): 18–22.

55. James Tyler Kent, *New Remedies, Clinical Cases, Lesser Writings, Aphorisms, and Precepts* (Chicago: Erhart and Karl, 1926), 211, quoted in Harris L. Coulter, *Science and Ethic in American Medicine, 1800–1914*, vol. 3 of *Divided Legacy: A History of the Schism in Medical Thought* (Washington, DC: McGrath Publishing, 1973), 329; see also Naomi Rogers, "American Homeopathy Confronts Scientific Medicine," in *Culture, Knowledge, and Healing: Historical Perspectives of Homeopathic Medicine in Europe and North America*, ed. Robert Jütte, Guenter B. Risse, and John Woodward (Sheffield: European Association for the History of Medicine, 1998), 32; and Rogers, *An Alternative Path: The Making and Remaking of Hahnemann Medical College and Hospital in Philadelphia* (New Brunswick, NJ: Rutgers University Press, 1998). On homeopathic appropriation of scientific medicine, see, e.g., B. Lyon Williams, *Homeopathy and the Doctors; or, A Plea for Scientific Medicine* (London: Henry Turner, 1870); and I. W. Heysinger, *The Scientific Basis of Medicine* (Philadelphia: Boericke & Tafel, 1897).

56. Wooster Beach, *The American Practice Condensed; or, The Family Physician, Being the Scientific System of Medicine, on Vegetable Principles, Designed for All Classes* (New York:

James M'Alister, 1847). On the Eclectic sect, see John S. Haller Jr., *Medical Protestants: The Eclectics in American Medicine, 1825–1939* (Carbondale: Southern Illinois University Press, 1994).

57. Quoted in John S. Haller Jr., *The People's Doctors: Samuel Thomson and the American Botanical Movement, 1790–1860* (Carbondale: Southern Illinois University Press, 2000), 245.

58. J. Redding, "Is Medicine a Science?" *Physio-Medical Journal* 8 (1882): 130, 148.

59. The best account of Christian Science healing is Rennie B. Schoepflin, *Christian Science on Trial: Religious Healing in America* (Baltimore, MD: The Johns Hopkins University Press, 2003). See also Rennie B. Schoepflin, "The Christian Science Tradition," in *Caring and Curing: Health and Medicine in the Western Religious Traditions*, ed. Ronald L. Numbers and Darrel W. Amundsen (New York: Macmillan, 1986), 421–46.

60. Carol Trowbridge, *Andrew Taylor Still, 1828–1917* (Kirksville, MO: Thomas Jefferson University Press, 1991), 139–41; see also Norman Gevitz, *The D.O.'s: Osteopathic Medicine in America* (Baltimore, MD: The Johns Hopkins University Press,1982).

61. Dennis Peterson and Glenda Wiese, *Chiropractic: An Illustrated History* (St. Louis: Mosby, 1995), 67 (science of magnetic healing), 83 (jail); Daniel David Palmer, *Text-Book of the Science, Art, and Philosophy of Chiropractic for Students and Practitioners* (Portland, OR: Portland Printing House, 1910), quoted in J. Stuart Moore, *Chiropractic in America: The History of a Medical Alternative* (Baltimore, MD: The Johns Hopkins University Press, 1993), 3. See also Walter I. Wardwell, *Chiropractic: History and Evolution of a New Profession* (St. Louis: Mosby Year Book, 1992), esp. "Chiropractic Philosophy, Science, Art," 179–210.

62. Steven C. Martin, "'The Only Truly Scientific Method of Healing': Chiropractic and American Science, 1895–1990," *Isis* 85 (1994): 207–27.

63. Adolph Lippe, "Valedictory Address, Delivered at the Eighteenth Annual Commencement of the Homeopathic Medical College of Pennsylvania, March 1, 1866," *Hahnemannian Monthly* 1 (1866): 308, quoted in John Harley Warner, "Orthodoxy and Otherness; Homeopathy and Regular Medicine in Nineteenth-Century America," in *Culture, Knowledge, and Healing: Historical Perspectives of Homeopathic Medicine in Europe and North America*, ed. Robert Jütte, Guenter B. Risse, and John Woodward (Sheffield: European Association for the History of Medicine, 1998), 19.

64. Louis Faugères Bishop, "The Evolution of Scientific Medicine," *Medical Times*, November 1898, 542–43. On the struggle between homeopaths and allopaths in New York, see Warner, "Ideals of Science," 454–78.

65. Charles A. L. Reed, "The President's Address," *Journal of the American Medical Association* 36 (1901): 1606, quoted in William G. Rothstein, *American Physicians in the Nineteenth Century: From Sects to Science* (Baltimore, MD: The Johns Hopkins University Press, 1972), 325. For a similar observation, see William Osler, *Aequanimitas* (Philadelphia: Blakiston's Son, 1932), 254–55 quoted in Rothstein, *American Physicians in the Nineteenth Century*, 325–26.

66. Abraham Flexner, *Medical Education in the United States and Canada: A Report to the Carnegie Foundation for the Advancement of Teaching* (New York: Carnegie Foundation, 1910), 157–61.

67. Regina Markell Morantz-Sanchez, *Sympathy and Science: Women Physicians in American Medicine* (New York: Oxford University Press, 1985), 144.

68. Porter, "Medical Science and Human Science in the Enlightenment," 57–60. In a particularly intriguing exercise, Lester S. King tries to decipher what the term "scientific medicine" meant to Benjamin Rush, who, as far as I know, never used the phrase; see King, *Transformations in American Medicine: From Benjamin Rush to William Osler* (Baltimore, MD: The Johns Hopkins University Press, 1991), 183; in a chapter devoted to "Changing Aspects of Scientific Medicine, 1800–1850." See also King, "Medicine Seeks to Be 'Scientific,'" *Journal of the American Medical Association* 241 (1983): 2469–74.

69. John Harley Warner and Janet A. Tighe, eds., *Major Problems in the History of American Medicine and Public Health* (Boston: Houghton Mifflin, 2001), 196.

70. Rudolf Virchow, "Standpoints in Scientific Medicine," trans. L. J. Rather, *Bulletin of the History of Medicine* 30 (1956): 437–39. In *Rudolf Virchow: Doctor, Statesman, Anthropologist* (Madison: University of Wisconsin Press, 1953) Erwin H. Ackerknecht refers to the organization as "the *Verein fuer wisenschaftliche Medizin*" (Society for Scientific Medicine) (11).

71. Claude Bernard, *An Introduction to the Study of Experimental Medicine*, trans. Henry Copley Greene (New York: Macmillan, 1927), 2, 205.

72. E. D. Adrian, "The Scientific Approach to Medical Research," in *Lectures on the Scientific Basis of Medicine*, vol. 1: *1951–52* (London: Athlone Press, 1953), 14–15.

73. N. S. Davis, "An Address on the Nature of the Science and Art of Medicine, and Their Relations to the Various Important Interests of the People," *Chicago Medical Journal & Examiner* 40 (1880): 453–54.

74. Lord Lister, "The Interdependence of Science and Medicine," *Scientific Monthly* 25 (1927): 193–212 (an address given to the British Association for the Advancement of Science in 1896).

75. Philip H. Pye-Smith, "Medicine as a Science and Medicine as an Art," *British Medical Journal*, August 4, 1900, 281.

76. Alexander Lambert, "The Adaptation of Pure Science to Medicine," *Journal of the American Medical Association* 42 (1904): 1669.

77. Bert Hansen, "America's First Medical Breakthrough: How Popular Excitement about a French Rabies Cure in 1885 Raised New Expectations for Medical Progress," *American Historical Review* 103 (1998): 373–418; Hansen, "New Images of a New Medicine: Visual Evidence for the Widespread Popularity of Therapeutic Discoveries in America after 1885," *Bulletin of the History of Medicine* 73 (1999): 629–78; see also Dan W. Blumhagen, "The Doctor's White Coat: The Image of the Physician in Modern America," *Annals of Internal Medicine* 90 (1979): 111–16.

78. Quoted in Frederic S. Lee, "The Relation of the Medical Sciences to Clinical Medicine," *Journal of the American Medical Association* 63 (1914): 2087.

79. Horst Oertel, "Science and Medicine," *Canadian Medical Association Journal* 6 (1916): 207.

80. "The Scientific Physician," *Boston Medical and Surgical Journal* 142 (1900): 701–2.

81. Kenneth L. Ludmerer, *Time to Heal: American Medical Education from the Turn of the Century to the Era of Managed Care* (New York: Oxford University Press, 1999), 337.

CHAPTER 9

Science and Technology

Ronald R. Kline

At the opening of Josiah Mason's Science College in Birmingham, England, in 1880, Thomas Huxley explained why he strongly favored education in science proper rather than its applications. "I often wish this phrase, 'applied science,' had never been invented," Huxley declared, "for it suggests that there is a sort of scientific knowledge of direct practical use, which can be studied apart from another sort of scientific knowledge, which is of no practical utility, and which is termed 'pure science.' But there is no more complete fallacy than this. What people call applied science is nothing but the application of pure science to particular classes of problems."[1]

Huxley's lament marks one position in a long-running discourse among scientists and engineers in Britain and the United States about the proper relationship between what was generally called "science and the useful arts" in the nineteenth century, "pure and applied science" from the 1880s to World War II, and "science and technology" from the 1930s to the present. The existence of multiple phrases in this discourse does not indicate a confusion in terminology nor a search for the correct terms to name timeless and spaceless referents. Instead, the multiplicity of key phrases, each of which conveyed a variety of meanings, indicates the historical complexity of debates about epistemology, the authority of science, and the boundaries marking the fields of science and engineering in the nineteenth and twentieth centuries.[2]

In calling certain words "keywords," historian Raymond Williams emphasizes their multiple, contested meanings. In regard to the term "culture," he says, "these variations, of whatever kind, necessarily involve alternative views of the activities, relationships and processes which this complex word indicates. The complexity, that is to say, is not finally in the word but in the problems which its variations of use signify." The history of such keywords as "culture" and "society" reveal that "earlier and the later senses [of the word] coexist, or become actual alternatives in which problems of contemporary belief and affiliation are contested." In studying British culture and society after World War II, Williams discovered that the changing meanings of the keywords "culture" and "society" were inextricably bound up with the very changes he wished to describe.[3]

English-speaking scientists and engineers were just as creative and combative in employing keywords in their debates. In contesting the relationships between their respective fields, they employed a variety of terms: "science," the "useful arts," "technology," "abstract science," "pure science," "practical science," "applied science," "fundamental science," "basic science," and "engineering science."[4] Rather than discuss the usages of all of these words, this chapter focuses on periods of contestation centered around three key phrases: "pure and applied science," "science and technology," and "engineering science." I draw mainly on public speeches made by physical scientists and engineers in the United States from the 1880s to the 1960s. My aim is not to answer the question, what was or is the relationship between what we now call "science and technology," but rather to explore how and why these discourse communities used the key phrases "pure and applied science," "science and technology," and "engineering science" to mark off the boundaries between their disciplines in this period.[5] I will also briefly discuss how changes in meaning of the keyword "technology" figured into those debates.

PURE AND APPLIED SCIENCE: COMPLEMENTARY BOUNDARY WORK

The phrase "pure and applied science," rather than "science and technology," dominated discussions among scientists and engineers about the relationship between systematic knowledge and the useful arts in the United States from the early 1880s to the 1930s.[6] The term "technology" was not part of these debates in the nineteenth century because at that time it chiefly meant the "scientific study of the practical or industrial arts," a meaning popularized by the names of such prominent technical colleges as the Massachusetts Institute of Technology, founded in 1865. This meaning had been used earlier in the century by Boston botanist and physician Jacob Bigelow's *Elements of Technology* (1829), which derived from a course of lectures Bigelow had given on the "Application of the Sciences to the Useful Arts," the topic of his Rumford Chair at Harvard. For the book, Bigelow had "adopted the general name, Technology, a word sufficiently expressive, which is found in some of the older dictionaries, and is beginning to be revived in the literature of practical men at the present day":

> Under this title it is attempted to include such an account as the limits of the volume permit, of the principles, processes, and nomenclatures of the more conspicuous arts, particularly those which involve applications of science, and which may be considered useful, by promoting the benefit of society together with the emolument of those who pursue them.[7]

Although Bigelow spoke of "technology" as applied in this passage, his usage of the term, comported with the contemporary English-language meaning, as the scientific study of the useful arts, which dates to the early seventeenth century. The earliest meaning of "technology," as the "systematic treatment of the arts" (primarily grammar and rhetoric), which dates to Aristotle's introduction of the term *technologia* in Greek philosophy, had become obsolete in English by the late seventeenth century.[8]

During the course of the nineteenth century in the United States, "technology," which initially referred to the study of all the practical arts, increasingly took on a masculine gender identity and an abstraction that lifted it above the messy complexities of craft culture. The word came to denote the abstract science of the useful arts built mainly by men, such as bridges, factories, and railroads, rather than those arts attributed to the realm of women, such as cooking, sewing, and cleaning.[9] A separate but related meaning, the practical arts collectively, began to appear at the turn of the twentieth century, primarily in the writings of the iconoclastic economist Thorstein Veblen. Influenced by a broad German discourse on *Technik* (the "rules, procedures, and skills" related to material culture), Veblen combined this meaning with the older one of knowledge to speak of "technology" as the "state of the industrial arts." Although social scientists in the United States adopted Veblen's usage in the early twentieth century, it was not common in American popular culture, nor among scientists and engineers, until the Great Depression of the 1930s.[10] Thus, to understand the debates from 1880 to 1930 about the relationship between what was later called "science and technology," we need to come to grips with the various meanings of the ubiquitous phrase "pure and applied science" in this period.

Although William Rogers used the term "applied science" in 1846 when he drew up the plans for the institution that became MIT, American scientists did not employ it regularly until the 1870s, when they frequently yoked it to "pure science," generally to mean purity of motive rather than purity of subject matter. At that time, prominent scientists transformed the prescription that science should be applied to the useful arts into a timeless, universal statement that all useful arts—past and present—were based on the application of science. By 1880, this interpretation had become part of the rhetorical arsenal of presidents and vice presidents of the American Association for the Advancement of Science (AAAS), who employed the terms "pure science" and "applied science" to define and promote their profession.[11]

In three areas of discourse—engineering, industrial research, and science policy—scientists and engineers expressed a wide spectrum of opinions about the relationship between science and innovation between 1880 and 1945. These extended from the idea that science is a necessary and sufficient source of knowledge for technical innovation on one end of the spectrum, to a complete independence of innovation from science on the other end.

The first area of discourse concerns the response of leading engineers to a pure-science ideal promoted by physicist Henry Rowland of The Johns Hopkins University. In a famous speech delivered before the AAAS in 1883, "A Plea for Pure Science," Rowland criticized newspapers for calling electrical innovation by the name of physics, was "tired of seeing our professors degrading their chairs by the pursuit of applied science instead of pure science," and upheld pure science as a pedagogical ideal that cultivated moral character. He objected to physics professors profiting from doing applied science instead of cultivating their proper field of pure science, which he held to be the source of all technical innovations.[12]

Rowland expanded on these themes the following year at the National Conference of Electricians, where he decided to show practical electricians how to apply science correctly by presenting a theory of the dynamo. Although electrical engineers later recognized the theory as fundamental to their field, most electricians at the conference, including Elihu Thomson, founder of an electrical firm, and Silvanus Thompson, a British professor of physics and electrical engineering, did not welcome the theory nor the condescending attitude of its author. Both men criticized Rowland for the better part of an afternoon, based on their considerable experience designing, building, operating, and writing about dynamos. At the start of the next day's session, Rowland replied that "every law of electricity necessary to be known is already known; it is only a question of the brain that has the power to evolve the perfect machine, and when we say that theory does not agree with practice, it means that we have not got brains enough to apply the theory to the facts and get at the result."[13]

This is a startlingly clear statement of the necessary and sufficient interpretation of the relationship between science and innovation, the furthermost point on one end of the spectrum. Machines could be derived in a complete and perfect form solely from Rowland's specialty, pure science, provided one had the brainpower to do it.

Among the contemporary critics of this ideology, engineer Robert Thurston created a sophisticated alternative based on the emerging ideals of his profession. A proponent of "school culture" over traditional "shop culture" in mechanical engineering education, Thurston developed an ideal of applied science composed of four broad usages of the term.

Thurston's speeches contain many instances of the meaning closest to Rowland's position—the application of scientific theories to the useful arts. He told the Virginia Mechanics Institute in 1894 that the unprecedented material progress in the nineteenth century "has come of the application, by the inventor and the mechanic, of the facts and laws of physical science to the useful purposes of their lives."[14] Yet in this speech and in many others, Thurston put equal emphasis on a second meaning of "applied science"—the application of Francis Bacon's inductive method to the useful arts. As a proponent of school culture, Thurston also used the term to signify a relatively autonomous body of knowledge and the practices of research, teaching, and innovation—the third and fourth meanings of the phrase. In this usage, "applied science" was a synonym for the knowledge gained from "scientific research in engineering,"[15] which he also called "engineering science" on at least two occasions.[16]

The four meanings of "applied science" formed complementary parts of Thurston's philosophy of science and innovation, an ideology best summarized by his claim that engineering was a union of science and the mechanic arts. In Thurston's ideal world, which was partially realized at his home institution, the Engineering College at Cornell University, engineering researchers applied scientific theories and methods to create a body of knowledge called "applied science," which engineers learned in college and practiced on the job. The four meanings meshed well with Thurston's campaign to raise the social status of engineering by arguing that it was a "learned profession" rather than a trade or a craft.[17] This wide spectrum of meanings is also evident in the writings of Charles Steinmetz, a prominent electrical engineer, and in speeches by three vice presidents of the engineering section of the AAAS in the late nineteenth century.

Many engineering leaders simplified Thurston's applied-science ideal after the turn of the century. Calling the engineer an "applied scientist" during a time of a growing prestige of science and a greater use of science in industry provided the basis for an ideology that claimed the engineer to be the agent of all technical change, a logical thinker free of bias, and a socially responsible professional. Presidents, as well as several vice presidents of the AAAS's engineering section, ignored the autonomous usage of "applied science" and moved its meaning toward the subservient end of the spectrum favored by Rowland. In this vein, Gano Dunn, president of the American Institute of Electrical Engineers in 1912, said that "engineering is not science, for in science there is no place for the conception of utility . . . Engineering is Science's handmaid following after her in honor and affection, but doing the practical chores of life."[18]

Despite the criticism of engineering researchers and their creation of broader meanings of the term "applied science," the dominant ideology among engineering leaders in the early twentieth century fell toward Rowland's end of the spectrum. These engineers, few of whom were researchers, reduced Thurston's complex ideal to the formulation that engineering was (simply) the application of science to the useful arts. The applied-science ideology was appealing as science began to amass a greater cultural authority in the Progressive Era, a phenomenon evident in the appeals to science by such new specialties as chemical engineering and highway engineering. By World War I, status-conscious engineering leaders had spent two decades promoting a simple applied-science interpretation of innovation that would probably not have displeased Rowland.

Our second area of discourse about pure and applied science centers on the new field of industrial research. Although some engineers had carved out an autonomy based (mostly) on an epistemologically dependent interpretation of the term "applied science," the growing number of industrial researchers in the early twentieth century blurred the boundaries between pure and applied science in a way that modified Rowland's pure-science ideal. The change in rhetoric is evident in debates about how to promote science and industrial research during World War I, when it became a matter of national defense.

The wartime discourse pitted those who preached the new gospel of industrial research against adherents of the older gospel of high culture and pure science. A vocal proponent of the latter ideal was astrophysicist George Ellery Hale. During his campaigns to revitalize the National Academy of Sciences and mobilize science during the war, Hale used the traditional arguments and historical figures of the pure-science ideal. In his efforts to bring industrial researchers into the academy to aid the war effort, he described a hierarchical, sharp, and rather condescending division of labor between pure and applied science when he appealed for help in this matter to his former MIT classmate, Willis Whitney, in 1915.

Fifteen years as head of industrial research at the General Electric Company had prepared Whitney to disagree with his friend. In a speech before the American Chemical Society in 1916, Whitney deplored the tendency of "some American scientists" to believe "that making a utility of the God-given discoveries of the truly beautiful phenomena of Nature was a prostitution to be deprecated, and that research could only be pure when it was sterile." In another speech given that year, he said "there are no sharp lines to be drawn through research to separate pure from applied, scientific from practical, useful from useless."[19] Charles Skinner, head of research at the Westinghouse Electric Company agreed with Whitney, while engineer John Carty, head of research at American Telephone & Telegraph (now a subsidiary of the modern-day AT&T), drew a distinction based on motive, rather than on method, in 1916. Several lab directors agreed with physicist Frank Jewett, Carty's replacement at AT&T, who said in 1917 that industrial labs should occasionally perform "pure scientific research."[20]

Both of these aspects of the gospel of industrial research—the blurring of boundaries between pure and applied science inside the lab and the recognition that these labs could do some pure research—weakened Hale's pure-science ideal (in much the same way that Thurston's applied-science ideal challenged Rowland). Yet many tenets of the pure-science ideal were not replaced by the gospel of industrial research.

Carty, for example, said in 1916 that the "natural home of pure science and of pure scientific research is to be found in the university," mainly because industrial labs would be compelled to keep its "pure science" secret.[21] Skinner initially agreed that a new lab at Westinghouse should perform fundamental research, but disagreement between him and the lab's director, physicist Perley Nutting, on its overall mission apparently led to Nutting's departure after the war.[22] Whitney also thought the universities were the natural home of pure science, yet he did not mention that, in his own lab, Irving Langmuir had performed fundamental research that had led to major improvements in electric lamps. After joining the National Research Council (NRC), Whitney adopted a pure-science rhetoric similar to Hale's and dropped his earlier statements about the closeness of pure and applied science in the laboratory.

During this period engineering researchers in industry and academia also took advantage of the increased prestige of research to argue for more recognition for their field. One way was to classify it as science. In 1916 Charles Steinmetz argued that F. W. Peek Jr.'s experiments on high-voltage lines, conducted in Steinmetz's engineering laboratory at GE, were entitled to be called "scientific." Steinmetz was infuriated that chemists recognized Langmuir's work on the lightbulb as "scientific" while physicists did not regard Peek's that way. Steinmetz also railed against scientific publications for not abstracting engineering research, particularly his earlier papers on magnetic hysteresis:

> Amongst the worst offenders in this unjustified exclusiveness are the physicists, while the chemists make a recommendable exception . . . Possibly the reason is, because applied chemistry is chemistry just as well as [is] theoretical chemistry, while applied physics goes under the name of engineering, and the average theoretical physicist is rather inclined not to recognize engineering as [being] scientific.

Although Steinmetz defined engineering as applied science in two other places in this speech, the above passages indicate that he used the term to mean an area of research and practice, rather than simply the application of scientific theories.[23]

Another way to raise the status of engineering research was to follow Thurston's example and work to have it recognized as an autonomous academic discipline, a path also taken by many European engineers at this time. One advocate of that ideal was Vladimir Karapetoff, an electrical engineering professor at Cornell. During a discussion at the American Institute of Electrical Engineers (AIEE) in 1917, he asked, "Are there only two kinds of research, industrial and pure scientific? Is there not an intermediate kind, which out of deference to my colleagues I will not call impure." This "semi-industrial research" was the "proper field" of the electrical engineering faculty. "We are not physicists nor mathematicians, so that pure research is closed to us . . . we teachers are forced into the middle path of work on the theory of electrical machines and other devices, a subject that is neither pure physics nor industrial research."[24]

This field, however, was often not given its due by key scientists. When physicist Robert Millikan, Hale's partner in mobilizing wartime science, gave a talk to the AIEE in 1918, he was at first deferential to engineering research. His experience coordinating the research of engineers and physicists on submarine detection during the war compelled him to say that the "distinctions between the man whom you commonly call the pure scientist and the man whom you call the applied scientist have absolutely disappeared." Yet when he got to the main purpose of his talk and asked engineers to provide more support for the NRC, he drew a sharp boundary between pure and applied science and made his pitch on the basis of a simple applied-science view of innovation. E. P. Hyde, a physicist-engineer in the audience remarked that the pure-science ideal he had learned at Johns Hopkins created "a veritable hell" for him during his early years at the National Bureau of Standards "because I was called upon to do something that was of some use to somebody."[25]

While the pure-science ideal of Hale and Millikan resembled the ideology of Rowland, the ideal was modified as the educational goals of the nineteenth century gave way to a growing utilitarianism associated with the commercial and political value of industrial research. In this discourse, "applied science" often referred to original research conducted in industrial labs by university-trained scientists and engineers using "scientific" methods. Motive was seen as a way to separate pure from applied in many of these labs, and universities did not have a monopoly on pure science. Yet the efforts to blur the boundaries between pure and applied and the growing tendency to refer to applied science as semi-autonomous research did not alter the pure-applied hierarchy, which was strengthened by the success of chemists and physicists in applying their knowledge to the war effort. By the end of the war, industrial scientists and engineers had (re) created a central aspect of Thurston's applied-science ideal in a new setting, which, ironically, reinforced the applied-science view of innovation. The "gospel of industrial research" had transformed, instead of replaced, the "gospel of high culture and pure science."

In our third area of discourse—science policy in the interwar years—academic scientists, industrial researchers, and engineers expressed a wider spectrum of beliefs about the relationship between what was increasingly called "science and technology" (see p. 236 below). The spectrum extended from a pure-science ideal that had been (slightly) transformed during World War I to a nascent ideology of autonomous engineering research. In these years, academic engineering research remained weak in comparison to academic science, status-conscious engineers still felt the need to identify their profession with science, and Hale's NRC exerted a wide influence in all of these fields.

Established on a permanent basis in 1918, the NRC was the major source of pure-science rhetoric during the interwar years. Its inner circle promoted an extreme version of this ideology mainly to lobby for financial support for pure science. In two addresses explaining the purpose of the reorganized NRC in 1919, Hale quoted Carty's wartime admonition for engineers to forward the work of "pure scientists" since they were the "advance guard of civilization."[26] Millikan put the matter squarely to a group of industrialists in 1929 when he claimed that *"Pure science begat modern industry."*[27] About as far from that position as the group strayed was when Gano Dunn defined engineering in 1930 as the "art of the economic application of science to the purposes of man."[28] By calling the engineer's creativity "art," the traditional definition of engineering, Dunn allowed his profession some autonomy within a hierarchical relationship to science.

Comments by the NRC inner circle's main contact with industry, Frank Jewett, show the continued influence of the gospel of industrial research on this pure-science ideal. Jewett recalled that his dissertation advisor, physicist Albert Michelson, complained that "I was prostituting my training and my ideals" by joining AT&T.[29] He told the Mellon Institute in 1937:

> Once the trail is blazed [by "fundamental science research"] there follow in succession the eras of development, first by other inventors, then by engineers who know more of science than the inventors but who rarely create essentially new knowledge, and finally the era in which development is mainly in the hands of research men and engineers working in intimate cooperation.[30]

Leaders of industrial research at GE advocated the NRC's pure-science ideal even more strongly than Jewett. It is not surprising that Irving Langmuir, the first industrial scientist to win a Nobel Prize, called his research on the lightbulb "purely scientific" four years before receiving the award. In 1937 Langmuir drew a sharp boundary between "fundamental research" and "engineering research" on the basis of a pure-applied hierarchy. Like Whitney, Langmuir recognized original research in the "applied" area of the GE lab but not in the company's engineering labs.[31] Industrial researchers at the Radio Corporation of America, the Mellon Institute, and AT&T took a more moderate position by acknowledging the difficult and fruitful work of "applied" science in their interwar speeches, but they all paid homage to research in "pure" or "fundamental" science, either in universities or in their own laboratories, as the basis of their success.

Engineering leaders also upheld the NRC's pure-science ideal, often to keep the engineer above the growing number of technicians in industry. Robert Ridgway, chief engineer of the board of transportation in New York City, told civil engineers in 1925 that "The pure scientist . . . discovers the fact and thus enables the engineer to apply it to a definite and useful purpose."[32] Vice presidents of the engineering section of the AAAS made similar statements. In 1928 Charles Richards, former dean of engineering at two Midwestern universities, repeated Thomas Huxley's argument from nearly fifty years before when he said that engineering "can not be classed as a fundamental science . . . It has often been classified as an applied science, although the term would seem to be a misnomer, for there is no other science than pure science."[33]

When some engineering leaders took a more moderate position, they often echoed Gano Dunn. Presidents of professional societies for electrical, mechanical, and civil engineering defined engineering as the "art" of applying science to useful ends. Civil engineer Harrison Eddy noted that "while there are wide differences in engineering activities, some lying within the definition of an art, and others within that of a science, there cannot be any doubt that Engineering is both an art and a science." But since Eddy's main purpose was to argue that engineering was a "learned profession," he followed this sentence by saying that "from both points of view Engineering depends in large measure upon the sciences."[34]

Most of the scientists and engineers who occupied the other end of the spectrum were committed to the practice and teaching of engineering research. Even the NRC's new inner circle in the 1930s included one such advocate, physicist Karl Compton. Compton's rhetoric changed after he became president of MIT in 1930. In speeches to engineering and scientific groups in the 1930s, Compton described MIT's successes in "engineering research," referred to "fundamental theories of engineering" as a necessary part of the MIT curriculum, and cited Dunn's definition of engineering as the "art" of applying science.[35]

Other advocates of engineering research in academia took rather weak rhetorical stances against the NRC's pure-science ideal, mainly by noting the similarities between scientific and engineering research. Chester Dawes at Harvard noted in 1929 that the "artificial dividing line between electrical engineering and physics is rapidly disappearing." Researchers generated "fundamental" knowledge in engineering, as well as in the physical sciences.[36] Vannevar Bush told the MIT electrical engineering department in 1935 that the "conception that the scientist should discover and the engineer apply is not sufficient." MIT electrical engineers had taken up the "research spirit of the scientist."[37]

Proponents of engineering research had been making similar claims since the war. In 1925, C. Edward Magnusson, director of the University of Wisconsin's experiment station, referred to "engineering fundamentals" and "basic laws of engineering."[38] Andrey Potter at Purdue spoke in 1930 of advancing the "frontier of engineering knowledge" and claimed that the future of industry rested "upon new engineering knowledge."[39]

But most colleges did not emphasize the type of original research on materials and machines pioneered by Thurston and Steinmetz. One prominent engineering educator, William Wickenden, recognized this state as vice president of the engineering section of the AAAS in 1938. "Engineering research pays dividends and they are soon realized, but over the decades and the centuries the richest returns come from the gratuitous and disinterested researches in pure science, from knowledge sought for its own sake alone."[40] The weak position of academic research before the war thus mitigated against engineers developing a strong rhetorical stance to counter the pure-science ideal. They did not take the alternative rhetorical path of engineering science until the 1950s (see p. 240 below).

Several important changes in terminology occurred in the interwar period. Although "engineering science" had not become an everyday phrase, the terms "basic science" and "fundamental science" began to replace "pure science" in the rhetoric of many scientists and engineers. Frank Jewett helped popularize "fundamental science" in the 1930s. In 1945 he told Bush that he objected to the term "pure science" because the "word 'pure' implies that all other kinds of research are 'impure.'"[41] Those who substituted these phrases for "pure science" probably shared Jewett's concerns about the connotations of the word "pure," especially since applied science had gained respectability through the work of industrial research laboratories.

The new terminology reflected changing practices in science and engineering. Both "fundamental science" and "basic science" appear to have been coined by industrial and engineering researchers, who were growing in numbers and influence during this period. They undoubtedly favored these terms over "pure science" because the words, which could refer to technological as well as academic science, did not belittle their research with the implication of impurity.

Although scientists and engineers often expressed opposing views and modified them substantially during this sixty-year period, they generally subsumed their differences under the flexible key phrase "applied science" in the process of drawing boundaries and promoting complementary self interests. Scientists tended to emphasize the epistemological dependency of applied science on pure science to argue for financial support for their research; engineers called themselves "applied scientists" and their field an "applied science" to raise their occupational status above that of artisans to the level of a "learned profession."[42]

SCIENCE AND TECHNOLOGY: FROM FIELDS OF STUDY TO SOCIAL FORCES

Of our three key phrases, "science and technology" is the most common one in use today. It permeates the English language to such an extent that reversing the order of the words, to technology and science, sounds foreign to our ears. The standard order, science and technology, reflects the common belief that technology depends upon science, especially when the phrase is used as a substitute for "pure and applied science." But the history of the phrase is more complex and revealing that a simple substitution.[43] Scientists and engineers used it early in the twentieth century to denote two separate fields of study, science *and* technology. The present meanings of "science and technology"—to denote the knowledge, practices, and artifacts produced by scientists, engineers, and inventors—did not become widespread in the United States until the 1930s and 1940s. The shift accompanied the changes discussed above in the meaning of the word "technology" in the early twentieth century, which was popularized during the Great Depression. Many social scientists and humanists at the time contributed to widespread use of the term "technology" during the debates over technological unemployment. Historian Charles Beard employed it widely to mean an agent of social change in a manner championed by Veblen. He also blurred the distinction between science and technology.[44] Eventually, "technology" came to refer to the products—the artifacts and systems—produced by the practical arts.[45]

The history of the keyword "technology" is reflected in the use of the phrase "science and technology" and similar locutions by physical scientists and engineers in the first half of the twentieth century. In these uses we see both an autonomous and a dependent relationship expressed between the individual keywords of this phrase.

In the early part of the last century, "science and technology" often referred to a combined field of study. In 1903, R. S. Woodward claimed that "In the educational transformation that has come about in the last three decades, our schools of science and technology have played an important role."[46] When A. H. Chamberlin referred to these as separate fields in 1909, he used the rare phrase "technology and science" to denote special technical schools outside of the state university system.[47]

The phrase "science and technology" could also refer to separate fields of study. In regard to research, Raymond Bacon of the Mellon Institute commented in 1914 that the "basis of this marked development in organic chemical industries is the combined working of science and technology."[48] The pure-science ideal held by leaders of the NRC did not prevent them from speaking about "researches in science and technology" and creating research "Divisions of Science and Technology" at its founding in 1919. Divisions were devoted to the physical sciences, engineering, chemistry and chemical technology, geology and geography, medical sciences, biology and agriculture, and anthropology and psychology.[49]

One prominent commentator equated technology with applied science and called it a field of industrial research. Alfred Flinn said in 1921 that "industrial research in technology, or applied science, demands practical experience in the industry as a preparation for successful work."[50] But others referred to a more dependent relationship between the two fields. In 1933, the venerable inventor Elihu Thomson praised MIT "as a center for scientific advance in all that concerns the application of the principles of science to practical needs, in the various branches of applied science, called Technology."[51]

This dependent meaning persisted into World War II, alongside the new discourse centered around changes in the meaning of the word "technology." Although Frank Jewett, the head of Bell Labs and a leading member of the Office of Scientific Research and Development in World War II, used "science and technology" a dozen times in a 1944 speech, he subordinated technology to science. "I am proposing to use the word 'technology' in its very broadest aspect to include not only the things we normally think of as technology, which are mainly the applications of the physical sciences to utilitarian ends, but also the application of the biological sciences." "All that we call 'technology,'" Jewett continued, "is nothing but the application of fundamental science discoveries and the employment of scientific methods for useful or desirable purposes."[52]

On the other hand, the prominent experimental physicist, Ernest Lawrence, claimed an interdependent relationship between science and technology in scientific instrumentation. In a speech published in 1937 he drew the audience's "attention to the great partnership of modern times—science and technology." Thirty years earlier, Veblen had referred to the "copartnership" between "science and technology" and argued extensively that "machine technology" conditioned the craft thinking that led to modern science.[53]

The change in the meaning of "technology" from a field of study and research to the practices and state of the industrial arts, initiated by Veblen in the social sciences, was reflected in the phrase "science and technology" as used by physical scientists and engineers. Speaking about science in the federal government, George Burgess, director of the National Bureau of Standards, said in 1924 that Leonardo da Vinci was "great in poetry, letters, painting, sculpture, architecture, science and technology, military and civil engineering."[54] Princeton astronomer John Stewart claimed in 1937 that "The first world war and the following booms and depressions in business were only incidents in the continuing drama of onrushing science and technology," which included improvements in airplanes, automobiles, ships, photography, and radio.[55] Hugh Taylor, a physicist at

Princeton, talked in 1944 about changes that "occurred in the science and technology of physics with advances in communications and transport, based upon electronics and aerodynamics, reaching a tremendous pace of growth only within the last few years."[56]

Many physical scientists and engineers joined the prevalent discourse of the turbulent 1930s and the early 1940s—punctuated by concerns about technological unemployment, a proposed moratorium on science, and the role of science in World War II—to expand the meaning of "science and technology" from that of a system of knowledge and practices to include that of a powerful agent (or agents) of social change, the meanings of "technology" developed by Veblen earlier in the century.[57] Engineering professor Andrey Potter remarked in 1933, at the depth of the depression, that "whether civilized man likes it or not, he is destined to live and work in an environment affected to an increasing extent by science and technology."[58] In a 1937 paper aptly entitled "Science and Society," F. R. Moulton, secretary of the AAAS, remarked that the "impact of new scientific discoveries on processes of wealth production has at times created serious economic and social maladjustments; and the fear has frequently been expressed that science and technology may even come to be the master rather than the servant of mankind—if, indeed, they may not lead to the destruction of society."[59]

In the fall of 1940, a year into World War II, Frank Jewett and Robert King, a vice president of AT&T, addressed concerns about the social implications of engineering in a speech at the bicentennial celebrations of the University of Pennsylvania:

> At a time like the present, we all realize that something is on trial . . . It would be very interesting to discover just what it is that stands before the bar. It may be fundamental science, it may be applied science, it may be science and technology in general, it may be religion, it may be domestic politics or world politics, or it may be that old and primeval scapegoat, human nature . . . Here in brief is the setting for any contemporary discussion of science and technology and their repercussions on the social order.

Jewett and King acknowledged that "the major problems and troubles of the day have, in considerable measure, a technological and scientific heritage. Both our civilization and our civilization-destroying engines are mechanistic." They dismissed a planned economy and other forms of government oversight as corrective measures, favoring instead a public-private partnership on the order of the National Defense Research Committee, to address social issues.[60] Jewett and King regarded science and technology as social forces, but ones that could be controlled through research organizations.[61]

During the war the phrase "science and technology" became common in the public rhetoric of scientists and engineers in the United States. It embraced a wide spectrum of meanings: fields of study; an epistemologically dependent relationship between knowledge and the useful arts; practices, artifacts, and systems; and a social force (or forces).[62] It thus had a flexibility similar to that of the older term "applied science"; as in the past, the flexibility enabled scientists and engineers to stake out boundaries between their fields. But in this case, there was a general agreement that measures had to be taken to address the new social force (or forces) denoted by the phrase "science and technology."

ENGINEERING SCIENCE: A PARADOX OF THE COLD WAR

Although scientists and engineers submerged their different interpretations of the relationships between their fields under the general rubrics of "applied science" and "science and technology," these debates broke out anew and in an intense manner when the phrase "engineering science" came into national prominence. During the early years of the cold war, scientists and engineering educators in the United States vigorously debated whether the field called engineering science, which the newly formed National Science Foundation (NSF) was chartered to support, existed, and, if it did, how should engineering colleges and the NSF support it?[63]

How there could be research in an "applied science" confused physicist Alan Waterman, the NSF's first director. In 1952 Waterman stated that "the fact that the National Science Foundation is directed to support basic research in engineering, which is specifically labeled as a science along with the mathematical, physical, medical, and biological sciences, creates what appears at first glance to be a paradox" since engineers themselves had defined engineering as "the art of the economic application of science to social purposes."[64] Waterman, his staff, and a number of prominent engineering educators struggled to resolve this apparent paradox for over a decade.

These debates were part of the cold war reconfiguration of engineering education in the United States. Guided by the belief that scientists had outperformed engineers in wartime laboratories, leading educators took advantage of a vast amount of federal funding to reform engineering education by including more science in the curriculum and by emulating scientific methods in engineering research. The NSF's use of the term "engineering science" encouraged colleges to develop research and teaching programs in this field in order to garner federal funds and to improve their status by substituting the ideal of engineering science for the ideal of engineering as applied science in the ideology of American engineering.

These debates are evident in the testimony of engineering groups during Senate hearings in 1945 on the proposed establishment of what became the NSF. The Engineering College Research Association, representing academia, and the Engineers Joint Council, representing the professional societies, complained that the congressional bills had not recognized that engineering research in such fields as aerodynamics, thermodynamics, hydrodynamics, and electronics could be "basic research" and thus eligible for funding by the proposed agency. The joint council statement, read by hydraulics expert Boris Bakhmeteff, defined engineering science as "fundamental knowledge of the laws of nature which permit the mastery of the resources and powers of nature." He recommended that "engineering scientists" be appointed to the foundation's board to ensure that "fundamental research in engineering science" was funded.[65]

Signed into law in mid-1950, the NSF Act authorized the new agency to "initiate and support basic scientific research in the mathematical, physical, medical, biological, engineering, and other sciences."[66] Bakhmeteff's panel took credit for the inclusion of engineering research in the act, and the NSF established the Division of Mathematical, Physical, and Engineering Sciences in 1951.

As mentioned above, Bakhmeteff drew on a long-standing, though minor, tradition of using "engineering science" to mean a science distinct to engineering that often originated in physics. This usage—which encompassed such fields as applied mechanics, strength of materials, fluid dynamics, thermodynamics, and aerodynamics—was popular with many engineering researchers, such as Bakhmeteff, Stephen Timoshenko, and Theodore von Karman, who emigrated to the United States from Russia and Europe in the 1920s and 1930s, and began teaching engineering sciences in American colleges. Several of these men had studied or taught at the University of Göttingen, where Felix Klein encouraged the research of Ludwig Prandtl and others in fields the Germans had called *Ingenieurwissenschaft* since the 1850s. The equivalent English phrase, "engineering science," had been used in British engineering education to mean a branch of study since the early 1860s. By the start of World War II, talk of engineering science as a relatively autonomous discipline was common among an elite group of educators in America, both European and native born.

In a 1948 booklet on engineering education, James Kip Finch, a colleague of Bakhmeteff's at Columbia University, quoted Bakhmeteff's Senate testimony to argue for support of engineering science and to counter the common, though erroneous idea, that "engineering is merely applied science."[67] Finch continued his attack in a 1951 book on the history of engineering by observing that this idea had been "exaggerated by the war experiences of some American scientific workers" and that it had an "adverse effect in securing support for research in engineering science." Finch argued against this "propaganda of natural science," as he called it, by showing that the engineering profession greatly antedated modern science. Even in later periods, natural science did not provide all the knowledge needed by engineers. Engineers then took over research into the "engineering sciences" and developed them "in directions, scope, and understandings which would not claim the interest, time, and efforts of workers in pure science." Finch aptly called the work of his colleagues the "engineering science movement" of the early 1950s.[68] Similar statements were made in 1951 by Morrough P. O'Brien, dean of engineering at the University of California at Berkeley who had helped bring European research on fluid dynamics to the United States in the 1930s, and Andrey Potter, dean of engineering at Purdue and a member of Bakhmeteff's Committee on Engineering Sciences of the Engineers Joint Council.

Despite these efforts to define the field of engineering science, establish a pedigree for it, and show its importance to engineering practice, the early directors of the NSF had difficulty understanding the concept. Paul Klopsteg, the first director of the Division of Mathematical, Physical, and Engineering Sciences and a professor of applied science on leave from Northwestern University, perceived an "incongruity" between the definition of engineering as the art of applying science to useful ends and the NSF's statutory obligation to support the science of engineering. Klopsteg attempted to resolve the dilemma by asserting that "research directed toward methods and procedures of applying knowledge toward useful ends is research as truly as that which seeks knowledge of science in the first instance."[69] In the same year, Waterman approved Klopsteg's approach as a way to resolve his perceived "paradox" of engineering science and said it was the basis for the NSF's "program for supporting basic research in engineering."[70] Waterman and Klopsteg—both of whom were trained as physicists—attempted to uphold a traditional pure-science ideal by defining the goal of engineering research to be improved means of *applying* "pure science" to practical problems, instead of the creation of "basic" knowledge useful to engineering.

At first the American Society of Engineering Education (ASEE) took Bakhmeteff's and Finch's strong position on the autonomy of research and teaching in engineering science, particularly in an influential report on engineering education, commonly called the Grinter report, after the committee's chair L. E. Grinter, dean of the graduate school at the University of Florida. The 1953 preliminary report addressed both research and teaching in engineering science. "The leaders of the engineering profession 25 years hence must be engineers who are at no loss in interpreting or themselves contributing to the extension of the fields of engineering science."[71] Following this statement of an engineering science ideal, the report identified nine engineering sciences (statics; dynamics; strength of materials; fluid flow; thermodynamics; electrical circuits, fields, and electronics; heat transfer; engineering materials; and physical metallurgy) that should be taught in the curriculum between the basic sciences and engineering design. The interim and final versions of the report, however, eliminated this preamble and presented only a shorter list of "engineering sciences" that should be taught. Many engineering educators climbed on the engineering science bandwagon after the publication of the Grinter report, especially when it formed the basis of accreditation changes.

Even though the NSF changed the name of its engineering program to Engineering Sciences in fiscal year 1953, the definition of this term was still problematical with many people at the agency. Raymond Seeger, acting director of the Mathematical, Physical, and Engineering Sciences Division, wrote in 1955:

> [One could] emphasize the adjective engineering, in which case engineering science is essentially a body of scientific knowledge useful for solving practical problems. On the other hand, one may stress the noun science so that engineering science may be regarded as genuinely basic science involving phenomena relating to engineering problems. In this sense, of course, physics itself might well be designated an engineering science.

Seeger recalled electrical engineering Professor Ernst Weber's definition of "engineering science as those parts of chemistry, mathematics and physics that are no longer of primary interest to chemists, mathematicians, and physicists."[72] But Seeger disliked this definition because it neglected the fact that more and more physicists were showing interest in the topic (probably because of NSF funding).

Seeger's ambivalence reflects the fact that the NSF had just established a committee to define engineering science and basic research in engineering. Harold Work of New York University reported on the joint efforts of the NSF and the Engineering College Research Association in this regard in 1956. Work thought that part of the difficulty facing the committee was due to the common belief that "basic research" was performed without any clear objective in mind other than seeking knowledge. "If this is accepted," said Work, "then basic engineering research faces a dilemma" because it has a clear objective. Work suggested focusing on what is sought, rather than how it is sought, in order to distinguish between basic research in science and engineering. Indicative of these definitional problems is that when the NSF gave a grant to the ASEE to study research needs in engineering, Eric Walker, the director of the study, decided that in order to "avoid controversy over definitions, the research will be merely characterized as research which engineers do."[73]

Earl Stevenson, chairman of the board of Arthur D. Little Company, was more pragmatic about the meaning of words. "Fortunately the difficulty of pushing and pulling words into an adequate definition has not held up a vigorous program of engineering research grants. Whatever it is, it is being supported!"[74] The disciplinary affiliation of a researcher largely determined if a proposal was put into the category of basic science or engineering science.

Yet turmoil over the definitions of engineering science and basic research continued, especially when the launch of *Sputnik 1* in October 1957 focused the country's attention on improving the quality of its engineering. In 1958 Gene Nordby, a civil engineer and the fourth NSF program director for engineering in seven years, reasoned that the transitional period toward scientific engineering "has resulted in confusion and conflict among members of the profession." "What is basic research in engineering and what is the impetus for the growing emphasis on the 'engineering sciences'?" Nordby asked. "In the classical sense, again, even the name 'engineering science' is a contradiction to the old-line practitioner of an art, and from one important viewpoint the title of 'applied science' is much more logical. Again by tradition, however, pure scientists who are applying their research to more practical problems hold a rather strong claim to that name. The engineer whose work may also be in this area is neither accepted as an applied scientist nor, perhaps more important, does he wish to surrender his identity as an engineer." Nordby's resolution to these dilemmas was to define engineering science as an "area which has been essentially 'abandoned' or neglected by one of the areas of pure science," a common definition.[75] One attractive feature of this definition—probably for both scientists and engineers—was that it resolved Waterman's epistemological problem by placing engineering science in the realm of (old) basic science.

The view that engineering must progress toward a scientific approach was echoed by another ASEE committee. Formed in 1956 because of the "mounting discussion of the section [of the Grinter report] devoted to the engineering sciences," the committee was chaired by prominent electrical engineers—W. L. Everitt, B. R. Teare, and Ernst Weber. These leaders from a "science-based" branch of engineering stated that "engineering science has its roots in basic science, but carries knowledge further toward applicability" through research based on scientific methods. The autonomy implied in this definition was weakened when the committee said that engineering science needed to follow developments in the basic sciences, yet wait until enough maturity had been reached in them to "permit the translation into applications."[76] More importantly for the practical needs of educators, the committee said the Grinter report should not be interpreted to mean that students had to take six individual courses on engineering science.

Some closure seems to have occurred on the question of engineering science at the end of this period. From 1958 onward, engineering educators and, increasingly, historians of technology recognized engineering science as a fairly well-defined set of research areas, formerly in the domain of physics, which were required courses for the accreditation of engineering colleges. Universities established programs in Engineering Science, the Society of Engineering Science was founded in 1961, and a handbook of engineering sciences was published in 1967.

Yet this closure may have precluded the wider usage of "engineering science" to mean systematic knowledge used by engineers (analogous to the term medical science) by restricting the field of engineering science to the epistemologically safe one of fields abandoned by leading-edge physicists and chemists. That is, this closure may have been appealing to both scientists and engineers because it did not significantly challenge the pure-applied hierarchy of the (commonly accepted) relationship between science and engineering.

Some texts, like those by Nordby and Work, also have traces of an opposition to the engineering science ideal by scientists and engineers that does not seem to have been as well represented in the engineering journals as the ideal of engineering science. In 1959, Nordby, no longer an NSF program director, complained:

> Members of the physical sciences—well reinforced by the traditionalists in the engineering ranks—define engineering as an applied science and say it is not concerned with "basic" research. Of course if this is the case, the physical scientists have placed a "halo" around their heads and are merely protecting their own bailiwick, usually because of their lack of understanding of the present trends of the engineering sciences. Nevertheless the other faction of our opposition [i.e., traditional engineers] is of more vital concern to me be-

> cause it will be difficult to win our battle for progress in engineering science research with a house divided.[77]

By calling postwar engineers who held the applied science ideology of engineering "traditionalists," Nordby raises an important issue for this period: the loss of status by some engineering educators to engineering scientists during the early cold war when they fell behind in the race to make engineering scientific in a more autonomous, research-oriented manner than in the past. These changes in the ideology of engineering—from a simple applied-science model to that of engineering science—are closely related to changes in teaching and research associated with the transformation of American engineering education in the cold war.

CONCLUSION

From about 1880 to 1960, three key phrases, "pure and applied science," "science and technology," and "engineering science," were at the center of boundary work conducted by physical scientists and engineers regarding the relationship between systematic knowledge and the useful arts that helped to define their fields. The uses and meanings of these phrases changed with the momentous social changes that occurred during this eighty-year period: the incorporation of engineering inside large firms, the growth of industrial research laboratories, the mobilization of science and engineering for two world wars, the specter of technological unemployment during the Great Depression, and the demand for more science in engineering education to help fight the cold war.

Some patterns are evident during this period. Scientists and engineers subsumed their differences about the sources of technical knowledge under the flexible rubric of "applied science," then "science and technology," to promote their complementary professional identities and to gain status and financial support. The move was made possible in regard to "applied science" because of the wide spectrum in meanings given to the term: from a pure-science ideal of applying scientific theories to the useful arts to an engineering-research ideal of applying scientific methods to produce technical knowledge. The spectrum was even wider for "science and technology"—from fields of study to social forces—and thus the epistemological debates over this term were not as intense. The intense debates over the paradox of "engineering science" were settled during the cold war, not by developing flexible meanings for the term, but by reducing its meaning to that of old basic sciences no longer of interest to physics and chemistry. By agreeing on this meaning scientists and engineers could preserve the epistemological hierarchy between their fields, while carving out autonomous areas of research in a mutually beneficial type of boundary work.

The settlement reached by scientists and engineers over these terms from 1880 to 1960 has lasted, to a great extent, to the present. Although "basic science" has largely replaced the term "pure science," the hierarchical meanings of "basic and applied science" are not that much different than the older "pure and applied science."[78] Many engineering researchers, if not practicing engineers, still refer to their field as an "applied science," an autonomous area of research. The term "engineering science," however, died out as the movement of bringing science into engineering education ran into complaints from accrediting agencies in the 1970s and 1980s that engineers were too scientific and no longer knew how to design. Then the requirement to teach a common core of engineering, called the engineering sciences, was gradually dropped at American universities.

The phrase "science and technology" has taken on broader meanings outside of science and engineering to the extent that it is no longer a marker of debates about the relationship between the referents of its individual terms. Unlike our other two phrases, "science and technology" has become a cultural keyword on a par with those studied by Raymond Williams. What most speakers do not realize, however, is that "science and technology" still carries vestiges of the contestations over its key terms. Using the phrase thus unwittingly reinforces a naive belief that technology is simply a result of applying science to the useful arts.

NOTES

Earlier versions of this chapter appeared as "Construing 'Technology' as 'Applied Science': Public Rhetoric of Scientists and Engineers in the United States, 1880–1945," *Isis* 86 (1995): 194–221 (University of Chicago Press, 1995); and "The Paradox of 'Engineering Science': A Cold War Debate about Education in the U. S.," *IEEE Technology and Society Magazine,* Fall 2000, 19–25.

1. Thomas H. Huxley, "Science and Culture," in *Collected Essays,* 9 vols. (1898; New York: Greenwood Press, 1968), 3:155.

2. On boundary work, see Thomas F. Gieryn, "Boundaries of Science," in *Handbook of Science and Technology Studies,* ed. Sheila Jasanoff et al. (London: Sage, 1995), 393–443.

3. Raymond Williams, *Keywords: A Vocabulary of Culture and Society,* rev. ed. (London: Oxford University Press, 1983), 22, 92.

4. Kline, "Construing 'Technology' as 'Applied Science.'"

5. On this approach to studying the word "technology," see Ruth Oldenziel, "Signifying Semantics for a History of Technology," *Technology and Culture* 47 (2006): 477–85.

6. This section is based on Kline, "Construing 'Technology' as 'Applied Science.'"

7. *Oxford English Dictionary,* s.v. "Technology"; Jacob Bigelow, *Elements of Technology: Taken Chiefly from a Course of Lectures Delivered at Cambridge, on the Application of the Sciences to the Useful Arts* (Boston: Hilliard, Gray, Little, and Wilkins, 1829), iii–v; Howard P. Segal, *Technological Utopianism in American Culture* (Chicago: University of Chicago Press, 1985), 78–81; and Eric Schatzberg, "*Technik* Comes to America: Changing Meanings of *Technology* before 1930," *Technology and Culture* 47 (2006): 491. Schatzberg notes that Bigelow did not use the term "technology" in the main part of the book, that Bigelow dropped it from the title of a second edition in 1840, and that the use of the term did not increase from 1829 to the establishment of MIT.

8. Carl Mitcham, *Thinking through Technology: The Path between Engineering and Philosophy* (Chicago: University of Chicago Press, 1994), 116–31.

9. Leo Marx, "The Idea of 'Technology' and Postmodern Pessimism," in *Does Technology Drive History? The Dilemma of Technological Determinism,* ed. Merritt Roe Smith and Leo Marx (Cambridge, MA: MIT Press, 1994), 237–57; Leo Marx, "*Technology:* The Emergence of a Hazardous Concept," *Social Research* 64 (1997): 965–68; and Ruth Oldenziel, "From Elite Profession to Mass Occupation," in *Making Technology Masculine: Men, Women, and Modern Machines in America, 1870–1950* (Amsterdam: Amsterdam University Press, 1999), 51–90.

10. Schatzberg, "*Technik* Comes to America," 494, 503.

11. See, e.g., the addresses of Julius E. Hilgard (1876), Edward C. Pickering (1877), and Ira Remsen (1879) in *Proceedings of the American Association for the Advancement of Science,* 25 (1876): 1–16; 26 (1877): 63–72; 28 (1879): 213–28.

12. Henry Rowland, "A Plea for Pure Science," in *The Physical Papers of Henry Augustus Rowland* (Baltimore, MD: The Johns Hopkins University Press, 1902), 594–613.

13. National Conference of Electricians, *Report of the Electrical Conference at Philadelphia in September 1884* (Washington, DC: US Government Printing Office, 1886), 86–107, 111.

14. Robert H. Thurston, "The Mechanic Arts and the Modern Educations; An Address Delivered before the Virginia Mechanics Institute . . ." (pamphlet, Richmond, VA, 1894), 14.

15. Robert H. Thurston, "On an Engineering Experiment Station," *Sibley Journal of Engineering* 10 (1896): 280.

16. Robert H. Thurston, "Our Progress in Mechanical Engineering," *Transactions of the American Society of Mechanical Engineers* 2 (1881): 431; and Thurston, "Needs and Opportunities of a Great Technical College," *Scientific American* 36 (Suppl.) (1893): 14587.

17. Robert H. Thurston, "Engineering as a Learned Profession," *Scientific American* 38 (Suppl.) (1894): 15467.

18. Gano Dunn, "The Relation of Electrical Engineering to Other Professions," *Transactions of the American Institute of Electrical Engineers* 31 (1912): 1031.

19. W. R. Whitney, "Incidents of Applied Research," *Journal of Industrial and Engi-*

neering Chemistry 8 (1916): 561; and Whitney, "Research as a National Duty," *Science* 43 (1916): 636.

20. Frank B. Jewett, "Industrial Research with Some Notes Concerning Its Scope in the Bell Telephone System," *Transactions of the American Institute of Electrical Engineers* 36 (1917): 845.

21. John J. Carty, "The Relation of Pure Science to Industrial Research," *Science* 44 (1916): 515.

22. Ronald Kline and Thomas Lassman, "Competing Research Traditions in American Industry: Uncertain Alliances between Engineering and Science at Westinghouse Electric, 1886–1935," *Enterprise and Society* 6 (2005): 601–45.

23. Charles P. Steinmetz, "Scientific Research in Its Relation to the Industries," *Journal of the Franklin Institute* 182 (1916): 711, 716, 717.

24. Vladimir Karapetoff, "Discussion," *Transactions of the American Institute of Electrical Engineers* 36 (1917): 890.

25. Robert A. Millikan, "Research in America after the War," *Transactions of the American Institute of Electrical Engineers* 37 (1918): 1725; and E. P. Hyde, "Discussion," *Transactions of the American Institute of Electrical Engineers* 37 (1918): 739.

26. George E. Hale, "The Responsibilities of the Scientist," *Science* 50 (1919): 144; and Hale, "The National Importance of Scientific and Industrial Research," *Bulletin of the National Research Council* 1 (1919): 6.

27. Robert A. Millikan, "The Relation of Science to Industry," *Science* 69 (1929): 28 (original emphasis).

28. Gano Dunn, "The Relationship between Science and Engineering," *Science* 71 (1930): 277.

29. Frank B. Jewett, "Problems of the Engineer," *Science* 75 (1932): 256.

30. Frank B. Jewett, "Communication Engineering," *Science* 85 (1937): 591.

31. Irving Langmuir, "Atomic Hydrogen as an Aid to Industrial Research," *The Collected Works of Irving Langmuir,* 12 vols. (New York: Pergamon Press, 1960–1962), 12:251–263; and Langmuir, "Chemical Research," in *The Collected Works of Irving Langmuir,* 12:309–34.

32. Robert Ridgway, "The Modern City and the Engineer's Relation to It," *Transactions of the American Society of Civil Engineers* 88 (1925): 1247.

33. Charles R. Richards, "The Functions of Section M—Engineering," *Science* 67 (1928): 6.

34. Harrison P. Eddy, "Trends in Engineering as a Profession in the United States of America," *Transactions of the American Society of Civil Engineers* 99 (1934): 1383.

35. Karl T. Compton, "Investment for Public Welfare," *Science* 83 (1936): 509; Compton, "Engineering in an American Program for Social Progress," *Science* 85 (1937): 277; and Compton, "Elihu Thomson the Scientist," *Science* 89 (1939): 188.

36. Chester L. Dawes, "Some Aspects of the Electrical Engineering Teaching Problem," *Journal of Engineering Education* 19 (1929): 640.

37. Vannevar Bush, "Electrical Engineering at the Massachusetts Institute of Technology," June 3, 1935, MIT Archives, collection MC 5, Box 3, Folder 244.

38. C. Edward Magnusson, "Engineering Research—An Essential Factor in Engineering Education," *Journal of the American Institute of Electrical Engineers* 44 (1925): 1243–45, 1244.

39. Andrey A. Potter, "Engineering Education," *Mechanical Engineering* 52 (1930): 504.

40. William E. Wickenden, "The Social Sciences and Engineering Education," *Science* 87 (1938): 154.

41. Frank Jewett to Bush, June 5, 1945, quoted in Daniel Kevles, *The Physicists: The History of a Scientific Community in Modern America* (New York: Knopf, 1977), 45.

42. For a criticism that this argument delegitimizes the rhetoric of American engineers by focusing on social interests, and my defense of the argument, see Paul Forman, "The Primacy of Science in Modernity, of Technology in Postmodernity, and of Ideology in the History of Technology," *History and Technology* 23 (2007): 1–152; and Ronald Kline, "Forman's Lament," *History and Technology* 23 (2007): 160–66.

43. As is implied in Kline, "Construing 'Technology' as 'Applied Science,'" 218.

44. Schatzberg, "*Technik* Comes to America," 509–11.

45. For early artifactual uses, which seemed to have occurred first in the social sciences, see Charles Beard, introduction to *The Idea of Progress*, by J. B. Bury (1932; New York: Dover, 1955), xxii, which comments on a narrow definition that includes machines; and Read Bain, "Technology and State Government," *American Sociological Review* 2 (1937): 860–74.

46. R. S. Woodward, "Education and the Work of Today," *Science* 18 (1903): 164. For a similar usage by an engineer, see Henry P. Talbot, "The Engineering Graduate: His Strengths and Weaknesses," *Science* 33 (1911): 842.

47. A. H. Chamberlin, "The Function and Future of the Technical College," *Science* 29 (1909): 725.

48. Raymond Bacon, "The Value of Research to Industry," *Science* 40 (1914): 875.

49. Paul Brockett, "Minutes of the Meeting of the Executive Board [of the NRC]," *Proceedings of the National Academy of Sciences* 5, no. 3 (March 1919): 92; and "The National Research Council," *Science* 49 (1919): 459.

50. Alfred Flinn, "The Relation of the Technical Schools to Industrial Research," *Science* 54 (1921): 508.

51. Elihu Thomson, *Elihu Thomson Eightieth Birthday Celebration at the Massachusetts Institute of Technology* (Cambridge, MA: Technology Press, 1933), 80. Since "Technology" was also the nickname of MIT, this statement was probably a play on words, as well.

52. Frank B. Jewett, "The Promise of Technology," *Science* 99 (1944): 2, 3.

53. Ernest O. Lawrence, "Science and Technology," *Science* 86 (1937): 295; and Thorstein Veblen, "The Place of Science in Modern Civilization," *American Journal of Sociology* 11 (1906): 597–98. On Veblen's analysis of the relationship between science and technology, the first explicit one published in the United States, see Schatzberg, "*Technik* Comes to America," 503–4.

54. George K. Burgess, "The Scientific Work which Our Government is Carrying on and Its Influence upon the Nation," *Science* 19 (1924): 116.

55. John Q. Stewart, "An Astronomer Looks at the Modern Epoch," *Scientific Monthly* 44 (1937): 403.

56. Hugh S. Taylor, "The Organization, Direction and Support of Research in the Physical Sciences," *Science* 99 (1944): 250.

57. Schatzberg, "*Technik* Comes to America," 504–7. On the social issues involving

science and technology in the 1930s and 1940s, see Carroll Pursell, "Government and Technology in the Great Depression," *Technology and Culture* 20 (1979): 162–74; Amy Sue Bix, *Inventing Ourselves out of Jobs? America's Debate over Technological Unemployment, 1929–1981* (Baltimore, MD: The Johns Hopkins University Press, 2000); and Paul Boyer, *By the Bomb's Early Light: American Thought and Culture at the Dawn of the Atomic Age* (New York: Pantheon, 1985).

58. A. A. Potter, "Whither Engineering?" *Transactions of the American Society of Mechanical Engineers* 55 (1933): 5.

59. F. R. Moulton, "Science and Society," *Science* 86 (1937): 388.

60. Frank B. Jewett and Robert W. King, "Engineering Progress and the Social Order," *Science* 92 (1940): 365, 366.

61. Attributing agency to science and technology, then proposing solutions to control them, was common among such later social critics of science and technology as Lewis Mumford and Langdon Winner, and tends to reinforce the problematic idea of technological determinism. See Merritt Roe Smith, "Technological Determinism in American Culture," in *Does Technology Drive History?* 1–35.

62. See, e.g., Arthur Compton, "What Science Requires of the New World," *Science* 99 (1944): 23–28 (11 usages); Lyman Chalkley, "Science, Technology, and Public Policy," *Science* 102 (1945): 289–92 (9 usages); and "Summary of the Report to the President on a Program for Postwar Scientific Research by Vannevar Bush, Director of OSRD," *Science* 102 (1945): 79–81 (used as an adjective on p. 80).

63. This section is based on Kline, "The Paradox of 'Engineering Science'".

64. Alan T. Waterman, "The National Science Foundation in Relation to Basic Engineering Research," *American Society of Naval Engineers—Journal* 64 (1952): 641–42.

65. U. S. Congress, Senate Committee on Military Affairs, *Hearings on science legislation (S. 1297 and related bills)* . . . (Washington DC: U.S. Government Printing Office, 1945–1946), 706–16.

66. James L. Penick Jr. et al., "A Historical Overview," in *The Politics of American Science, 1939 to the Present*, rev. ed., ed. James L. Penick Jr. et al. (Cambridge, MA: MIT Press, 1972), 31.

67. James K. Finch, *Trends in Engineering Education* (New York: Columbia University Press, 1948), 68.

68. James K. Finch, *Engineering and Western Civilization* (New York: McGraw-Hill, 1951), 96, 315, 316

69. Paul E. Klopsteg, "Engineering Research in the Program of the National Science Foundation," *Journal of Engineering Education* 42 (1942): 264.

70. Waterman, "The National Science Foundation in Relation to Basic Engineering Research," 641–42.

71. L. E. Grinter et al., "Summary of Preliminary Report [of] Committee on Evaluation of Engineering Education," *Journal of Engineering Education* 44 (1953): 143–144.

72. R. J. Seeger, "Physics Is Not Engineering," *Journal of Engineering Education* 46 (1955): 127.

73. Harold K. Work, "Research Needs of Engineering and the Engineering Sciences and How the Needs Shall Be Met," *Journal of Engineering Education* 47 (1956): 101.

74. Earl P. Stevenson, "The Scientist-Engineer," *Journal of Engineering Education* 47 (1956): 150.

75. Gene M. Nordby and Robert N. Faiman, "Engineering Sciences—Still Another Look," *Journal of Engineering Education* 48 (1958): 527–29.

76. "Report on the Engineering Sciences, 1956–1958" special issue of *Journal of Engineering Education* 49 (1958): 37.

77. Gene M. Nordby, "The Role of the Federal Government in Supporting Basic Engineering Research," *Journal of Engineering Education* 49 (1959): 323.

78. As seen in the prevalence of the linear-model of innovation; see Benoit Godin, "The Linear Model of Innovation: The Historical Construction of an Analytical Framework," *Science, Technology, and Human Values* 31 (2006): 639–67. For a debate on the prevalence of the linear model, see David Edgerton, "The 'Linear Model' Did not Exist," in *The Science-Industry Nexus: History, Policy, Implications*, ed. Karl Grandin, Nina Wormbs, and Sven Widmalm (Sagamore Beach, MA: Science History Publications, 2004), 31–57; and David Hounshell, "Industrial Research: Commentary," in *The Science-Industry Nexus*, 59–65.

CHAPTER 10

Science and Religion

Jon H. Roberts

In recent years the effort to arrive at an understanding of the relationship between science and religion has generated enormous interest within Anglo American culture. Courses in the subject have proliferated and are now taught at hundreds of colleges and universities throughout the United States and Great Britain. The volume of inquiry and the need for scholarly outlets through which to disseminate the results of that inquiry have led to the publication of several journals and numerous newsletters dedicated to "science and religion." Interdisciplinary organizations and think tanks have arisen to foster innovative ways of relating science and religion, and organizations such as the Templeton Foundation and the Discovery Institute regularly provide these institutions, as well as numerous individuals, with sizable sums of money for their efforts. A seemingly endless stream of books and magazine articles dealing with everything from the religious implications of the big bang to Catholic perspectives on bioethics pours off the presses. Thinkers who have advanced their careers by pontificating on the relative merits of science and religion have attained near-celebrity status. There is even discussion in some circles about the desirability of creating a new discipline out of the study of science and religion. "Science and religion" has become hot.

THE PREHISTORY OF "SCIENCE AND RELIGION"

It was not always so. Indeed, prior to about the middle of the nineteenth century, the trope "science and religion" was virtually nonexistent. To be sure, there was a good deal of discussion of the appropriate relationship between nature and the Bible, the two "books" that God had given humans for their edification. But because virtually all of the participants in that discussion agreed that the Author of those two books would ensure that no real contradiction in their testimony could occur, they tended to limit their focus to the relative value and perspicuity of God's two modes of revelation. They did not grapple with the more global issue of the interaction of "science and religion."[1]

The pairing of science and religion did not become prominent until the definition of both terms attained recognizably modern form. "Religion" was the first to do so. Within the Christian tradition, with which this paper deals, elite members of the church hierarchy had long devoted attention to theology, but prior to the seventeenth century, most adherents of the faith had treated religion primarily as a life of piety and communal devotion and a set of ritual practices. But with the proliferation of beliefs that occurred in the wake of the Reformation and the discovery of a plethora of "heathen" beliefs during the course of Europeans' voyages into hitherto uncharted parts of the world, doctrinal claims assumed new importance, and by the end of the Enlightenment "beliefs" had become central to the way that people envisioned the nature of religion.[2]

The nature and meaning of "science" also changed in important ways. During the seventeenth and eighteenth centuries, students of the natural world called natural philosophers and natural historians sought to associate their efforts with theological inquiries and made God-talk central to their endeavors. Isaac Newton, for example, maintained that whatever truths might be gleaned from the Scriptures, "the proof of a Deity and what are his Properties belong to experimental philosophy." John Winthrop IV, an American who made something of a name for himself in the British colonies as a practitioner of natural philosophy, echoed Newton's claim, declaring in 1753 that "the consideration of a DEITY is not peculiar to *Divinity*, but belongs also to *natural Philosophy*." Natural historians, by calling attention to the multitude of ways in which living things attested to the existence of a rational and benevolent Deity, also gave their studies a distinctly theological cast. In the face of such commitments, titles juxtaposing "natural philosophy and religion" or "natural history and religion" were conspicuous by their absence.[3]

The boundary between the natural sciences and religion in Great Britain and North America, the geographical foci of this chapter, remained extremely porous well into the nineteenth century. This has prompted some historians to suggest that the production and dissemination of science and religion were emanating from a "common context." There are certainly some grounds for adopting that position. Many Christians believed that scientific inquiry disclosed "statements, specifications, facts, details, that will illustrate the wonderful perfections of the infinite Creator"; therefore, they thought it reasonable to regard "the calm investigation of science, stamped with the seal of Christian charity," as "the best of all swords and all shields." Moreover, some Christians valued the sciences for the light they could shed on the meaning of biblical passages relating to Creation, Noah's Flood, and other incidents in sacred history. Although efforts were sometimes made to show that the conclusions of individual sciences could be reconciled with revealed or natural theology, there was little sense that religion and science constituted fundamentally different enterprises. Rather, the view that prevailed in the mid-nineteenth century was that articulated by Samuel Harris, a clergyman who eventually became a professor of theology at Yale, when he declared in 1852 that "every science runs into theology; every science borders on theology, and the explorer cannot traverse it without presently crossing over into the theological domain."[4]

Notwithstanding the prevalence of that view, however, as early as 1750 an ever-growing number of investigators in the realms of natural philosophy and natural history began making determined efforts to pursue their inquiries untrammeled by concern with the testimony of the biblical narrative. They also began to substitute natural laws and agencies for the supernatural in accounting for the history, structure, and operation of nature. Hostility to religion motivated few of these investigators, and for much of the nineteenth century most continued to ascribe events to supernatural intervention when they found it impossible to explain them adequately in other ways. Nevertheless, the trend was clear: "men of science" increasingly came to assume that "it is the aim of science to narrow the domain of the supernatural, by bringing all phenomena within the scope of natural laws and secondary causes." Charles Darwin may not have sufficiently appreciated the options available to an omnipotent God when he asserted that the doctrine of special creation was no explanation at all. However, the thrust of his complaint highlighted the direction in which thinking about the nature of scientific explanation was moving. Although the rate at which the norms and practice of scientific investigation became naturalistic was slow and uneven, by 1875 many natural scientists clearly preferred to confess their ignorance rather than invoke the supernatural in discussing natural phenomena. What later came to be called "methodological naturalism" emerged as the reigning norm within the Anglo American scientific community.[5]

The elimination of God-talk from scientific discourse constitutes a defining feature of modern science. In view of the important role that theological categories had played within both natural philosophy and natural history, it is not coincidental that at the same time students of nature were attempting to detach themselves from such categories, they were also beginning to employ new terms in describing their vocation. The changes that occurred were neither abrupt nor universal; as late as 1830 John Herschel entitled his ambitious treatise on scientific methodology *A Preliminary Discourse on the Study of Natural Philosophy*. Still, by the time that Herschel published that treatise, the term "natural science" was rapidly replacing "natural philosophy," and by the end of the century the label "scientist," coined in the 1830s, was being widely used.[6]

THE EMERGENCE OF THE TROPE

During the late nineteenth century the term "science" underwent a significant constriction. Although some thinkers continued to employ the venerable definition of science as systematized knowledge, most began to restrict their use of the word to the empirical investigation of natural (and sometimes social) phenomena. In 1874 the conservative Princeton theologian Charles Hodge, who three years earlier had described theology as a science on the grounds that it took on the task of arranging biblical data systematically, complained that the term "science" was "becoming more and more restricted to the knowledge of a particular class of facts, and of their relations, namely the facts of nature or of the external world." Two years later the liberal Protestant clergyman James Thompson Bixby similarly observed that a "great revolution in thought" had occurred; whereas "science" had once referred to "systematized knowledge of any kind," it now referred simply to "*physical* knowledge." As a result, he opined, the use of terms such as "the science of religion" or "the science of God" now seemed "to involve a figurative extension of the word beyond its proper sphere."[7]

By the second half of the nineteenth century, a wide variety of Anglo American intellectuals had concluded that science and theology had become distinctly different enterprises. Recognition of this fact underlay the Baptist clergyman Samson Talbot's acknowledgement of 1872 that "mere physical science . . . begins and ends with nature." Even Samuel

Harris, who had contended in 1852 that study of the natural world was "interlinked at every point with the study of the Creator," had come to believe by 1883 that "empirical science no more takes cognizance of God than a mechanic investigating a watch takes cognizance of the man who made it."[8]

As more and more people conceived of religion and science as different conceptual entities, the phrase "science and religion"—or variants such as "science and the Bible" and "science and theology"—came into common use. Some books and articles assessing the relationship suggested that "angry conflict" characterized those relationships. One of the most well-known expressions of that view appeared in John William Draper's *History of the Conflict between Religion and Science* (1874). Draper, a physician and professor of chemistry at New York University whose scholarly interests had shifted to history in the latter stages of his career, was commissioned to write the book as a contribution to The International Scientific Series published by Appleton and Company. It proved to be enormously popular, selling more copies than any other title in the series.[9]

Although Draper reified both religion and science, he devoted little attention to defining either term. Possibly because he himself believed that God presided over a universe pervaded by natural law, he made no effort to claim that all varieties of religion had been in conflict with science. He praised the Islamic instigators of the "Southern Reformation" for promoting the development of science and described modern science and the Protestant Reformation as "twin-sisters." While concerned that some Protestants appeared bent on severing the family ties, he expressed the hope that such "misunderstandings" could be resolved and that a "cordial union" between Protestantism and science could be restored.[10]

Draper's organizing principle was "the conflict of two contending powers, the expansive force of the human intellect on one side, and the compression arising from traditionary faith and human interests on the other." In illuminating the nature of that conflict, Draper revealed that the primary object of his hostility was actually religious authoritarianism. Although he denounced such despotism when it took the form of biblicism just as vehemently as when it appeared within ecclesiastical structures, he regarded Roman Catholicism as the primary culprit—"partly because its adherents compose the majority of Christendom, partly because its demands are the most pretentious, and partly because it has commonly sought to enforce those demands by the civil power." Draper upbraided the Roman hierarchy for claiming "that blind faith is superior to reason; that mysteries are of more importance than facts." His distaste for Catholicism, born in his Protestant background and nurtured by the promulgation of the "Syllabus of Errors" (1864) and the doctrine of papal infallibility (1869), led Draper to throw down the gauntlet: "Roman Christianity and Science are recognized by their respective adherents as being absolutely incompatible; they cannot exist together; one must yield to the other; mankind must make its choice—it cannot have both."[11]

That the title of Draper's volume was only tangentially related to its actual content suggests just how evocative the trope "religion and science" had become by the 1870s. In 1876 James T. Bixby observed:

> The conflict now going on between the physical discoveries and theories of these latter days, and the forms of faith which have hitherto ruled the mind of Christendom, is one of the most noticeable phenomena of the intellectual movement of the times. The constant discussions from pulpit and platform, the numerous essays, pamphlets, and books, in which these two opponents are arrayed one against the other, and attack, defense, or effort at reconciliation made, allow no intelligent man or woman to remain unaware of the controversy.[12]

In the face of the growing tension between religion and science, Anglo American opinion leaders who had no desire to dispense entirely with either category developed a variety of strategies to harmonize the two realms of human experience. At the institutional level, a number of colleges and universities established professorships dedicated to that task. Efforts to establish appropriate lines of demarcation, however, extended well beyond these institutional efforts. In fact, the late nineteenth and early twentieth centuries witnessed the creation of what one commentator called "whole libraries" devoted to reconciling religion and science. That estimate is confirmed by the data contained in figures 10.1 and 10.2, which reveal that what started as a trickle of books and articles addressing "science and religion" before 1850 became a torrent in the 1870s.[13]

During the late nineteenth and early twentieth centuries, Christian thinkers sympathetic to the scientific enterprise commonly emphasized that "natural science is neither Christian nor anti-Christian, neither theistic nor atheistic, any more than the multiplication table." They recognized, however, that the work of scientists raised important interpretive issues, such as the relationship between science and biblical testimony. In dealing with that issue, most Christians who embraced the conclusions of modern science shared the views of the Scottish clergyman James S. Candlish, who suggested that "the general principle on which we must ultimately fall back in all cases is, that the Bible contains a revelation of religious truth and not of science at all, and in all its references to the

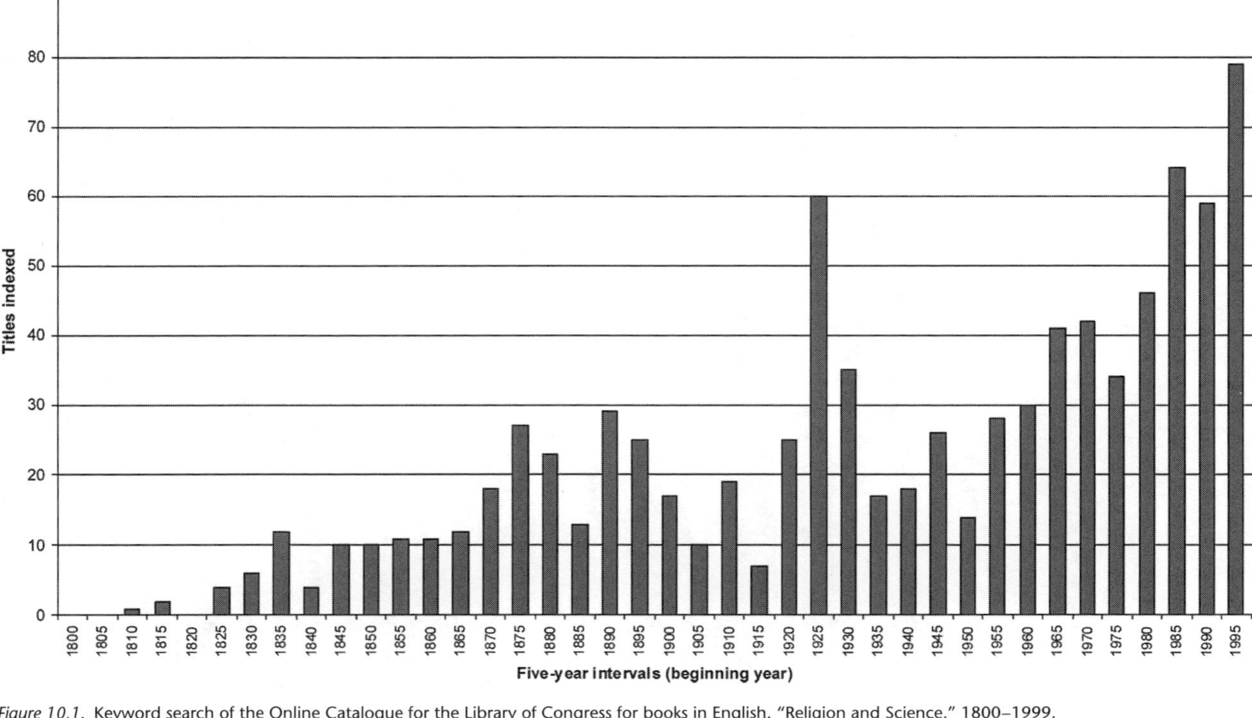

Figure 10.1. Keyword search of the Online Catalogue for the Library of Congress for books in English, "Religion and Science," 1800–1999.

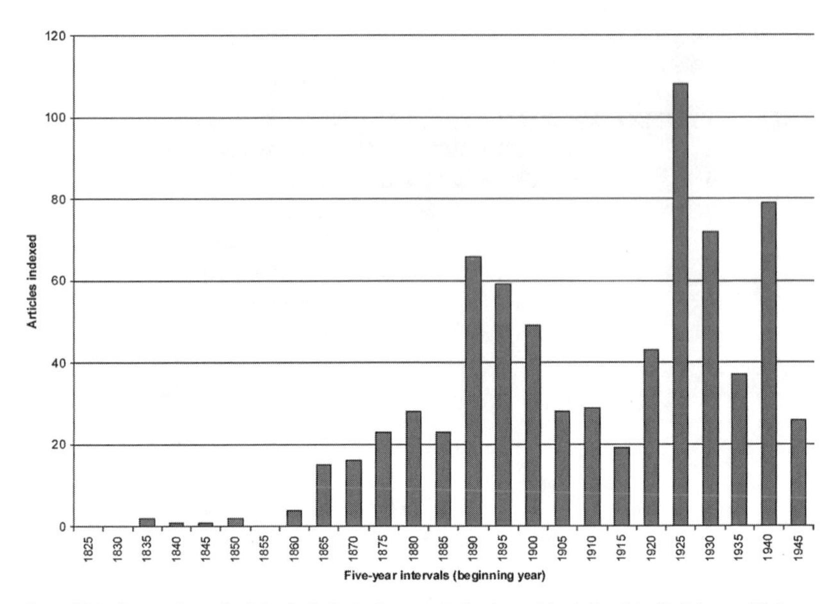

Figure 10.2. Keyword search of the Periodicals Contents Index for articles in English, "Religion and Science," 1825–1949.

physical world speaks according to the appearances of things and the current ideas of the times." Acceptance of a "canon of interpretation" that made science and biblical theology complementary rather than overlapping enterprises enabled thinkers to exempt themselves from a position that "makes Faith appear as a defendant, continually obliged to Science for permission to live."[14]

A few Christians seeking to reconcile science and theology continued to ascribe "otherwise inexplicable effects" in the realm of natural events to supernatural agency. Most, however, concluded that "the natural and supernatural are not two mutually exclusive spheres." Those Christians typically embraced a division of labor that envisioned science as an enterprise designed to answer "how" questions—in particular, to arrive at the "rules" that described the operation of the cosmos—and religion as a source of conceptual resources for answering "why" questions. According to this arrangement, the law-like behavior of natural phenomena disclosed by science simply revealed "the method by which God works." Proponents of this view, who included numerous clergy, theologians, and scientists, commonly held that all force was simply the manifestation of immanent divine "energy" and that as a result, "wherever in nature efficient power operates, that power is God." Hence, whereas science described phenomena, the task of actually explaining them—of uncovering the "Supreme Efficiency" that actually "lies at the bottom and works from behind the phenomena"—fell to theology. Science and religion thus became, as the Andover Seminary theologian George Harris asserted, "different modes of viewing the same facts."[15]

Some people resisted the idea that theology played an important role in understanding the natural world. A small but culturally influential group of "scientific naturalists," for example, sought to establish the cultural hegemony of science, promote the professionalization of scientific inquiry, enhance the intellectual and social prestige of scientists, and promote the secularization of society by extending scientific inquiry to "the whole of Nature." From this perspective, the Irish physicist John Tyndall, who trumpeted "the impregnable position of science," demanded that "all religious theories, schemes, and systems which embrace notions of cosmogony, or which otherwise reach into its domain, must, in so far as they do this, submit to the control of science." Some members of this camp differentiated between theology, a cognitive enterprise that typically obstructed the advance of knowledge, from the larger religious impulse, which nurtured the moral, aesthetic, and emotional realms of human life—"in those deeps of man's nature which lie around and below the intellect, but not in it." On occasion, scientific naturalists treated science itself as a form of religion. Thomas H. Huxley, for instance, entitled one of his works *Lay Sermons* and alluded to "the church scientific."[16]

SEPARATING THE SPHERES

Although few Anglo American intellectuals shared the scientific naturalists' conviction that "there is but one kind of knowledge and but one method of acquiring it," many did concur with their view that the seemingly "ceaseless conflicts" between science and religion were the result of "the imperfect separation of their spheres and functions." Thus, in the late nineteenth century there developed a sustained tendency among Anglo American thinkers to look to realms other than nature when discussing the foundations and impulses underlying religion. Although they continued to affirm that "nature is wonderfully suggestive of God," they acknowledged that "external arguments for the being of God" had become less compelling. In the face of this situation, many Christians in Great Britain and the United States—especially those who viewed themselves as liberals—began to draw more heavily than ever before on the work of German theologians and philosophers such as Friedrich Schleiermacher and

Immanuel Kant, who had located the wellsprings of faith in a sense of contingency, or dependence, or in the possession of a conscience and "moral intuitions"—in short, in the feelings associated with the human condition. Some even affirmed the existence of a special "religious faculty"—a "religious consciousness"—that enabled human beings to transcend the realm of the senses. For those Christians, "religious experience" constituted "the supreme evidence of the faith." In contrast to science, which sprang from intellectual curiosity about the interaction of natural phenomena, religion constituted "a mode of thought, of feeling, and of action determined by the consciousness of our relations to God."[17]

Reinforcing the conviction that theology addressed different aspects of human experience than scientific inquiry was the work of the German theologian Albrecht Ritschl. Ritschl, who made a sharp distinction between the realms of nature and spirit, held that while religion and the natural sciences were epistemologically on the same level, they appealed to different elements of human experience. Whereas the sciences of nature were "disinterested" cognitive enterprises centered on discovering facts relating to natural phenomena and formulating theories based on those facts, religion involved the use of the feelings and the will in addition to the intellect in making "independent value judgments." Such judgments, Ritschl believed, differed "in kind from those of theoretical science."[18]

At the turn of the twentieth century, Ritschl powerfully influenced Anglo American liberal Protestants. William Adams Brown, a professor of theology at Union Theological Seminary who embraced the Ritschlian perspective, contrasted science, which provided us with "better understanding of nature," with religion, which dealt with a realm of values—"the ideals of truth, justice, beauty, and goodness"—that were grounded in the personhood of God. Similarly, Gerald Birney Smith, a theologian at the University of Chicago, suggested that whereas science sought to interpret the natural world "in terms of exact cause and effect, so as to be able to control the processes of nature mechanically," religion dealt with "spiritual meanings," which he described as "those aspects of reality which justify man in his desire to establish relations of trust and love and moral confidence between himself and the world-process."[19]

This way of thinking, which had the effect of highlighting the salience of inner experience in the lives of Christians while minimizing the significance of doctrinal beliefs, received strong support at the turn of the twentieth century from the work of practitioners in the emerging field of the psychology of religion. Using questionnaires and autobiographies to "carry the well-established methods of science into the analysis and organisation of the facts of the religious consciousness," those behavioral scientists concluded that in the lives of most people, theology played a relatively minor role. In 1922, Edward L. Schaub of Northwestern University, summarizing the research of the previous quarter century, emphasized that psychologists of religion had "set into clearer light the fact that religion is not creed but life in its loftiest reach; and they have likewise shown that all man's conscious powers—cognitive, emotional, and conative—enter into its structure."[20]

By the end of the nineteenth century, a number of ideological currents had converged to convince many liberals that religion should be distinguished from doctrinal theology. This context served as the backdrop for Andrew Dickson White's evolving interpretation of the history of the relationship between science and religion. In 1869, White, who taught history at the University of Michigan before becoming Cornell's first president, presented a lecture at New York's Cooper Union titled "The Battlefields of Science." Arguing that "interference with Science in the supposed interest of religion" had led to "the direst evils both to Religion and Science," he contended that religion and science should "stand together as allies, not against each other as enemies." In his short book *The Warfare of Science* (1876), White continued to depict "religion" and "science" as the central parties in the "warfare," and he admonished religious leaders to see that the "readjustment of religion to science be made as quietly and speedily as possible."[21]

By 1896, the year that White's massive, lavishly documented two-volume *A History of the Warfare of Science with Theology in Christendom* appeared, he had subtly but significantly altered his views. He now refused to characterize the issue as a struggle between science and religion. Rather, he announced that he had only the highest regard for religion, which he envisioned in terms laid down by Matthew Arnold (the recognition of "a Power in the universe, not ourselves, which makes for righteousness") and, more broadly, as "the love of God and of our neighbor." From this perspective, he denounced the notion that "religion and science are enemies" as the "most mistaken of all mistaken ideas." The villain, he asserted, was not religion but "Dogmatic Theology." Only after science had cleared away the debris of "mediaeval" theological dogma could "the stream of 'religion pure and undefiled' . . . flow on broad and clear, a blessing to humanity."[22]

As a number of critics at the time and since have suggested, White exaggerated both the incidence and the severity of conflict between partisans of theology and practitioners of science. He also oversimplified the complexities of the interaction by implying that the "warfare" found scientists on one side and partisans of dogmatic theology on the other. Still, the distinction that he made between doctrinal theology and experiential religion exemplified a strategy that became increasingly common among liberal Christians, including many scientists, in the United States and Great Britain during the half century after 1890. That distinction enabled its proponents to detach the power and meaning of religion from its ability to provide an understanding of the natural world. It also allowed them to concede the realm of cognitive claims regarding natural phenomena to people who specialized in the sciences without feeling that in so doing, they were fatally weakening the position of religion. Noting that the primary meaning and value of religion lay in its "power to release faith and courage for living, to produce spiritual vitality and fruitfulness; and by that she ultimately stands or falls," Harry Emerson Fosdick emphasized that Christians possess a "standing-ground and a message" independent of the verdict of science. Although it was important "not to be unscientific," it was fatuous to assume that "the highest compliment that could be paid to Almighty God was that a few scientists believed in him."[23]

Liberals may have recognized that the realm of doctrine constituted the source of most of the difficulties Christians had in relating their worldview to scientific inquiry, but few of them wanted to dispense with theology altogether. They therefore redoubled their efforts to demarcate appropriate boundaries between that discipline and science. Proponents of the idea that "religious experience" rather than efforts to explain natural phenomena lay at the heart of the religious impulse reasoned that "theological knowledge is founded in experience as really as physical science is." From this perspective, some thinkers concluded that the primary bond linking religion and science lay in a common commitment to a rigorously empirical methodology, increasingly called the "scientific method." A forthright expression of this approach appeared in the work of liberal theologians who espoused what they called "empirical theology." In 1919 one of the most well-known proponents of that position, the Yale theologian Douglas Clyde Macintosh, published a work demanding that theology itself "become genuinely scientific." Macintosh maintained that just as practitioners of "the empirical sciences assume the existence, and the possibility of empirical knowledge, of the objects they undertake to investigate," so theologians were entitled to "posit the existence of God" and then use the data of religious experience to discern in more detail "just what attributes and relations can be ascribed to that religious Object." Shailer Mathews, a liberal who served as dean of the University of Chicago Divinity School, did not place himself within the camp of empirical theologians, but he did define the "modernist" version of evangelicalism that he embraced as "the use of the methods of modern science to find, state and use the permanent and central values of inherited orthodoxy in meeting the needs of a modern world."[24]

Although some liberals endorsed the notion that science and theology employed a common methodology, a larger and ever-growing number of Anglo American intellectuals during the early twentieth century held that those two realms of inquiry dealt with "separate spheres" of experiential data. In practice, they typically consigned the structure and operation of natural phenomena to scientific investigation while reserving the realm of values and ultimate causes for theological inquiry and speculation. This had the effect of placating the many participants in the discussion who believed that scientists should be given hegemony in dealing with natural phenomena while simultaneously conceding that subjective and metaphysical issues belonged to the sphere of religion.

MILITANT NAYSAYERS AND PACIFIST SEPARATISTS

Many conservative Christians, however, renounced the idea of placing science and religion within separate spheres. Defining "science" as it was found in the dictionary, they maintained that "true science" comprised certain knowledge based on inductively ascertained and verified facts and principles; it did not necessarily embrace the theories and practices espoused by the scientific community of their day.[25]

This conception of science, which nineteenth-century conservatives had sometimes associated with Baconianism, enabled them to draw several inferences that proved decisive in shaping their conception of the relationship between science and religion. First, it prompted a number of conservatives to include theology itself among the sciences. Charles Hodge, for example, insisted that *all* knowledge gleaned through induction, not just data gleaned through the senses, constituted "scientific" knowledge and should thus be taken into account in attempting to describe reality. Hodge maintained that since the Bible contained the theologian's "store-house of facts," and since the theologian employed the same inductive method of discerning the teachings of the Bible that the natural scientist used in disclosing the truths of nature, it was appropriate to include theology among the sciences. The equation of Christian doctrines with "facts" prompted the theologically conservative J. Gresham Machen to assert that "we ought to try to lead scientists and philosophers to become Christians not by asking them to regard science and philosophy as without bearing on religion, but on the contrary by asking them to become more scientific and more philosophical through attention to all, instead of to some, of the facts."[26]

If the "Baconian" conception of science embraced by conservatives was sufficiently broad to include theology within its purview, it proved to be sufficiently narrow to justify their imposing strict limitations on what passed muster as "science" in the investigation of nature. Those Christians commonly insisted that some theories put forward under the banner of modern science were unduly speculative. William Jennings Bryan assailed Darwinism on the grounds that it "is not science at all; it is guesses strung together." Conservatives such as Bryan employed their definition of science to reject the notion that all physical events could be described in naturalistic terms. Convinced that "science can not grasp all the processes of the universe," they concluded that it was necessary to invoke the supernatural activity of God to bridge "*many impassable gulfs*" in explaining the workings of nature.[27]

During the 1920s, militant conservatives calling themselves "fundamentalists" used their conception of science as a springboard for attempting to ban the teaching of human evolution in the public schools. Those efforts, which yielded legislation in five states and many local school districts, led to increased interest in the topic of religion and science. Some commentators unsympathetic to the conservative agenda invoked the now-familiar image of conflict in making sense of the cultural landscape. The journalist Frederick Lewis Allen, for example, described "the conflict between religion and science" as "one of the most momentous struggles of the age." For such people, convinced that religion contributed nothing of real worth for humanity, there seemed to be little reason to try to end hostilities. The British biologist C. H. Waddington, for example, maintained that science alone could "provide mankind with a way of life which is, firstly, self-consistent and harmonious, and, secondly, free for the exercise of that objective reason on which our material progress depends." Logical positivists insisted that the methods of science constituted the only valid way of verifying claims about experience. Theological statements might *seem* to be about experience, but they were unverifiable and hence meaningless.[28]

Although such denigration of religion did not reflect the attitude of the vast majority of Anglo Americans who addressed the relationship between science and religion during the first half of the twentieth century, commentators in that period increasingly leaned toward the notion that science and religion dealt with different, albeit complementary, areas of human experience. Catholic Neo-Thomists, for example, emphasized that whereas scientific problems focused on description, theology addressed ultimate metaphysical questions. Most proponents of the notion that science and religion occupied "separate spheres" advanced the claim that whereas the interaction of natural phenomena constituted the appropriate realm of science, the province of religion comprised the private, subjective domain of meaning and values. Adoption of that perspective led partisans of that view to suggest that any conflict that persisted between theology and science resulted from careless trespassing.[29]

Robert A. Millikan, a Nobel Laureate in physics, provided a particularly influential exposition of the separate-spheres doctrine. In 1923 he secured the signatures of more than thirty eminent scientists, clergymen, and theologians to a widely circulated "Joint Statement upon the Relations of Science and Religion." According to this statement, science and religion "meet distinct human needs" and thus "supplement rather than displace or oppose each other." Whereas the "purpose of science is to develop, without prejudice or preconception of any kind, a knowledge of the facts, the laws, and the processes of nature," the role of religion "is to develop the consciences, the ideals, and the aspirations of mankind."[30]

Proponents of "neo-orthodoxy" sought to sever completely the concerns of theology from the influence of the natural sciences. Emerging from liberal Protestantism during the late 1920s, neo-orthodoxy claimed the support of a sizable number of theologians and well-educated clergymen for the next half century. Convinced that religion centered on the redemptive activity of a "wholly other" God, the neo-orthodox joined the Swiss theologian Karl Barth in maintaining that God's revelation in the Bible, grounded in Jesus, provided the appropriate foundation of Christianity. E. G. Homrighausen, a Reformed clergyman who went on to obtain a professorship at Princeton Seminary, asserted that whereas natural science was "a blind leader" that "cannot redeem us from our hopeless and meaningless existence," the Christian Gospel "offers man that which can save him from frustration, meaninglessness, defeat, limitation, and sin."[31]

The conviction that religion and science consisted of separate enterprises continued to find support in the second half of the twentieth century. Theologians and clergy who found sustenance in the insights of existentialism, for example, maintained that in contrast to science, which addressed all phenomena, including human beings, as objects of analysis and maintained indifference toward the human condition, religion centered on the "existential encounter . . . of the inward self with God's Word in faith." On a different front, some partisans of analytic philosophy, an approach to linguistic analysis similar to that of logical positivism but devoid of its hostility to religious discourse, espoused the notion that theological and scientific approaches to human experience were quite different. For example, Donald D. Evans, an influential professor of philosophy at the University of Toronto, invoked linguistic analysis to defend the claim that "there are fundamental differences between religion and science." Evans held that whereas science consists of a body of impersonal and empirically testable assertions, religion is "self-involving," both expressing a commitment to a relationship with the divine and interpreting particularly meaningful experiences as manifestations of "divine revelation."[32]

THE RE-MERGING OF THE SPHERES

Although the view that science and theology should remain separate commanded the allegiance of a sizable number of opinion leaders in the twentieth century, many religious thinkers insisted that the natural world served as an appropriate object for theological reflection. In the period between the two world wars the Cambridge University philosophical theologian F. R. Tennant emphasized that scientific descriptions of natural phenomena could serve as useful resources for the construction of a theistic interpretation of reality. A few scientists also espoused that idea. Probably the most popular and influential of these was Sir James Jeans, a Fellow of the Royal Society who taught applied mathematics at Princeton and Cambridge before deciding to use his wealth to support a career as an independent scholar. In work published during the early thirties, Jeans, noting that the "new physics" of relativity theory and quantum mechanics differed significantly from the materialistic, deterministic, and mechanistic cosmos envisioned by late-nineteenth-century scientific naturalists, argued that the cosmos was "more like a great thought than like a great machine." Jeans believed that a "universal mind," which he also called "the Great Architect of the Universe," had not only designed the natural world but retained control of it.[33]

During the third quarter of the twentieth century, a growing number of Christian intellectuals dissented from the notion that science and theology could be meaningfully detached. Many of these thinkers possessed professional credentials in both areas. For example, in 1953 Arthur F. Smethurst, an English cleric who possessed a PhD in geology, denounced the idea of imposing "an 'iron curtain' between the world of nature and the realm of the spirit" on the grounds that this would "produce not only distrust but actual misunderstanding and confusion of thought between the two." Religion, he insisted, "by its very character must be concerned with the whole of reality" and thus incorporate the revelations of science within its purview. William Grosvenor Pollard, a physicist who became also an Episcopalian clergyman, emphasized that science and religion were both incomplete human enterprises. Whereas science was "fully competent to deal with any element of experience which arises from an object in space and time," it was necessary to invoke the existence of a transcendent, "non-conceptual" realm of experience that he described as "supernature" to account for why the natural world possessed the properties it did. Ian Barbour, a Carleton College professor who received graduate training in both physics and theology, warned that Christianity must not "disparage or neglect the natural order." After all, he reminded his readers, "if Christianity is radically interiorized, nature is left devoid of meaning and the stretches of cosmic history before man's appearance are unrelated to God."[34]

For their part, conservative Protestants believed that it was imperative to bring theological perspectives to bear on the natural world and typically criticized claims that investigation of the natural world should remain the exclusive preserve of the scientific community. In 1954 Bernard Ramm, a conservative evangelical theologian committed to the idea that "*a positive relationship must exist between science and Christianity*," affirmed in his widely read work, *The Christian View of Science and Scripture*, that the natural world served as a resource for scientists and biblical theologians alike. Ramm acknowledged that biblical descriptions of nature ran the gamut from "*popular*" and "*prescientific*" to "*non-postulational*," but he insisted that theologians had the right to draw on biblical perspectives in affirming that divine purpose and "a personal, meaningful, valuational core" lay at the very heart of the cosmos. He concluded that "the emphasis in science is on the visible universe and in theology the emphasis is on the invisible universe, *but it is one universe*."[35]

Not all conservatives followed Ramm in seeking to harmonize science and religion by abandoning a literal reading of the Bible. In 1961 the biblical theologian John Whitcomb, Jr., teamed up with the engineer Henry M. Morris to write *The Genesis Flood*. Affirming a fundamentalist commitment to "the verbal inerrancy of Scripture," Whitcomb and Morris insisted that since the province of science was limited to "present processes" and use of the scientific method depended on "experimental reproducibility," evolutionary theory and uniformitarianism were no more rigorously scientific than the scriptural narrative. Hence, people trying to understand creation and subsequent natural history were free to invoke biblical testimony. While they conceded that their own position did not qualify as science, this did not prevent them from adducing as much scientific evidence as they could for the "new scheme of historical geology" that served as the basis for the young-earth creationist movement.[36]

RECENT DIALOGUE

The scientifically inflammatory rhetoric of young-earth creationists and other religionists skeptical of the claims of modern science has played an important role in generating increased interest in the relationship between science and religion during the past few decades. Since 1970, religious conservatives—from fundamentalists and Pentecostals to Missouri Lutherans and Seventh-day Adventists—have aggressively contested the notion that the vocabulary of modern science, standing alone, provides an adequate description of the natural world. For those Christians, radically demarcating science from religion holds little appeal.

Many late-twentieth-century creationists continued to equate "science" with the idea of systematic knowledge based on generalization from observed facts. This led two of them in the middle of the 1980s to assail the theory of evolution by suggesting that it could "be beaten to a pulp with the dictionary." Other creationists, however, have appealed to a somewhat broader conception of science as a set of disciplines that arrange their data within theoretical structures, or "models." They have insisted that using that conception, creationism is just as scientific as evolutionism. In defending that view, "scientific" creationists have appealed to the work of philosophers of science such as Karl Popper and Thomas Kuhn to suggest that creationism constitutes a "research program" no less rigorously "scientific" than Darwinism. Since the middle of the 1970s, this way of framing the issue has prompted many creationists to demand "equal time" for their views in public education.[37]

In 1982 the federal judge William R. Overton ruled against an Arkansas law mandating "balanced treatment for creation-science and evolution-science" in the public schools on the grounds that "creation science is not science." Overton's justification for this view was that creation science "is not explanatory by reference to natural law, is not testable, and is not falsifiable" and is "inspired by the Book of Genesis." Notwithstanding that ruling, which even some opponents of creationism have acknowledged was "reached for all the wrong reasons and by a chain of argument which is hopelessly suspect," advocates of creationism have continued to defend the scientific pedigree of their views. However, most also assume that, as one creationist has recently put it, "the Bible is an essential part of the practice of science for a Christian. It provides the context for the study of God's creation."[38]

Other than scientific creationists, the religious conservatives in the Anglo American world who have played the most prominent role in discussing the relationship between science and religion are proponents of what has come to be known as intelligent design theory. Although most intelligent design advocates avoid appealing to the Bible, they share the creationists' conviction that theological beliefs should be brought to bear on the scientific enterprise. Lying at the heart of the intelligent design movement, which attained visible form in the 1990s, is a rejection of the notion that "methodological naturalism" should serve as the reigning norm of scientific investigation. Advocates of intelligent design have steadfastly maintained that legitimate "demarcation" criteria cannot be found to justify excluding God from the realm of scientific inquiry. Thus, at a 2005 hearing of the Kansas Board of Education, they "proposed changing the definition of science to include explanations other than 'natural.'" The recourse of proponents of intelligent design to a "theistic science" is predicated on what they regard as the commonsense notion that "Christians ought to consult all they know or have reason to believe in forming and testing hypotheses, explaining things in science and evaluating the plausibility of various scientific hypotheses, and among the things they should consult are propositions of theology."[39]

Religious believers convinced of the inadequacy of modern scientific perspectives have not monopolized aggressive rhetoric. During the past few decades the level of militance among proponents of scientific naturalism has also escalated. In contrast to the first half of the twentieth century, when scientists tended to stress the compatibility of science with religion, a number of more recent scientific luminaries, such as the physicists Steven Weinberg and Stephen Hawking, the biologists Richard Dawkins and Francis Crick, and the philosopher Daniel Dennett, have dismissed, even scorned, the beliefs of orthodox Christians. Although some of the critics have continued to employ such figures of speech as "discovering the mind of God" in describing their work, they clearly regard the notion of a supernatural or transcendent realm of reality as an unedifying fiction. The Harvard biologist Richard Lewontin has suggested that a "commitment to materialism" constitutes an important element of the apparatus of modern science. "We are forced by our *a priori* adherence to material causes," he wrote in 1997, "to create an apparatus of investigation and a set of concepts that produce material explanations, no matter how counterintuitive, no matter how mystifying to the uninitiated."[40]

Some scientific naturalists have claimed that science can account for religion itself. The biologist Edward O. Wilson not only acknowledged that religious impulses had conferred evolutionary advantage on human beings but predicted that those feelings would remain powerful into the foreseeable future. He also suggested, however, that "theology is not likely to survive as an independent intellectual discipline."[41]

In recent years most participants in the contemporary science-and-religion dialogue have resisted efforts to compartmentalize science and theology. Some of those thinkers have invoked the work of philosophers, sociologists, and historians of science who have challenged the notion that science provides an objective understanding of nature or possesses a privileged means of generating natural knowledge to deny sharp epistemic distinctions between theology and science. For example, Nancey Murphy, a professor of Christian philosophy at Fuller Theological Seminary, has drawn on the postmodernist claim that science, like other cultural endeavors, never fully escapes from the social context of its birth and hence in important ways remains local and time-bound, as well as the work of the philosopher of science Imre Lakatos, to collapse the longstanding epistemological divide between science and theology. Contending that both enterprises engage in "research programs" that employ "canons of probable reasoning" and that "the purpose of theology, like that of science, broadly speaking, is knowledge," Murphy has concluded that "theology is methodologically indistinguishable from the sciences." This has led her to endorse "theology's scientific status."[42]

Other participants in the contemporary science-and-religion dialogue cite scientific knowledge in support of theological claims. Many of the sponsors of this project have backgrounds in scientific investigation. John Polkinghorne, a particle physicist and Fellow of the Royal Society who surrendered his chair at the University of Cambridge to enter the Anglican priesthood (and subsequently became a highly respected figure in science-and-religion circles), refuses to treat science and theology in terms of a simple "dichotomy." It is more reasonable, suggests the knighted physicist priest, to affirm their interaction, with "theology explaining the source of the rational order and structure which science both assumes and confirms in its investigation of the world; [and] science by its study of creation setting conditions of consonance which must be satisfied by any account of the Creator and his activity." Restricting religion to the realm of human values strikes Polkinghorne as "narrowly parochial." Thus he and other interactionist activists of the "new natural theology" look for reasons for belief in God lying outside that realm. The confidence that scientific investigation can provide theologically valuable evidence has led the Anglo Australian physicist Paul Davies, author of such books as *God and the New Physics* and *The Mind of God: The Scientific Basis for a Rational World,* to assert that "science offers a surer path to God than religion."[43]

Critics outside the community of self-described science-and-religion scholars have inclined more toward maintaining the absolute independence of science from religion. For the past quarter century the National Academy of Sciences has defended the view that "religions and science answer different questions about the world." Whereas science asks about how the universe operates, religion inquires into "whether there is a purpose to the universe or a purpose for human existence." The eminent paleontologist Stephen J. Gould, an ardent scientific naturalist and sometime president of the American Association for the Advancement of Science, echoed this position in his book *Rocks of Ages: Science and Religion in the Fullness of Life.* There Gould insisted that science and religion each possess "a legitimate magisterium, or domain of teaching authority—and these magisteria do not overlap." In contrast to science, which deals with "the empirical universe: what is it made of (fact) and why does it work this way (theory)," the "net of religion extends over questions of moral meaning and value." Given this division of labor, Gould concluded that determining the relationship between the two "magisteria" presented the "great nonproblem of our times."[44]

One of the attractions of the separate-spheres model for scientists is that it gives them a monopoly in dealing with the world encompassed in time and space while consigning religion to the more subjective and private world of value and meaning. More difficult to understand is the appeal of this position for a number of theologians and lay people ostensibly sympathetic to the cause of religion, such as Don Cupitt, Dean of Emmanuel College, Cambridge. The controversial Cupitt, once dubbed a Christian Buddhist, assigns science the role of "understanding and controlling" the realm of natural phenomena and religion the task of generating "an order of meanings and values for us to live by."[45]

However they have described the relationship between science and religion, most modern intellectuals have joined their predecessors in assuming that both enterprises possess distinctive elements and in exploring their interrelationship. Most have also assumed that science and religion provide valuable resources for fostering the full flowering of human life and society. The enormous popularity of books seeking to "reconcile" science and religion would seem to suggest that many readers in the West remain committed to the idea that the creation of a boundary between science and religion need not, and should not, contribute to their mutual estrangement.

NOTES

I am indebted to Frederick Gregory, Peter Harrison, Bernard Lightman, and Michael Shank for providing me with very helpful responses to earlier drafts of this paper. Ron Numbers has read and commented helpfully on so many drafts that I feel somewhat guilty that I have not listed him as a coauthor. As always, I have benefited enormously from the love and support of my wife, Sharon (ILYS), and my son, Jeff.

1. Kenneth J. Howell, *God's Two Books: Copernican Cosmology and Biblical Interpretation in Early Modern Science* (Notre Dame, IN: University of Notre Dame Press, 2002); Peter Harrison, *The Bible, Protestantism, and the Rise of Natural Science* (New York: Cambridge University Press, 1998), 268–69; and Ronald L. Numbers, "Reading the Book of Nature through American Lenses," in *The Book of Nature in Early Modern and Modern History*, ed. Klaas van Berkel and Arjo Vanderjag (Leuven: Peeters, 2006), 261–74.

2. Peter Harrison, "'Science' and 'Religion': Constructing the Boundaries," *Journal of Religion* 86 (2006): 92–93; Peter Harrison, *"Religion" and the Religions in the English Enlightenment* (New York: Cambridge University Press, 1990), 1–14.

3. Isaac Newton [1706], quoted in Andrew Cunningham, "How the *Principia* Got Its Name: Or, Taking Natural Philosophy Seriously," *History of Science* 29 (1991): 384; John Winthrop IV [1753], quoted in Charles Edwin Clark, "Science, Reason, and an Angry God: The Literature of an Earthquake," *New England Quarterly* 38 (1965): 353 (original emphasis). More generally, see Harrison, "'Science' and 'Religion,'" 81–91.

4. Robert M. Young, "Natural Theology, Victorian Periodicals and the Fragmentation of a Common Context," in *Darwin's Metaphor: Nature's Place in Victorian Culture* (Cambridge: Cambridge University Press, 1985), 126–50; [Orville Dewey], "Diffusion of Knowledge," *North American Review* 30 (1830): 312; Review of *The Indications of the Creator*, by George Taylor, *Knickerbocker* 39 (1852): 85; Samuel Harris, "The Harmony of Natural Science and Theology," *New Englander* 10 (1852): 16.

5. William North Rice, "The Darwinian Theory of the Origin of Species," *New Englander* 26 (1867): 608. See also Charles Darwin, *On the Origin of Species by Means of Natural Section, or the Preservation of Favoured Races in the Struggle for Life*, facsimile ed. (1859;Cambridge, MA: Harvard University Press, 1964), 413, 481–83; Ronald L. Numbers, "Science Without God: Natural Laws and Christian Beliefs," in *When Science and Christianity Meet*, ed. David C. Lindberg and Ronald L. Numbers (Chicago: University of Chicago Press, 2003), 272–79, 282.

6. Numbers, "Science Without God," 265–85; Sydney Ross, "*Scientist:* The Story of a Word," *Annals of Science* 18 (1962): 65–85.

7. Charles Hodge, *What Is Darwinism?* (New York: Scribner, Armstrong and Company, 1874), 127; James T. Bixby, *Similarities of Physical and Religious Knowledge* (New York: D. Appleton and Company, 1876), 16–17, 67–68 (quotes, 67; original emphasis).

8. Samson Talbot, "Development and Human Descent," *Baptist Quarterly* 6 (1872): 146; Harris, "The Harmony of Natural Science and Theology," 17.

9. Joseph Le Conte, *Religion and Science: A Series of Sunday Lectures on the Relation of Natural and Revealed Religion, or the Truths Revealed in Nature and Scripture* (1873; New York: D. Appleton and Company, 1874), 231. My discussion of Draper and his work draws heavily on Donald Fleming, *John William Draper and the Religion of Science* (Philadelphia: University of Pennsylvania Press, 1950). Another especially useful treatment of Draper's discussion of the conflict between religion and science is James R. Moore, *The Post-Darwinian Controversies: A Study of the Protestant Struggle to Come to Terms with Darwin in Great Britain and America 1870–1900* (Cambridge: Cambridge University Press, 1979), 22–29.

10. John William Draper, *History of the Conflict between Religion and Science* (1874; New York: D. Appleton and Company, 1875), 68–118, 158–59, 352, 353–54, 363–64.

11. Ibid., vi, x–xi, 362, 363; Moore, *Post-Darwinian Controversies*, 25–26.

12. Bixby, *Similarities of Physical and Religious Knowledge*, 7. See also James Martineau, "Nature and God," [1860], in *Essays, Philosophical and Theological* (Boston: William V. Spencer, 1866), 122.

13. E. L. Youmans, "Draper and His Critics," *Popular Science Monthly* 7 (1875): 231; Jon H. Roberts and James Turner, *The Sacred and the Secular University* (Princeton, NJ: Princeton University Press, 2000), 32. I am indebted to Ronald L. Numbers, for suggesting this quantitative approach, and to Stephen E. Wald, who compiled the data and created the figures that I have included in this chapter.

14. James Woodrow, *An Examination of Certain Recent Assaults on Physical Science* [reprinted from the *Southern Presbyterian Review*, July 1873] (Columbia, SC: Presbyterian Publishing House, 1873), 27; J. S. Candlish, "Reformation Theology in the Light of Modern Knowledge," *Presbyterian Review* 8 (1887):, 231; George P. Fisher, "The Alleged Conflict of Natural Science and Religion," *Princeton Review* n.s., 12 (1883): 37–38; Bixby, *Similarities of Physical and Religious Knowledge*, 14.

15. M. H. Valentine, "The Influence of the Theory of Evolution on the Theory of Ethics," *Lutheran Quarterly Review* 28 (1898): 218; George T. Ladd, "The Origin of the Concept of God," *Bibliotheca Sacra* 34 (1877): 35; James McCosh, *The Religious Aspect of Evolution*, rev. ed. (New York: Charles Scribner's Sons, 1890), 58; Myron Adams, *The Continuous Creation: An Application of the Evolutionary Philosophy to the Christian Religion* (Boston: Houghton, Mifflin and Company, 1889), 96; L. Bell, "Unbelief, Half-Belief, and a Remedy," *New Englander* n.s., 7 [43] (1884): 72–73 (quote, 72); [Alexander] Winchell, "Huxley and Evolution," *Methodist Quarterly Review* 4th ser., 29 (1877): 305; Ladd, "Origin of the Concept of God," 34; George Harris, *Moral Evolution* (Boston: Houghton, Mifflin and Company, 1896), 188–89 (quote, 189). See also Peter J. Bowler, *Reconciling Science and Religion: The Debate in Early-Twentieth-Century Britain* (Chicago: University of Chicago Press, 2001), 226–27; Jon H. Roberts, *Darwinism and the Divine in America: Protestant Intellectuals and Organic Evolution, 1859–1900* (Madison: University of Wisconsin Press, 1988), 137–43.

16. [E. L. Youmans], "Purpose and Plan of Our Enterprise," *Popular Science Monthly* 1 (1872): 113; [John Tyndall], *Advancement of Science: The Inaugural Address of Prof. John Tyndall, D.C.L., LL.D. F.R.S., Delivered before the British Association for the Advancement of Science, at Belfast, August 19, 1874, with Portrait and Biographical Sketch* (New York: Asa K.

Butts & Co., 1874), 83; Thomas H. Huxley, quoted in Ruth Barton, "Evolution: The Whitworth Gun in Huxley's War for the Liberation of Science from Theology," in *The Wider Domain of Evolutionary Thought*, ed. David Oldroyd and Ian Langham (Dordrecht: D. Reidel Publishing Company, 1983), 264; Thomas H. Huxley, "On the Reception of the 'Origin of Species,'" in *The Life and Letters of Charles Darwin, Including an Autobiographical Chapter*, ed. Francis Darwin, 3 vols. (London: John Murray, 1887), 2:186. See also Bernard Lightman, *The Origins of Agnosticism: Victorian Unbelief and the Limits of Knowledge* (Baltimore, MD: The Johns Hopkins University Press, 1987), 128–34; Frank M. Turner, "The Victorian Conflict between Science and Religion: A Professional Dimension," *Isis* 69 (1978): 356–76.

17. Thomas H. Huxley, "On the Advisableness of Improving Natural Knowledge" [1866], *Methods and Results*, in *Collected Essays by T. H. Huxley* (1893; New York: Hideschein, 1970), 1:41; Herbert Spencer, *First Principles* (London: Williams and Norgate, 1862), 106; Newman Smyth, *The Religious Feeling: A Study for Faith* (New York: Scribner, Armstrong and Company, 1877), 148, 34–35; Lewis F. Stearns, "Reconstruction in Theology," *New Englander* n.s., 5 [41] (1882): 86–87 (quote, 86); W. P. King, "Christian Faith and the New Apologetics," *Methodist Review Quarterly* 62 (1913): 338; B. F. Cocker, *The Theistic Conception of the World: An Essay in Opposition to Certain Tendencies of Modern Thought* (New York: Harper & Brothers, 1875), 345. For the lukewarm reception given to the "metaphysics" of scientific naturalism, see Bernard Lightman, "Victorian Sciences and Religions: Discordant Harmonies," *Osiris* 2d ser., 16 (2001): 351; Bowler, *Reconciling Science and Religion*, 18, 27, 122–23.

18. Albrecht Ritschl, *The Christian Doctrine of Justification and Reconciliation: The Positive Development of the Doctrine*, ed. H. R. Mackintosh and A. B. Macaulay, 3 vols., 3d ed. (New York: Charles Scribner's Sons, 1900), 3:16, 398, 205, 225.

19. William Adams Brown, *Beliefs That Matter: A Theology for Laymen* (1928; New York: Charles Scribner's Sons, 1930), 73–74, 77; Gerald Birney Smith, "Systematic Theology and Christian Ethics," in *A Guide to the Study of the Christian Religion*, ed. Gerald Birney Smith (Chicago: University of Chicago Press, 1916), 546–47.

20. Edwin Diller Starbuck, *The Psychology of Religion: An Empirical Study of the Growth of Religious Consciousness* (New York: Charles Scribner's Sons, 1900), 1; Edward L. Schaub, "The Present Status of the Psychology of Religion," *Journal of Religion* 2 (1922): 377.

21. "First of the Course of Scientific Lectures—Prof. White on 'the Battle-Fields of Science,'" *New York Daily Tribune*, December 18, 1869; Andrew Dickson White, *The Warfare of Science*, 2d ed. (London: Henry S. King & Co., 1877), 148–49.

22. Andrew Dickson White, *A History of the Warfare of Science with Theology in Christendom*, 2 vols. (New York: D. Appleton and Company, 1896), 1:ix, xii, 410, v–vi, 12.

23. Harry Emerson Fosdick, "The Dangers of Modernism," *Harpers Magazine* 152 (1926): 408; Harry Emerson Fosdick, "Beyond Modernism," *Christian Century* 52 (December 4, 1935): 1552; Harry Emerson Fosdick, "Yes, But Religion Is an Art!" *Harper's Magazine* 162 (1931): 132. See also William Newton Clarke, *An Outline of Christian Theology* (1898; New York: Charles Scribner's Sons, 1922), 1, 4, 54–55; Eugene W. Lyman, "What is Theology? The Essential Nature of the Theologian's Task," *American Journal of Theology* 17 (1913): 330–32; Henry S. Pritchett, *What Is Religion? And Other*

Student Questions: Talks to College Students (Boston: Houghton, Mifflin and Company, 1906), 52–53. For critical reaction to White's work by contemporaries, see Glenn C. Altschuler, *Andrew D. White—Educator, Historian, Diplomat* (Ithaca, NY: Cornell University Press, 1979), 209–16.

24. Samuel Harris, *The Philosophical Basis of Theism* (New York: Charles Scribner's Sons, 1883), 15–16 (quote, 16); Douglas Clyde Macintosh, *Theology as an Empirical Science* (New York: Macmillan, 1919), 5, 25–28; Shailer Mathews, *The Faith of Modernism* (New York, Macmillan, 1924), 23 (the original was in italics), 35–36.

25. Hodge, *What Is Darwinism?* 126; Anonymous, "What is 'Science,' and What the 'Inductive Method[']," *Bible Student and Teacher* 7 (1907): 107; and W. B. Riley, "Skepticism in Our Schools," *Bible Student and Teacher* 6 (1907): 452; Ronald L. Numbers, *The Creationists: From Scientific Creationism to Intelligent Design*, expanded ed. (Cambridge, MA: Harvard University Press, 2006), 65.

26. Hodge, *What Is Darwinism?* 126–31; Charles Hodge, *Systematic Theology*, 3 vols. (1871; New York: Scribner, Armstrong and Company, 1873), 1–17 (quote, 10); J. Gresham Machen, "The Relation of Religion to Science and Philosophy," *Princeton Theological Review* 24 (1926): 64.

27. William Jennings Bryan, *In His Image* (New York: Fleming H. Revell Company, 1922), 94; Alfred Fairhurst, *Atheism in Our Universities* (Cincinnati, OH: Standard Publishing Company, 1923), 91 (original emphasis).

28. Numbers, *The Creationists*, 69–87; Judith V. Grabiner and Peter D. Miller, "Effects of the Scopes Trial," *Science* 185 (1974): 832–37; Edward J. Larson, *Trial and Error: The American Controversy Over Creation and Evolution*, 3rd ed. (New York: Oxford University Press, 2003), 28–92; Frederick Lewis Allen, *Only Yesterday: An Informal History of the Nineteen-Twenties* (New York: Harper & Brothers, 1931), 195; C. H. Waddington, *The Scientific Attitude*, 2nd ed. (Middlesex, UK: Penguin Books, 1948), 170.

29. Étienne Gilson, *God and Philosophy* (New Haven, CT: Yale University Press, 1941), 119; George H. Richardson, "Scientist and Theologian," *American Church Monthly* 24 (1928): 336.

30. The joint statement, along with a list of those who signed the document, is reprinted in "Science and Religion," *Science* n.s., 57 (1923): 630–31. See also Alfred North Whitehead, *Science and the Modern World* (1925; New York: Free Press, 1967), 12, 185.

31. E. G. Homrighausen, *Christianity in America: A Crisis* (New York: Abingdon Press, 1936), 80–81, 62. In my description of the neo-orthodox treatment of science I have drawn on the following works: Ian G. Barbour, *Issues in Science and Religion* (Englewood Cliffs, NJ: Prentice-Hall, 1966), 116–19; and Langdon Gilkey, *Naming the Whirlwind: The Renewal of God-Language* (Indianapolis, IN: Bobbs-Merrill Company, 1969), 82–91.

32. Carl Michalson, *The Rationality of Faith: An Historical Critique of the Theological Reason* (New York: Charles Scribner's Sons, 1963), 26; Langdon B. Gilkey, "The Concept of Providence in Contemporary Theology," *Journal of Religion* 43 (1963): 181–82 (quote, 182); Donald D. Evans, "Differences Between Scientific and Religious Assertions," in *Science and Religion: New Perspectives on the Dialogue*, ed. Ian G. Barbour (New York: Harper & Row, 1968), 101, 111–25 (quote, 125), 102, 107, 109. See also Ian T. Ramsey, *Religious Language* (London: SCM Press, 1957), 11–18, 185–86, 26. Useful discussions

of the role of existentialism in postwar religious belief include Barbour, *Issues in Science and Religion*, 119–21; and David E. Roberts, *Existentialism and Religious Belief*, ed. Roger Hazelton (New York: Oxford University Press, 1957).

33. F. R. Tennant, *Philosophical Theology*, 2 vols. (Cambridge: Cambridge University Press, 1930), 2:1–2, 63; James Jeans, *The Mysterious Universe*, rev. ed. (1932; New York: Macmillan, 1933), 186–87, 17–53, 175, 165. See also Henry P. Van Dusen, "Trends in Contemporary Theism," *Religion in Life* 1 (1932): 183–84.

34. Arthur F. Smethurst, *Modern Science and Christian Beliefs* (New York: Abingdon Press, 1955), xvi–xvii, 71–72; William Grosvenor Pollard, *Physicist and Christian: A Dialogue Between the Communities* (New York: Seabury Press, 1961), 54, 106–10; Barbour, *Issues in Science and Religion*, 453–54.

35. Bernard Ramm, *The Christian View of Science and Scripture* (Grand Rapids, MI: Wm. B. Eerdmans Publishing Company, 1954), 33, 47, 76, 34–35, 37 (original emphasis).

36. John C. Whitcomb Jr. and Henry M. Morris, *The Genesis Flood: The Biblical Record and Its Scientific Implications* (Philadelphia, PA: Presbyterian and Reformed Publishing Co., 1961), xx, 116–211 (quote, 119); Henry M. Morris and John C. Whitcomb Jr., "Reply to Reviews," *Journal of the American Scientific Affiliation* 16 (June 1964): 59. For a more extended account of the work of Whitcomb and Morris, see Numbers, *The Creationists*, 208–38.

37. Marshall Hall and Sandra Hall, quoted in Numbers, *The Creationists*, 274, cf. 268–69. See also Henry M. Morris and Gary E. Parker, *What Is Creation Science?* rev. ed. (El Cajon, CA: Master Books, 1987), 9; Henry M. Morris, *History of Modern Creationism* (San Diego, CA: Master Books, 1984), 24.

38. [William R. Overton], "McLean v. Arkansas: Opinion of William R. Overton, U. S. District Judge, Eastern District of Arkansas, Western Division (Dated 5 January 1982)," in *Creationism, Science, and the Law: The Arkansas Case*, ed. Marcel C. La Follette (Cambridge, MA: MIT Press, 1983), 68, 60, 58; Larry Laudan, "Commentary on Ruse: Science at the Bar—Causes for Concern," in La Follette, *Creationism, Science, and the Law*, 161–66 (quote, 161): Wayne Frair and Gary D. Patterson, "Creationism: An Inerrant Bible and Effective Science," in *Science & Christianity: Four Views*, ed. Richard F. Carlson (Downers Grove, IL: InterVarsity Press, 2000), 28–29.

39. J. P. Moreland, introduction to *The Creation Hypothesis: Scientific Evidence for an Intelligent Designer*, ed. J. P. Moreland (Downers Grove, IL: InterVarsity Press, 1994), 12–13; J. P. Moreland, "Theistic Science & Methodological Naturalism," in *The Creation Hypothesis*, 49; Stephen C. Meyer, "The Scientific Status of Intelligent Design: The Methodological Equivalence of Naturalistic and Non-Naturalistic Origins Theories," in *Science and Evidence for Design in the Universe: Paper Presented at a Conference Sponsored by the Wethersfield Institute, New York City, September 25, 1999* (San Francisco: Ignatius Press, 2000), 151–211; Yudhijit Bhattacharjee, "Kansas Gears Up for Another Battle Over Teaching Evolution," *Science* 308 (2005): 627. For a useful historical overview of the intelligent design movement, see Numbers, *The Creationists*, 373–98.

40. Richard Lewontin, "Billions and Billions of Demons," *New York Review of Books*, January 9, 1997, 31. See also Steven Weinberg, *The First Three Minutes: A Modern View of the Origin of the Universe* (New York: Basic Books, 1977), 154; Richard Dawkins, *River Out of Eden: A Darwinian View of Life* (New York: Basic Books, 1995), 132–33. See also

Karl Giberson and Mariano Artigas, *Oracles of Science: Celebrity Scientists versus God and Religion* (New York: Oxford University Press, 2007). For the use of religious metaphors, see, for example, Stephen W. Hawking, *A Brief History of Time: From the Big Bang to Black Holes* (New York: Bantam Books, 1998), 175; Steven Weinberg, *Dreams of a Final Theory* (New York: Pantheon Books, 1992), 242.

41. Edward O. Wilson, *On Human Nature* (Cambridge, MA: Harvard University Press, 1978), 172, 192.

42. Nancey Murphy, *Theology in the Age of Scientific Reasoning* (Ithaca, NY: Cornell University Press, 1990), 16, 51, 198. See also Michael C. Banner, *The Justification of Science and the Rationality of Religious Belief* (Oxford: Clarendon Press, 1990).

43. John Polkinghorne, *One World: The Interaction of Science and Theology* (Princeton, NJ: Princeton University Press, 1986), 62, 97, 82; John Polkinghorne, *The Faith of a Physicist: Reflections of a Bottom-Up Thinker; The Gifford Lectures for 1993–4* (Princeton, NJ: Princeton University Press, 1994), 5; John Polkinghorne, *Belief in God in an Age of Science* (New Haven, CT: Yale University Press, 1998), 10, xiii; Paul Davies, *God and the New Physics* (1983; New York: Simon & Schuster, 1984), ix, 229. See also A. R. Peacocke, *Science and the Christian Experiment* (London: Oxford University Press, 1971), 5; Arthur Peacocke, *Theology for a Scientific Age: Being and Becoming—Natural, Divine and Human*, rev. ed. (Minneapolis, MN: Fortress Press, 1993), 21.

44. National Academy of Sciences, Working Group on Teaching Evolution, *Teaching About Evolution and the Nature of Science* (Washington DC: National Academy Press, 1998), 58; Stephen Jay Gould, "Nonoverlapping Magisteria," *Natural History* 106 (March 1997): 19; Stephen Jay Gould, *Rocks of Ages: Science and Religion in the Fullness of Life* (New York: Ballantine, 1999), 92.

45. Don Cupitt, *The Sea of Faith* (New York: Cambridge University Press, 1984), 79, 32. See also Nicholas Lash, "Production and Prospect: Reflections on Christian Hope and Original Sin," in *Evolution and Creation*, ed. Ernan McMullin (Notre Dame, IN: University of Notre Dame Press, 1985), 278–79; Steven Goldberg, *Seduced by Science: How American Religion Has Lost Its Way* (New York: New York University Press, 1999), 1; and George A. Lindbeck, *The Nature of Doctrine: Religion and Theology in a Postliberal Age* (Philadelphia, PA: Westminster Press, 1984), 69.

CHAPTER 11

Science, Pseudoscience, and Science Falsely So-Called

Daniel P. Thurs and Ronald L. Numbers

On July 1, 1859, Oliver Wendell Holmes, a fifty-year-old Harvard Medical School professor and littérateur, submitted to being "phrenologized" by a visiting head-reader, who claimed the ability to identify the strength of character traits—amativeness, acquisitiveness, conscientiousness, and so forth—by examining the corresponding bumps on the head. Holmes found himself among a group of women "looking so credulous, that, if any Second Advent Miller or Joe Smith should come along, he could string the whole lot of them on his cheapest lie." The astute operator, perhaps suspecting the identity of his guest, offered a flattering assessment of Holmes's proclivities, concluding with the observation that his subject "would succeed best in some literary pursuit; in teaching some branch or branches of natural science, or as a navigator or explorer." Holmes found the event so revealing of contemporary gullibility that shortly thereafter he drew on his experience in writing his *Atlantic Monthly* feature "The Professor at the Breakfast-Table."[1] He began by offering a "definition of a *Pseudo-science*," the earliest explication of the term that we have found:

> A Pseudo-science consists of a nomenclature, with a self-adjusting arrangement, by which all positive evidence, or such as favors its doctrines, is admitted, and all negative evidence, or such as tells against it, is excluded. It is invariably connected with some lucrative practical application. Its professors and practitioners are usually shrewd people; they are very serious with the public, but wink and laugh a good deal among themselves. . . . A

> Pseudo-science does not necessarily consist wholly of lies. It may contain many truths, and even valuable ones.

Holmes repeatedly punctuated his account by denying that he wanted to label phrenology a pseudoscience; he desired only to point out that phrenology was *"very similar"* to the pseudosciences. On other occasions Holmes categorically included phrenology—as well as astrology, alchemy, and homeopathy—among the pseudosciences.[2]

Before pseudoscience became available as a term of reproach, critics of theories purporting to be scientific could draw on a number of older deprecatory words: humbuggery, quackery, and charlatanism. But no phrase enjoyed greater use than "science falsely so-called," taken from a letter the apostle Paul wrote to his young associate Timothy, advising him to "avoid profane and vain babblings, and oppositions of science falsely so called" (1 Timothy 6:20). Although "science" had appeared in the original Greek as *gnōsis* (meaning knowledge generally), English translators in the sixteenth and seventeenth centuries chose "science" (a synonym for knowledge). By the mid-eighteenth century (at the latest), the phrase was being applied to disagreeable natural philosophy. In 1749, for example, the philosophically inclined Connecticut minister Samuel Johnson complained that "it is a fashionable sort of philosophy (a science falsely so-called) to conceive that God governs the world only by a general providence according to certain fixed laws of nature which he hath established without ever interposing himself with regard to particular cases and persons."[3]

Even after the appearance of the term pseudoscience, science falsely so-called remained in wide circulation, especially among the religious. In the wake of the appearance of Charles Darwin's *On the Origin of Species* (1859) and other controversial works, a group of concerned Christians started the Victoria Institute, or Philosophical Society of Great Britain; they dedicated it to defending "the great truths revealed in Holy Scripture . . . against the opposition of Science, falsely so called." The religiously devout Scottish anti-Darwinist George Campbell, the eighth Duke of Argyll, dismissed Thomas H. Huxley's views as "science falsely so called." After Huxley's good friend John Tyndall used his platform as president of the British Association for the Advancement of Science in 1874 to declare all-out war on theology, one Presbyterian wit dubbed Britain's leading scientific society "The British Association for the Advancement of 'Science, Falsely So-Called.'" Ellen G. White, the founding prophet of Seventh-day Adventism, condemned "science falsely so-called"—meaning "mere theories and speculations as scientific facts" opposed to the Bible—over a dozen times.[4]

Few epithets, however, have drawn as much scientific blood as pseudoscience. Defenders of scientific integrity have long used it to damn what is not science but pretends to be, "shutting itself out of the light because it is afraid of the light." The physicist Edward Condon captured the transgressive, even indecent, implications of the term when he likened it to "Scientific Pornography."[5] Such sentiments might make it appear that tracing the history of pseudoscience would be an easy enough task, simply requiring the assembly of a rogue's gallery of obvious misconceptions, pretensions, and errors down through the ages. But, in fact, writing the history of pseudoscience is a much more subtle matter, especially if we eschew essentialist thinking.

If we want to tell the history of pseudoscience, we have to come to grips with the term's fundamentally rhetorical nature. We also need to take a historically sensitive track and focus on those ideas that have been rejected by the scientific orthodoxy of their own day. But here, too, problems arise. For most of the history of humankind up to the nineteenth century, there has been no clearly defined orthodoxy regarding scientific ideas to run afoul of, no established and organized group of scientists to pronounce on disputed matters, no set of standard scientific practices or methods to appeal to. However, even in the presence of such orthodoxy, maintaining scientific boundaries has required struggle. Rather than relying on a timeless set of essential attributes, its precise meanings have been able to vary with the identity of the enemy, the interests of those who have invoked it, and the stakes involved, whether material, social, or intellectual. The essence of pseudoscience, in short, is how it has been used.

English-speakers could have paired "pseudo," which had Greek roots, and "science," which entered English from Latin by way of French, at any time since the medieval period. However, pseudo-science (almost universally written with a hyphen before the twentieth century) did not become a detectable addition to English-language vocabularies until the early 1800s. Its greater circulation did not result from a sudden realization that false knowledge was possible. Instead, it involved larger shifts in the ways that people talked, including a greater tendency to append pseudo to nouns as a recognized means of indicating something false or counterfeit. This habit was apparent as early as the seventeenth century, but became particularly common during the nineteenth.[6]

Even more significant, increased usage paralleled important changes in the concept of science. Pseudoscience appeared at precisely the same time during the early part of the 1800s that science was assuming its modern meaning in English-speaking cultures to designate knowledge of the natural world. The more the category of science eclipsed and usurped significant parts of those activities formerly called natural philosophy and natural history, the more rhetorical punch pseudoscience packed as a weapon against one's enemies. By the same token, even taking account of its many possible meanings, pseudoscience gave people the ability to mark off scientific pretense and error as especially worthy of notice and condemnation, making science all the more clear by sharpening the outlines of its shadow and opening the door to attestations of its value in contrast with other kinds of knowledge. In this sense, pseudoscience did not simply run afoul of scientific orthodoxy—it helped to create such orthodoxy.

Pseudoscience began its career in the English-speaking world rather modestly. While certainly in general use during the early and mid-1800s, it was still somewhat rare, particularly in contrast to the latter decades of the century and the 1900s. This is visible, for instance, in full-text searches for the term in American magazine articles from 1820 to 1920.[7] In general, the results of such a survey show no usage at first, then a steadily increasing appeal, with a slight surge in the 1850s and a dramatic increase in the 1880s, leading to a comparatively high and somewhat constant level of use around the turn of the century. By this time pseudoscience was becoming an international term of opprobrium. The French used the same word as the English did; however, other nationalities coined cognates: *Pseudowissenschaften* in German, *pseudoscienza* in Italian, *seudociencia* in Spanish, *pseudovetenskap* in Swedish, *pseudowetenschap* in Dutch, and *псевдонаука* in Russian.[8] Americans, however, seemed to have been fondest of the term, which makes the American context particularly interesting for examining the rise and evolution of pseudoscience.

During the 1830s and 1840s a wide variety of novel ideas appeared on the American intellectual landscape that some people thought strange. These novelties included religious groups such as Mormons and Millerites and social-reform movements associated with women's rights and abolitionism. A host of new scientific and medical approaches, ranging from the do-it-yourself botanical cures of Thomsonianism to the minute doses of homeopathy, also circulated. Probably the most emblematic of the crop of -isms, -ologies, and -athies that flourished in the antebellum soil was phrenology. In its most popular form, which linked the shape of the skull (and the cerebral organs underneath) to the details of an individual's personality, phrenological doctrine was spread across the nation by a cadre of devoted lecturers and head readers and by a thick stack of cheap literature.[9]

For many of its skeptics, phrenology provided one of the primary examples of pseudoscience and, sometimes with special emphasis, "*pseudo* science." As we indicated above, the very first reference we have found to "pseudo-science" appeared in 1824 and was directed obliquely at phrenology.[10] Still, even amidst antebellum debate over phrenological ideas, invocations of pseudoscience remained fairly few. Phrenology was identified as pseudoscience more frequently during the latter portions of the nineteenth century than it was in its heyday. This was partly because in the early years a strong scientific orthodoxy remained more hope than actuality. Historian Alison Winter has argued that individual scientific claims during this period had to be established without the help of an organized community of practitioners with shared training, beliefs, and behaviors.[11] The same was true of attempts to exile ideas from science. Even among the most notable members of the American scientific scene, there was less-than-universal agreement about the status of novel ideas. One of the most prominent men of science of the period, Yale's Benjamin Silliman, was publicly friendly to phrenology, albeit it in one of its more scholarly forms.

The most important explanation for the relative rarity of charges of pseudoscience in the early parts of the nineteenth century was that science remained a somewhat amorphous term. It had, by the second quarter of the 1800s, largely taken the place of earlier names, such as natural philosophy and natural history, for the study of the natural world. But enough of its former connection to reliable and demonstrable knowledge in general remained that science went well beyond the natural and included a huge swath of areas, from theology to shorthand. Such enormous extent was supported by contemporary methodological standards that made it much easier to include fields within porous scientific boundaries than to exclude them. This fuzziness actually made science difficult to use in many cases, and Americans appealed to "the sciences" collectively or to individual sciences, such as chemistry or geology, more frequently than during later eras.[12] It was also difficult to know exactly how to describe scientific transgression. Popular rhetoric had not quite settled on pseudoscience as the means to do that. Many Americans also denounced "pseudo-chemists," "pseudo-induction," "pseudo-observation," and, in the case of a new religious phenomenon with scientific pretensions, "pseudo-spiritualism." In Britain the mathematician Augustus de Morgan contributed the term "paradoxer" to describe those whose ideas "deviated from general opinion, either in subject-matter, method or conclusion."[13]

The fuzziness of science meant that many ways of categorizing knowledge with false pretensions to truth did not have any direct link with the scientific at all. Defenders of scientific integrity routinely denounced "mountebanks" and "pretenders." Even some of the American practitioners who agitated most for a more controlled and organized scientific community, including the secretary of the Smithsonian Institution, Joseph Henry, and the director of the United States Coast Survey, Alexander Dallas Bache, talked about expelling "charlatans" and "quacks" from the scientific world.[14] Indeed, by far the most common term for error in the guise of science was quackery. Quackery was particularly common in discussions of medicine but the scope of the category was not tightly linked to the medical or scientific alone. The definition offered by English physician Samuel Parr, as paraphrased by the American anti-quackery crusader David Meredith Reese, identified "every practitioner, whether educated or not, who attempts to practise imposture of any kind" as a quack. And though Reese noted in his *Humbugs of New-York* (1838) that the "epithet is often restricted within narrow limits, and is attached ordinarily only to those ignorant and impudent mountebanks, who, for purposes of gain, make pretensions to the healing art," he included abolitionism within the sphere of humbuggery.[15]

Likewise, many Americans offered broad definitions of the pseudoscientific that did little more than identify it with pretensions to an unwarranted scientific status. Astrology and alchemy were two widely used historical examples of pseudoscience because they resembled actual sciences in name, but were not, or so it was universally assumed, truly scientific. When it came to explaining pretension, some people focused on the "servile sophistry of pseudo-science" and on the perversion of legitimate scientific terminology.[16] An 1852 article in *Harper's New Monthly Magazine* decried the "dextrous use of some one long new-coined term."[17] Such verbiage was another widely recognized sign of pseudoscience, as was the tendency to theorize too quickly or leap to conclusions "in lieu of proof."[18] In other cases, observers attributed false claims cloaked in science to ethical lapses. To such leading American men of science as Henry and Bache, moral fiber was an essential possession of the true scientific practitioner.[19]

This lack of specificity paralleled the fuzziness of science itself. At the same time, there were some emerging ideas about the nature of pseudoscience, quackery, charlatanism, or mountebankery applied specifically to scientific matters that were beginning to draw lines between science and not-science in new and ultimately modern-sounding ways. For instance, one antebellum characteristic of pretense in the scientific or medical worlds was the improper influence of the hope for material gain. Both David Meredith Reese and Oliver Wendell Holmes pointed to the corrupting influence of money in the creation and diffusion of quackery and pseudoscience.[20] Such claims helped to create a zone of pure scientific knowledge or medical practice.

By far the most important distinction emerging between science and not-science was the one between scientific and popular knowledge. This distinction played a profoundly important role in the control of scientific knowledge and practice by aspiring leaders of science and medicine. Since such so-called professionalization was more advanced in Britain than in the United States, one of the earliest public links between pseudoscience and "popular delusions" or the "slowness in the capacity of the popular mind" occurred in an evaluation of homeopathy in the Edinburgh-based *Northern Journal of Medicine*. American pens inscribed less strident connections between pseudoscience and untutored interest in scientific subjects, thanks in part to their incompatibility with the majority of antebellum public rhetoric; but it was evident beneath the surface. In a letter to Joseph Henry in 1843, John K. Kane, the senior secretary of the American Philosophical Society, worried that "all sorts of pseudo-scientifics" were on their way to try to win the recently vacated position as head of the Coast Survey by way of enlisting Henry's help to encourage Bache to make a bid for the position. Henry himself lamented in a letter to Bache the recent "avalanch of pseudo-science" and wondered how the "host of Pseudo-Savants" could "be controlled and directed into a proper course."[21]

The last quarter of the nineteenth century witnessed a marked growth in references to pseudoscience, both in American periodicals and in English-speaking culture generally. Method remained a powerful means of identifying pseudoscience, particularly the tendency to leap to conclusions "in advance of the experimental evidence which alone could justify them." Likewise, the presence of potential profit continued to signal the presence of pseudoscience. One correspondent in an 1897 issue of *Science* noted that the recent discovery of x-rays had "already been made a source of revenue by more than one pseudo-scientist."[22] At times, a moral note appeared in broadsides against pseudoscience. Daniel G. Brinton's presidential address at the annual meeting of the American Association for the Advancement of Science (AAAS) in 1894 strongly linked pseudoscience to the fundamental dishonesty and snobbery of "mystery, concealment, [and] occultism." Mary Baker Eddy's Christian Science found itself both denounced as a pseudoscience (and a science falsely so-called) and divided by "pseudo Scientists," who questioned the prophet.[23]

In addition, observers could point to an enduring set of historical touchstones in their attempts to delineate the pseudoscientific, including astrology, alchemy, and what most critics assumed were the thoroughly discredited notions of phrenology.[24] Of course, not everyone agreed on the last point. Phrenology remained a topic of popular interest. The number of phrenological publications even experienced something of a small surge in the 1870s. Still, in the culture as a whole, phrenology had lost a certain claim to respectability and plausibility. One author in 1887 caught the mood of phrenology's fall from grace in the attempt to salvage the field under the banner of a "new phrenology," while casting aspersions on the "old phrenology," the latter leading to the former just like astrology and alchemy lead to astronomy and chemistry. They were ultimately distinguished not by the elimination of supernatural or superstitious but by the fact that the new phrenology brought together the study of the brain with "mathematics, physics, chemistry, and other sciences where the old phrenology was isolated in this respect."[25] Implicit in this move, as well as the increasingly prominent "outsider" status of phrenology, was a much stronger sense of orthodoxy than prevailed earlier in the century, not simply in a patchwork fashion but one that extended and bound together the whole range of science into one whole.

Indeed, orthodoxy had made substantial gains over the 1800s, thanks in no small measure to the champions of what came to be known as professionalized scientific practice. Particularly in Britain, a core group of young men of science had begun actively to campaign for a more restricted and tightly controlled scientific community, often through the capture of key institutions, such as the British Association for the Advancement of Science. The American inheritors of Bache and Henry scored successes as well, though American scientific institutions tended to be weaker than British ones. The work of aspiring professionals also had an intellectual edge. Two of the most notable British professionalizers of the period, John Tyndall and Thomas Henry Huxley, combined institutional activity with a vocal defense of naturalism in science, including support for Charles Darwin's ideas about evolution. Darwinian evolution proved a valuable tool in the hands of such aspiring professionals and other people sympathetic to the idea of a science surrounded by strict boundaries because it provided a potent symbol of the natural distinguished from the supernatural. When belief in an evolutionary account of life was used as a litmus test for separating scientific insiders from outsides, it also became a means of freeing science from what some maintained was unacceptable theological interference.[26]

Emerging distinctions between science and religion became one of the primary fault lines in the growing gap between science and not-science. But it was not the only one. Other potent divisions in American public culture included those between pure and applied science and between the generation of scientific knowledge and its popularization. All these new rhetorical habits enhanced the value of ejecting ideas and people from the scientific fold, and therefore the invocation of pseudoscience.

They also paralleled a number of important shifts in widespread ways of talking about pseudoscience. An 1896 letter to the editor of *Science* from one reader criticized "'practical' or pseudo-science" men, by which he meant those uneducated in theory.[27] Equations between pseudoscience and popular science also became far more common than earlier in the century. As men of science (or "scientists" as they were increasingly being called) began to privilege research over the diffusion of science—which in their opinion often simplified and degraded pure scientific knowledge—popularization often found itself associated with pseudoscience. An 1884 article in the *New York Times* leveled as one of its primary complaints against the "pseudo-science" of phrenology that books and pamphlets about it were "within reach of everybody."[28]

The growing sense of "popular science" as something distinct from science itself that informed such characterizations of science and not-science went hand in hand with evolving ideas about the nature of the public that was consuming popularized science. One of its chief characteristics in late-nineteenth-century depictions was its credulity. A variety of commentators noted, particularly in the years before 1900, that the scientific discoveries of the previous hundred years had been so dramatic and extensive that ordinary people had become "ready to accept without question announcements of inventions and discoveries of the most improbable and absurd character" or that the "general public has become somewhat overcredulous, and untrained minds fall an easy prey to the tricks of the magazine romancer or to the schemes of the perpetual motion promoter."[29] According to some observers, "an army of pseudo-scientific quacks who trade upon the imperfect knowledge of the masses" had grown up alongside legitimate purveyors of orthodoxy. In this sense, the enormous and bewildering power of science was itself to blame for the spread of pseudoscientific ideas. But the "evil influence of a sensational press" also played a deleterious role.[30]

Such concerns reflected genuine worries, but they also paralleled the growth of the commercialized mass media as a new cultural force, as well as the creation of a new kind of mass public such publications made possible. By the early 1900s, many Americans had adopted the habits and methods of the advertising industry to promote everything from public-health campaigns to religious revivals. An article in an 1873 issue of the *Ladies' Repository* decried the supposed tendency of many magazine readers to "swallow any bolus that speculative doctors of chances may drop into their gullets," particularly when an article began "'Dr. Dumkopf says.'"[31] The geologist and science-popularizer Joseph LeConte similarly complained about mistakes "attested to by newspaper scientists, and therefore not doubted by newspaper readers."[32] For champions of science, such conditions could appear to be prime breeding grounds for error. One observer worried about "plausibly written advertisements," particularly for medicines, that falsely invoked science and thus threatened to dupe and unwary public with pseudoscientific claims.[33] The battle between science and pseudoscience in the press occasionally produced a call to arms among professional practitioners. In 1900, the president of the AAAS recognized in his annual address that while scientists' "principle business is the direct advancement of science, an important, though less agreeable duty, betimes, is the elimination of error and the exposure of fraud."[34]

As the invocation of pseudoscience began to proliferate in the emerging gaps between the scientific and popular, it also started to appear with greater frequency along another emerging distinction, namely the one between the physical and social sciences. Encouraging this boundary was an increasing tendency to link science with physical nature, a move that often left the study of humans in a kind of limbo. One character in a serialized story originally published in the British *Contemporary Review* and later reprinted for American audiences in *Appleton's Monthly* lamented that, though potentially important, social science was "at present not a science at all. It is a pseudo-science."[35] Sometimes, the harshest critics were social-science practitioners themselves, a fact that might indicate some small measure of anxiety about the status of their field and the need to root out unacceptable practices. In 1896, the American economist-cum-sociologist Edward A. Ross claimed that ethics was a pseudoscience "like theology or astrology" because it sought to combine the mutually exclusive perspectives of the individual and the group. Albion W. Small, holder of the first American chair in sociology (at the University of Chicago), asserted, with a nod to the growing boundary between the scientific and popular, that his own field was "likely to suffer long from the assumptions of pseudo-science" because "sociologists are no more immune than other laymen against popular scientific error." Still others charged that social pseudoscience, particularly in economics, was misusing the tools of physical science, concealing "its emptiness behind a breastwork of mathematical formulas."[36]

Still, it was discussion of science and religion that seemed to generate the most invocations of pseudoscience during the late 1800s. A large amount of pseudoscientific rhetoric spilled out of the contemporary debate over evolution and its implications. In 1887, Thomas Henry Huxley himself published an article entitled "Scientific and Pseudo-Scientific Realism," the first salvo in an extended exchange between Huxley and George Campbell, the eighth Duke of Argyll, an outspoken opponent of

Darwinism. After a reply by Campbell, Huxley penned a second article simply called "Science and Pseudo-Science," to which the duke responded with an essay entitled "Science Falsely So Called." All these appeared in the British magazine *The Nineteenth Century*. Huxley's contributions were reprinted across the Atlantic in the *Popular Science Monthly*; his second article was additionally reprinted in the *Eclectic Magazine*.[37] The primary issue in this exchange was the status of natural law and its relationship with the divine. The Duke of Argyll claimed that natural laws were directly ordained and enforced by God. To Huxley, by contrast, any suggestion that the uniformity of natural law implied the existence of divine providence was illegitimate and pseudoscientific.

Huxley, however, was in the minority. Many practicing men of science were wary of supernatural interference in the natural world by the late 1800s, but few of them used the term "pseudoscience" to tar such belief. Instead, the vast majority of cries of pseudoscience came from the opponents of evolution. One author in the conservative *Catholic World* depicted the Huxleys, Tyndalls, and Darwins of the world as the "modern Cyclops, who in forging their pseudo-sciences examine nature, but only with one eye."[38] Other defenders of orthodox science and traditional religion, Catholic and Protestant, decried the "materialistic or pseudo-scientific skepticism of the day" or denounced the "pseudo-scientific sect" of Darwinian evolutionists.[39] Sometimes, antievolutionists refined their attacks by equating the pseudoscientific nature of evolution with "foreign tendencies which are alien from science or philosophy," including materialism and atheism.[40] Other critics sometimes claimed, without convincing evidence, that evolution did not meet the approval of most men of science, or at least the most prominent ones, thereby invoking a sense of orthodoxy for their cause.

Between the late nineteenth and early twentieth centuries, the mechanisms both for enforcing and communicating orthodoxy in scientific matters grew to new heights in the United States and in Britain. Graduate training, specialized journals, and membership in exclusive organizations all helped to establish a substratum of agreed-upon practices, facts, and concepts that most scientists learned to share. The increased professionalization of science paralleled more stringent boundaries around the scientific, including distinctions between scientists and laypeople and between legitimate scientific knowledge and scientific error and misunderstanding. Less permeable divisions were also emerging within science among the various disciplines, shaped by the need to master the expanding volume of specialized knowledge in any one area of work. Such an environment provided considerable encouragement for invocations of pseudoscience.

It remained a prominent feature of talk about such areas as the social sciences and popular ideas. Over the course of the early 1900s, the term also began to take important new features that would eventually dominate the rhetoric of pseudoscience, particularly during the last third of the century.

One area of continuity in early-twentieth-century discussions of pseudoscience involved method (see chapter 12). In order to show that textual criticism was a pseudoscience, a 1910 article in the journal of the Modern Language Association of America made what was by that time a venerable equation between the pseudoscientific and the violation of inductive reasoning.[41] Over the next several decades, ideas about scientific methodology changed in some significant ways, particularly in much more positive assessments of the role of theory. But though the methods of science changed, pseudoscience remained in violation of them. By far the most important shift in ideas about the methodology of science, however, was the emerging concept of the "scientific method." The proliferation of "scientific method" in public discussion implied a growing sense that science operated in special ways distinct enough to require its own name. It was, in short, a product of stronger boundaries around science, just like pseudoscience. Appropriately, a stricter view of "the established methods of science" was even more intimately linked with the pseudoscientific than during the previous century. A 1926 article in *California and Western Medicine* depicted nonorthodox medical ideas as an "attack upon the scientific method not alone in medicine but in all fields of knowledge."[42] In a more neutral mood, it was also possible to see a uniquely scientific method as what actually joined science and pseudoscience. The anthropologist Bronislaw Malinowski called magic "a pseudo-science" not because he thought it was illegitimate but because it had practical aims and was guided by theories, just like science.[43]

Talk about scientific method grew especially intense in the 1920s in discussions of the scientific status of the social sciences, which continued to find themselves depicted as scientific outsiders. A 1904 article in the *American Journal of Sociology* complained about charges leveled by "workers in other sciences" that sociology was pseudoscientific. In *The Public and Its Problems* (1927), the influential philosopher and psychologist John Dewey pointed out that no methodology could eliminate the distinction between "facts which are what they are independent of human desire" and "facts which are what they are because of human interest and purpose . . . In the degree in which we ignore this difference, social science becomes pseudo-science." Some critics described psychology as "the pseudoscience of thought." Others found in Sigmund Freud an example of the archetypical "pseudo-scientist."[44] As social scientists reacted to the ferment in their own rapidly changing disciplines, they sometimes painted the ideas of opponents in pseudoscientific tones. Advocates of Franz Boas's new ideas about cultural anthropology sometimes claimed that "the old classical anthropology . . . is not a science but a pseudo-science like medieval alchemy." By midcentury one critic, exasperated by "the deference paid to the pseudo-sciences, especially economics and psychology," declared that "if all economists and psychiatrists were rounded up and transported to some convenient St. Helena, we might yet save something of civilization."[45]

In discussions of race in particular, the concept of pseudoscience proved a useful tool for redrawing scientific maps for a new century. In a 1925 letter, W. E. B. Du Bois claimed that talk about racial disparity was "not scientific because science is more and more denying the concept of race and the assumption of ingrained racial difference." However, he lamented that "a cheap and pseudo-science is being sent broadcast through books, magazines, papers and lectures" asserting "that yellow people and brown people and black people are not human in the same sense that white people are human and cannot be allowed to develop or to rule themselves."[46] Over twenty years later, another critic of racial stereotypes asked "what greater evidence of the use of pseudo-science can we ask than that afforded by Nazi doctrines of the 'superiority' of *Das Herrenvolk*?"[47]

The particular basis on which certain ideas or entire disciplines were accused of being psuedoscientific varied considerably. Whatever their specifics, assertions about scientific status—pro and con—have drawn from the greater physicalization of science since the late nineteenth century. Skeptics typically pointed out that the social sciences did not measure up to the objectivity visible in such disciplines as biology or physics. Supporters emphasized new methods for overcoming potential subjectivity and prejudice in the study of human beings, often through the incorporation of mathematics. In an article discussing the differences between history and science, the philosopher Ferdinand C. S. Schiller asserted that "the *essential* characteristics of scientific knowledge, which distinguish it from pseudo-science, divination, guesswork, metaphysics, verbiage, and nonsense, are *prediction and control*."[48] Some social scientists, such as the psychologist John B. Watson, claimed that their fields promised a measure of control over human behavior. That promise often appealed to business leaders and members of the emerging advertising industry, though such interest in the practical side of social science was not immune to pseudoscientific aspersions. The author of a 1920 *Scientific Monthly* article noted interest among industrialists in intelligence and personality testing to "help solve the labor problem of industry." Sadly, however, their demand had resulted in the revival of such examples of "pseudo-science" as "phrenology, physiognomy, character reading and character analysis."[49]

Meanwhile, the conceptual boundary between science and the general public also generated a large share of the rhetoric of pseudoscience. That boundary had been enhanced by the specialization and increasingly technical nature of scientific knowledge. As the space between the scientific and popular grew, textbooks for the expanded public-school system and the developing profession of science journalism helped to establish sanctioned crossings of scientific boundaries and to spread more or less accurate reflections of scientific orthodoxy to the general public. Still, despite the work of dedicated popularizers, some observers claimed to detect a "great public interest in pseudo-science." In 1927, A. W. Meyer, then professor of anatomy at Stanford University, noted Sir William Osler's claim that there was no shortage of pseudoscience in medicine. Meyer added his own observation that public credulity seemed to be peaking, driven, he suggested, by the stresses of modern life, particularly those related to war and religion. Another author wrote with some alarm that "pseudo and unscientific cults are springing up and finding it easy to get a hold on the popular mind." As evidence, he cited West Virginia's legal recognition in 1925 of chiropractic and naturopathy.[50] Though such depictions could hardly brighten the heart of admirers of science, beliefs about the popularity of pseudoscience could also inspire and justify popularization. As early as 1921, an editor of *The Month* magazine had raised alarm about a "plague" of pseudoscientific ideas and asserted that "the follies and pretenses of pseudo-science are best met by vigorous and persistent exposure."[51]

Just as during the late 1800s, worries persisted that the media's handling of science was as much a cause of problems as of solutions. Popular treatments of science were often prime culprits in creating and spreading pseudoscientific misconceptions. As we have seen, W. E. B. Du Bois attributed widespread racial pseudoscience to "books, magazines, papers and lectures."[52] One of the new forms of literature emerging during the early 1900s that seemed especially likely to spread scientific error was the genre of the "scientific romance," exemplified by such authors as Jules Verne and H. G. Wells. Writing in the *Saturday Review*, one skeptic worried that many scientific novels offered fairy stories rather than "the anticipations of a future science founded upon the positively ascertained facts of the present" and aimed to amuse "boys and maidens" rather than educate serious-minded seekers after science. "On the whole," he advised his readers, "it is better when we want science to read science; and when we want fiction not to read a composite thing in which the science diverts us from the fiction, and the fiction is not more imaginary than the pseudo-science." He expressed the vain hope that the scientific romance would not have a bright future.[53] Over the course of the twentieth century, science fiction became a hugely popular genre of literature, as well as radio and television shows and movies. And though the "father of science fiction," Hugo Gernsback, did his best to turn it into a means of popularizing orthodox science, it continued to be a source of worry for those focused on the eradication of pseudoscientific ideas.[54]

One of the consequences of depictions of pseudoscience that made it essentially popular in character (and that sometimes also tended to view all popular knowledge as potentially pseudoscientific) was the perception of the United States, "always prolific in strange types," as "especially favored in the development of all sorts of pseudo-scientific cults and anti societies."[55] In many ways, that perception was facilitated by the association of American culture with democracy, heightened by early American public rhetoric and the impressions of a stream of European visitors during the early 1800s. However, well before the American Revolution, many Europeans and colonials had been aware of the dangers of democracy, including its potential enshrinement of the lowest common denominator and the consequent loss of order, disrespect for orthodoxy, and willingness to entertain novel, utopian, or simply weird notions. By the early 1900s, and particularly after the Great War, those concerns were coupled with increasing worries about the unreliability of the public among American intellectuals, including Walter Lippmann. By the second half of the twentieth century, assertions that such unorthodoxies as belief in UFOs or support for scientific creationism were uniquely American had become fairly common, even when those assertions were totally untrue. In fact, despite claims to the contrary by skeptics and critics, both the UFO phenomenon and scientific creationism have become global in extent.[56]

The increasingly sharp boundaries of science made the scientific appear more distinct and separate from what it was not, but they also cast a shaper shadow. Playing off descriptions of science as something unified and set apart were portrayals of pseudoscience that made it resemble a shadow science. On occasion such pseudoscience became a kind of semi-coherent, though still deeply flawed, collection of pseudo-disciplines with their own practitioners, sources of support, and methods of working. In 1932, the journalist and acidic wit H. L. Mencken suggested that every branch of science had an evil twin, "a grotesque Doppelganger" that transmuted legitimate scientific doctrines into bizarre reflections of the truth. Though Mencken did not gather these doppelgängers together into a single pseudoscientific horde, others did address the "assumptions and pretensions of the hydra-headed pseudo-science." The sense of a wide variety of pseudoscientific ideas originating from an essential core culminated in a tendency to see pseudoscience not simply as the scattered errors of true science but as an example of "anti-science," a concept that would become widespread among those worried about popular misconception during the rest of the century.[57]

The rhetorical trends of the previous century and a half laid the foundations for an explosion of talk about pseudoscience in America and elsewhere during the last third of the 1900s. After a small increase during the 1920s, usage of the term rose dramatically in the pages of English-language print media from the late 1960s onward, dwarfing the previous level of public invocation.[58] Such talk was especially aimed at a series of unorthodoxies that appeared to erupt into popular culture after World War II, including the astronomical theories of Immanuel Velikovsky, sightings of UFOs and their occupants, and reports of extrasensory perception (ESP). Such notions were not necessarily more transgressive than, say, phrenology, but they did occur against the backdrop of the greater establishment of science, including the massive infusion of material support for research from the federal government, particularly the military. That establishment helped to create an environment that encouraged protecting the boundaries of science against invasion. Scientists now had much more to lose. The considerable increase in support for scientific work also helped to establish a heightened sense of scientific orthodoxy. A highly developed system of graduate education provided the required credentials to would-be scientists and socialized them into certain shared practices, beliefs, and pieces of knowledge. Professional journals, policed by the peer-review process, ensured a similar synchronization, at least on basic matters of fact, theory, and method.

An enhanced sense that there was a scientific orthodoxy, as well as mechanisms for ensuring one, resulted in a much more strongly bounded concept of science. Against this backdrop Edward Condon, who had himself constructed a career within the military-industrial-academic establishment, denounced pseudoscience as "scientific pornography." Where consensus broke down, charges of pseudoscience could mobilize a communal distaste for transgression against one's rivals. Social scientists, ethnoscientists, sexologists, sociobiologists, and just about anyone working in psychology and psychiatry were susceptible to be labeled pseudoscientists. Emerging new fields, clamoring for respect and support, were often met with pseudoscientific accusations. While skeptics portrayed the so-called search for extraterrestrial intelligence, or SETI, as beyond the pale of scientific respectability, proponents such as the Cornell astronomer Carl Sagan asserted that SETI had moved from "a largely disreputable pseudoscience to an interesting although extremely speculative endeavor within the boundaries of science."[59]

Alongside the construction of a much stronger and more organized scientific establishment emerged a more structured means of communicating scientific consensus to the general public, whether they were reading textbooks in the expanding public-school system, scanning the daily newspaper, or watching the evening news on television. But some observers inside and outside the scientific community continued to worry that popular treatments of science were as much a source of pseudoscientific ideas as they were a means to combat them. Complaints about the media's handling of scientific topics provided a constant drumbeat. Even more insidious were the publications unfettered by the sorts of respectability that often kept mainstream media in check. Condon's 1969 denunciation of scientific pornography laid responsibility squarely on "pseudo-science magazine articles and paper back books," which sold by the tens of thousands and even millions.[60]

Since the early 1950s concerned citizens had been agitating for the formation of "one organization that could represent American science in combating pseudo-science." After decades of delay, champions of science finally banded together in 1976 to police the public sphere. Aroused by the popularity of Immanuel Velikovsky's *Worlds in Collission* (1950), Eric von Däniken's *Chariots of the Gods* (1968), and Charles Berlitz's *The Bermuda Triangle* (1974)—to say nothing of Uri Geller's spoon-bending and Jeane Dixon's prophesying—a group of skeptics under the leadership of philosopher Paul Kurtz formed the Committee for the Scientific Investigation of Claims of the Paranormal (CSICOP), renamed the Committee for Skeptical Inquiry (CSI) in 2006, and began publication of a pseudoscience-busting journal that they soon called the *Skeptical Enquirer*. One supporter, Carl Sagan, who became a crusader against pseudoscience during the last third of the twentieth century, asserted in the journal that "poor popularizations of science establish an ecological niche for pseudoscience" and worried that there was a "kind of Gresham's Law by which in popular culture the bad science drives out the good."[61]

Inspired by CSICOP and the slightly older Association française pour l'information scientifique, publisher of *Science et pseudo-sciences*, similar groups sprang up around the world. By 1984 there were organizations of skeptics in Australia, Belgium, Canada, Ecuador, Great Britain, Mexico, the Netherlands, New Zealand, Norway, and Sweden. During the next quarter century the anti-pseudoscience movement spread to Argentina,

Brazil, China, Costa Rica, the Czech Republic, Denmark, Ecuador, Finland, Germany, Hungary, India, Israel, Italy, Japan, Kazakhstan, Korea, Peru, Poland, Portugal, Russia, South Africa, Spain, Sri Lanka, Taiwan, Venezuela, and elsewhere. Many of the societies in these lands published their own magazines or newsletters. Concern often paralleled the eruption of some activity regarded as pseudoscientific, such as the Falun Gong in China and the Indian government's plan to introduce "Vedic astrology" as a legitimate course of study in Indian universities.[62]

Beyond equating the pseudoscientific and popular, late-twentieth-century invocations of pseudoscience continued along the lines set down during the 1800s and early 1900s. The transgression of proper methodology remained a primary means of identification. Sagan claimed in 1972 that the reason some people turned toward UFOs or astrology was "precisely that they are often beyond the pale of established science, that they often outrage conservative scientists, and that they seem to deny the scientific method."[63]

New methodological authorities also appeared. In the 1930s the Viennese philosopher Karl Popper began writing about "falsifiability" as a criterion *"to distinguish between science and pseudo-science."* He knew "very well," he later said, "that science often errs, and that pseudo-science may happen to stumble on the truth," but he nevertheless thought it crucial to separate the two.[64] Michael Ruse's invocation of falsifiability to distinguish between science and religion in a high-profile creation-evolution trial in Little Rock, Arkansas, in the early 1980s prompted Larry Laudan, another well-known philosopher of science, to charge his colleague with "unconscionable behavior" for failing to disclose the vehement disagreements among experts regarding scientific boundaries in general and Popper's lines in particular. By emphasizing the nonfalsifiability of creationism in order to deny its scientific credentials, argued Laudan, Ruse and the judge had neglected the "strongest argument against Creationism," namely, that its claims had already been falsified. Laudan dismissed the demarcation question itself as a "pseudo-problem" and a "red herring." Ruse, in rebuttal, rejected Laudan's strategy as "simply not strong enough for legal purposes." Merely showing creation science to be "bad science" would have been insufficient in this case, because the constitution does not ban the teaching of bad science in public schools.[65]

Just as they had during the late 1800s, differences between science and religion loomed large in characterizations of pseudoscience, although in precisely the opposite way than they had before. Rather than signaling the overaggressive separation of scientific and religious concerns, Americans during the second half of the twentieth century more often linked pseudoscience to the illegitimate mixture of science and religion. Charges of pseudoscience aimed at a wide variety of targets, from creationism to UFOs to federal standards for organic food, all of which were denounced as involving religious motivations rather than scientific ones. By the turn of the twenty-first century, particularly in the context of a number of politically charged debates involving science, there was also a pronounced tendency to see pseudoscience as arising from the intrusion of political concerns onto scientific ground. In controversies over global warming, stem-cell research, intelligent design, and even the demotion of Pluto from planetary status (because the final decision was made by vote), partisans depicted their opponents as proceeding from political motivations and thus distorting pure science. In reaction to the comments of President George W. Bush that appeared to open the possibility of including information on intelligent design in public schools, critics charged him with raising a "pseudoscience issue" and "politicizing science," which "perverted and redefined" the true nature of scientific knowledge. Assertions of a "Republican war on science" from the Left of the political spectrum echoed this sentiment.[66]

The most dramatic development in portrayals of pseudoscience after midcentury was the emergence of what we might call "Pseudoscience" with a capital "P" and without a hyphen. This reflected a growing sense during the 1920s and 1930s that pseudoscientific beliefs were not simply scattered errors to be exorcized from the boundaries of science, but rather a complex system of notions with their own set of boundaries, rather like an "alternative" version of science. The loss of the hyphen was a subtle indication of this transition, insofar as it weakened a seeming dependence on the scientific and suggested something more than simply false science. From the late 1960s on, many skeptical scientists and popularizers explicitly depicted links among a large collection of unusual topics, including "everything from PK (psychokinesis, moving things by will power) and astral projection (mental journeys to remote celestial bodies) to extraterrestrial space vehicles manned by web-footed crews, pyramid power, dowsing, astrology, the Bermuda triangle, psychic plants, exorcism and so on and so on." In the midst of public debate over intelligent design, journalist John Derbyshire blasted the "teaching of pseudoscience in science classes" and asked "why not teach the little ones astrology? Lysenkoism? . . . Forteanism? Velikovskianism? . . . Secrets of the Great Pyramid? ESP and psychokinesis? Atlantis and Lemuria? The hollow-earth theory?" Though lists differed, they often revolved around a similar core of unorthodoxies, including what one author characterized in 1998 as the "archetypical fringe theory," namely, belief in UFOs.[67]

Supporters of science also wrote encyclopedic condemnations, most notably Martin Gardner's pioneering *In the Name of Science* (1952), subsequently published as *Fads & Fallacies in the Name of Science*, which gathered a variety of subjects together under the general banner of pseudoscience. Indeed, universalized depictions of pseudoscience became a convenient and clearly articulated target for those dedicated to crusading against antiscience in all its forms. From the 1970s on, CSICOP's *Skeptical Enquirer* proved to be one of the most important locations in which pseudoscience was forged, elaborated on, and stridently denounced. In 1992 a likeminded organization in Southern California, the Skeptics Society, began publishing a second major magazine devoted to "promoting science and critical thinking," *Skeptic*, published and edited by the historian of science Michael Shermer. To focus on medical matters, CSICOP in 1997 helped to launch *The Scientific Review of Alternative Medicine and Aberrant Medical Practices*, followed five years later by a sister journal, *The Scientific Review of Mental Health Practice: Objective Investigations of Controversial and Unorthodox claims in Clinical Psychology, Psychiatry, and Social Work.*[68]

But pseudoscience did not just provide a nicely packaged enemy; it also provided an object for more neutral study. As early as 1953 the History of Science Society proposed adding a section on "Pseudo-Sciences and Paradoxes (including natural magic, witchcraft, divination, alchemy and astrology)" to the annual *Isis* critical bibliography. Later, history and philosophy of science (HPS) departments featured the study of pseudoscience. "Another important function of HPS is to differentiate between science and pseudo-science," announced the University of Melbourne. "If HPS is critical of the sciences, it is even more so when dealing with pseudo-sciences and the claims they put forth to defend themselves." In the late 1970s scholarly studies of pseudoscience—by scientists as well as historians, philosophers, and sociologists of science—began appearing in increasing numbers.[69] Sociologists of science associated with the so-called "strong programme" in the sociology of knowledge at the University of Edinburgh were especially influential in encouraging this development. Among the cardinal tenets of this initiative was impartiality "with respect to truth and falsity" of scientific claims; in other words, the same type of explanations would be applied to "true and false beliefs" alike. Thus encouraged, reputable historians of science devoted entire books to such topics as phrenology, mesmerism, parapsychology, and creationism. By the early twenty-first century, Michael Shermer was able to bring out a two-volume encyclopedia of pseudoscience.[70]

Despite attempts to situate it in a less negative context, pseudoscience almost always remained a term of denunciation. Still, it did capture something real. People interested in unusual topics had begun to link them together. An examination of one extensive bibliography suggests a growing tendency, particularly during the 1960s and 1970s, to combine multiple unorthodoxies into a single volume. This practice had its roots in the 1920s, particularly in the work of the former journalist and failed novelist Charles Fort. In his *Book of the Damned* (1919) and in several subsequent volumes, Fort cataloged stories about a wide range of unusual phenomena, including strange aerial objects; rains of frogs, fish, and other unusual things; psychic events; accounts of spontaneous human combustion; and other phenomena he claimed had been ignored or "damned" by orthodox science.[71] His efforts ultimately inspired the formation of the Fortean Society in 1932, as well as the publication of a number of self-described Fortean magazines that continued the compilation of strange phenomena, and a range of aspiring UFOlogists and parapsychologists. Fort's ideas also bled into popular culture. Many of the topics covered by Fort appeared, sometimes in nearly identical terms, in episodes of the *X-Files* during its nine-year run on television (1993–2002). Devotees of the unusual have typically avoided the term pseudoscience in favor of "alternative," "forbidden," or "weird" science. They have also emphasized what one observer has labeled a "kinder, gentler science," more accessible than mainstream science.[72] In recent decades critics of alternative science have created their own synonyms for pseudoscience, including anti-science, cargo-cult science, and junk science.[73] But all such rhetoric, along with the grandparent of them all—pseudoscience—remains closely connected to the preservation of scientific boundaries and the protection of scientific orthodoxy.

NOTES

1. Oliver Wendell Holmes, "The Professor at the Breakfast-Table," *Atlantic Monthly* 1859, 232–43; Hjalmar O. Lokensgard, "Oliver Wendell Holmes's 'Phrenological Character,'" *New England Quarterly* 13 (1940): 711–18, reprints the phrenologist's reading of Holmes's head. "Second Advent Miller" refers to William Miller, who had predicted the Second Coming of Christ in 1844; "Joe Smith" was Joseph Smith, the founder of the Church of Jesus Christ of Latter-Day Saints (Mormons).

2. Holmes, "The Professor at the Breakfast-Table," 241–42; Lokensgard, "Holmes's 'Phrenological Character,'" 713. On Holmes's identification of astrology, alchemy, and homeopathy as pseudosciences, see Holmes, *Homœopathy, and Its Kindred Delusions* (Boston: William D. Ticknor, 1842), 1.

3. Theodore Hornberger, "Samuel Johnson of Yale and King's College: A Note on

the Relation of Science and Religion in Provincial America," *New England Quarterly* 8 (1935): 391. For a brief historical survey, see Ronald L. Numbers, "Pseudoscience and Quackery," in *The Oxford Companion to United States History*, ed. Paul S. Boyer (New York: Oxford University Press, 2001), 630–31.

4. Ronald L. Numbers, *The Creationists: From Scientific Creationism to Intelligent Design*, expanded ed. (Cambridge, MA: Harvard University Press, 2006), 162 (Victoria Institute); Duke of Argyll, "Science Falsely So Called," *Nineteenth Century* 21 (1887): 771–74; David N. Livingstone, "Darwinism and Calvinism: The Belfast-Princeton Connection," *Isis* 83 (1992): 411 (BAAS); *Comprehensive Index to the Writings of Ellen G. White* (Mountain View, CA: Pacific Press, 1963), 3:2436; Ellen G. White, *The Great Controversy between Christ and Satan* (Mountain View, CA: Pacific Press, 1888), 522 (mere theories). See also Ronald L. Numbers, "Science Falsely So-Called: Evolution and Adventists in the Nineteenth Century," *Journal of the American Scientific Affiliation* 27 (March 1975): 18–23.

5. Daniel G. Binton, "The Character and Aims of Scientific Investigation," *Science* 4 (1895): 4 (light); Edward Condon, "UFOs I Have Loved and Lost," *Bulletin of the Atomic Scientists* 25 (December 1969): 6–8.

6. *Oxford English Dictionary*, s.v. "pseudo."

7. American Periodical Series Online, http://www.proquest.com/products_pq/descriptions/aps.shtml.

8. Pierre Larousse, *Grand dictionnaire universel du XIXe siècle*, 17 vols. (Paris: Administration du Grand dictionnaire universel, 1866–1879.); Émile Littré, ed., *Dictionnaire de la langue française*, 4 vols. (Paris: Libraire Hachette, 1873–74). We are grateful to Camilo Quintero for his assistance in searching foreign-language dictionaries and reference works.

9. On phrenology in America, see Daniel Patrick Thurs, "Phrenology: A Science for Everyone," in *Science Talk: Changing Notions of Science in American Culture* (New Brunswick, NJ: Rutgers University Press, 2007), 22–52.

10. For an example of "*pseudo* science," see "Sir William Hamilton on Phrenology," *American Journal of Insanity* 16 (1860): 249. See also untitled review, *Medical Repository of Original Essays and Intelligence*, new ser. 8 (1824): 444.

11. Alison Winter, *Mesmerized: Powers of Mind in Victorian Britain* (Chicago: University of Chicago Press, 1998), esp. 306–43.

12. Thurs, "Phrenology: A Science for Everyone," 24.

13. "'Scientific Agriculture,'" *Country Gentleman* 7 (1856): 93; "Editor's Table," *Harper's New Monthly Magazine* 4 (1852): 839 (pseudo-spiritualism); Christine Garwood, *Flat Earth: The History of an Infamous Idea* (London: Macmillan, 2007), 70 (paradoxer).

14. Hugh Richard Slotten, *Patronage, Practice, and the Culture of American Science: Alexander Dallas Bache and the U.S. Coast Survey* (Cambridge: Cambridge University Press, 1994), 28.

15. David Meredith Reese, *Humbugs of New-York* (New York: Weeks, Jordan, 1838), 110–11.

16. Andrew Ure, *A Dictionary of Arts, Manufactures, and Mines* (New York: Appleton, 1853), 368.

17. "Editor's Table," 841.

18. Review of *Homeopathy Unmasked*, 391.

19. Slotten, *Patronage, Practice, and the Culture of American Science*, 29.

20. Reese, *Humbugs of New-York*, 111; Holmes, "The Professor at the Breakfast-Table."

21. John K. Kane to Joseph Henry, November 20, 1843, in *The Papers of Joseph Henry* (Washington, DC: Smithsonian Institution Press, 1992), 5:451; Joseph Henry to Alexander Dallas Bache, April 16, 1844, ibid., 6:76–77.

22. George M Sternberg, "Science and Pseudo-Science in Medicine," *Science* 5 (1897): 199–206. See also J. B. Stallo, "Speculative Science," *Popular Science Monthly* 21 (1882): 145–64.

23. Daniel G. Brinton, "The Character and Aims of Scientific Investigation," *Science* 1 (1895): 3–4; C. F. Nichols, "Divine Healing," *Science* 19 (1892): 43–44; William Leon Brown, *Christian Science Falsely So Called* (New York: Fleming H. Revell, 1911); "Separation of the Tares and the Wheat," *Christian Science Journal* 6 (1889): 546 (pseudo scientists). See also Rennie B. Schoepflin, "Separating 'True' Scientists from 'Pseudo' Scientists," in *Christian Science on Trial: Religious Healing in America* (Baltimore, MD: The Johns Hopkins University Press, 2003), 82–109.

24. Augustine F. Hewitt, "The Warfare of Science," *Catholic World* 911(1891): 679; Daniel G. Brinton, *Myths of the New World: A Treatise on the Symbolism and Mythology of the Red Race of America* (New York: Henry Holt, 1876), 298; J. H. M'Clelland, *The Mind: The Annual Address Delivered before the Homeopathic Medical Society of Pennsylvania* (Pittsburgh: Wm. G. Johnston, 1876), 22; "Character in Finger Nails," *New York Times*, December 5, 1884.

25. "The Old and New Phrenology," *The Open Court* 1 (1887): 435.

26. See Frank M. Turner, "The Victorian Conflict between Science and Religion: A Professional Dimension," *Isis* 69 (1978): 356–76.

27. B. E. Fernow, "Pseudo-Science in Meteorology," *Science* 3 (1896): 706.

28. "Character in Finger Nails," *New York Times*, December 5, 1884.

29. R. S. Woodward, "Address of the President," *Science* 12 (1900): 14.

30. "Time Wasted," *Science* 6 (1897): 969.

31. E. F. Carr, "A Theory, an Extravaganza," *Ladies' Repository* 12 (1873): 125.

32. Joseph LeConte, "Rough Notes of a Yosemite Camping Trip—III," *Overland Monthly*, 2nd ser. 6 (1885): 636.

33. Sternberg, "Science and Pseudo-Science in Medicine," 202.

34. Woodward, "Address of the President," 14.

35. W. H. Mallock, "Civilization and Equality," *Contemporary Review* 40 (1881): 660; reprinted in *Appletons' Journal*, new ser. 11 (1881): 526–38.

36. Edward A. Ross, "Social Control," *American Journal of Sociology* 1 (1896): 513–35, on 513 and 530; Albion W. Small, "The Scope of Sociology," *American Journal of Sociology* 6 (1900): 42–66, on 59; Review of *Mathematical Economics* by Wilhelm Launhardt, *Science* 1 (1886): 309.

37. See Thomas Henry Huxley, "Scientific and Pseudo-Scientific Realism," *Nineteenth Century* 21 (1887) 191–204, repr. in *Popular Science Monthly* 30 (1887): 789–803; Thomas Henry Huxley, "Science and Pseudo-Science," *Nineteenth Century* 21 (1887): 481–98, repr. in *Popular Science Monthly* 31(1887): 207–24, and in *Eclectic Magazine*, new. ser., 45 (1887): 721–31. For Campbell's articles, see Duke of Argyll, "Professor Huxley on Cannon Liddon," *Nineteenth Century* 21 (1887): 321–39; Duke of Argyll, "Science Falsely So Called," 771–74.

38. "Socialism and Communism in 'The Independent,'" *Catholic World* 28 (1879): 812–13.

39. Review of *Inductive Inquiries in Philosophy, Ethics, and Ethnology*, by A. H. Dana, *Presbyterian Quarterly and Princeton Review*; "Darwin on Expressions," *Littell's Living Age*, new. ser. 2, July 5, 1873,: 561 (pseudo-scientific skepticism).

40. Augustine F. Hewit, "Scriptural Questions," *Catholic World* 44 (1887): 660–61.

41. Frederick Tupper Jr., "Textual Criticism as Pseudo-Science," *PMLA* 25 (1910): 176.

42. A. B. MaCallum, "Scientific Truth and the Scientific Spirit," *Science* 43 (1916): 444; Peter Frandsen, "Anti-Scientific Propaganda," *California and Western Medicine* 25 (1926): 336.

43. Bronislaw Malinowski, *Magic, Science and Religion* (Garden City, NY: Doubleday, 1954), 87.

44. Albion W. Small, "The Subject-Matter of Sociology," *American Journal of Sociology* 10 (1904): 281; John Dewey, *The Public and Its Problems* (New York: Henry Holt, 1927), 7; L. S. Hearnshaw, "A Reply to Professor Collingwood's Attack on Psychology," *Mind* 51 (1942): 165; Rose Macaulay, *Keeping Up Appearances* (London: W. Collins Sons, 1928), 213.

45. Edward Sapir, quoted in Leslie A. White, "Evolutionism in Cultural Anthropology: A Rejoinder," *American Anthropologist* 49 (1947): 407; Hugh Ross Wiliamson, "Pseudo-Science," *National and English Review* 139 (1952): 48.

46. Letter dated February 9, 1925, in *The Correspondence of W. E. B. Du Bois*, ed. Herbert Aptheker (Amherst: University of Massachusetts Press, 1973–1978), 1: 303.

47. Wilton Marion Krogman, "Race Prejudice, Pseudo-Science and the Negro," *Phylon* 8 (1947): 14.

48. F. C. S. Schiller, "Are History and Science Different Kinds of Knowledge," *Mind*, new ser. 31 (1922): 462 (original emphasis).

49. Arthur Frank Payne, "The Scientific Selection of Men," *Scientific Monthly* 11 (1920): 545.

50. "Irresponsible Pseudo-Science," *Bookman* 81 (1932): 576; A. W. Meyer "Reflections on Credulity," *Scientific Monthly* 24 (1927): 530, 532; Frandsen, "Anti-Scientific Propaganda," 336.

51. Editor, "The Plague of Pseudo-Science," *Month* 137 (1921): 531.

52. Letter dated February 9, 1925, in *Correspondence of W. E. B. Du Bois*, 1:303.

53. "Science in Romance," *Saturday Review* 96 (1905): 414–15.

54. See Mark Richard Siegel, *Hugo Gernsback: Father of Modern Science Fiction* (San Bernadino, CA: Borgo Press, 1988).

55. Editor, "The Plague of Pseudo-Science," 531; Frandsen, "Anti-Scientific Propaganda," 336.

56. Ronald L. Numbers, "Creationism Goes Global," in *The Creationists*, 399–431.

57. H. L. Mencken, "Nonsense as Science," *American Mercury* 27 (1932): 509–10; Peter Guthrie Tait, "Religion and Science," in *Life and Scientific Work of Peter Guthrie Tait*, ed. Cargill Gilston Knott (Cambridge: Cambridge University Press, 1911), 293 (hydra-headed), an essay originally published in 1888. For a link between pseudo- and anti-science, see, e.g., Frandsen, "Anti-Scientific Propaganda," 336–38.

58. This was true of articles with "pseudoscience" in their titles as indexed by Readers' Guide and the Periodicals Contents Index (now Periodicals Index Online). These searches were conducted in January 2003.

59. C[arl] Sagan, "The Recognition of Extraterrestial Intelligence," *Proceedings of the Royal Society of London B* 189 (1975): 143.

60. Edward Condon, "UFOs I Have Loved and Lost," *Bulletin of the Atomic Scientists* 15 (December 1969): 6–8.

61. Samuel A. Miles, letter to the editor, *Science* 114 (1951): 554; Paul Kurz, "A Quarter Century of Skeptical Inquiry: My Personal Involvement," *Skeptical Inquirer* 25 (July/August): 42–47; Carl Sagan, "The Burden of Skepticism," *Skeptical Inquirer* 12 (Fall 1987): 46. In 1974 Philip H. Abelson, editor of *Science*, had called attention to the problem in "Pseudoscience," *Science* 184 (1974): 1233.

62. "The State of Belief in the Paranormal Worldwide," *Skeptical Inquirer* 8 (Spring 1984): 224–38; "International Committees," *Skeptical Inquirer* 9 (Fall 1984): 97; "Show-and-Tell Time Exposes Pseudo-Science," *People's Daily Online*, March 26, 2000, http://english.peopledaily.com.cn/english/200003/26/eng20000326N120.html; R. Ramachandran, "Degrees of Pseudo-Science," *Frontline* 18 (March 31–April 13, 2001); T. Jayaraman, "A Judicial Blow," *Frontline* 18 (June 9–22, 2001). For a current list of international organizations, see http://www.csicop.org/resources/international_organizations.

63. Carl Sagan, introduction to *UFO's—A Scientific Debate*, ed. Carl Sagan and Thornton Page (Ithaca, NY: Cornell University Press, 1972), xiii.

64. Karl Popper, *Conjectures and Refutations: The Growth of Scientific Knowledge* (London: Routledge & Kegan Paul, 1963), 33 (original emphasis); Karl Popper, *Logik der Forschung* (Vienna: Springer, 1934; published in English as *The Logic of Scientific Discovery* [New York: Basic Books, 1959]). For other philosophical discussions of pseudoscience, see R. G. Collingwood, *An Essay on Metaphysics* (Oxford: Clarendon Press, 1940); and Imre Lakatos, "Lecture One: The Demarcation Problem," in Imre Lakatos and Paul Feyerabend, *For and Against Method*, ed. Matteo Motterlini (Chicago: University of Chicago Press, 1999), 20–31.

65. Numbers, *The Creationists*, 277–68; Larry Laudan, "The Demise of the Demarcation Problem," in *Physics, Philosophy, and Psychoanalysis: Essays in Honor of Adolf Grünbaum*, ed. R. S. Cohen and Larry Laudan (Dordrecht, Holland: D. Reidel, 1983), 111–27, repr. in *But Is It Science? The Philosophical Question in the Creation/Evolution Controversy*, ed. Michael Ruse (Buffalo, NY: Prometheus Books, 1988), 337–50; Michael Ruse, "Pro Judice," in *But Is It Science?* 357.

66. Gerry Wheeler, quoted in Claudia Wallis, "The Evolution Wars," *Time*, August 15, 2005, 28; Jonathan Alter, "Monkey See, Monkey Do: Offering ID as an Alternative to Evolution is a Cruel Joke," *Newsweek*, August 15, 2005, 27; Robert George Sprackland, "A Scientist Tells Why 'Intelligent Design' Is NOT Science," *Educational Digest* 71 (January 2006): 33; Chris Mooney, *The Republican War on Science* (New York: Basic Books, 2005).

67. John Pfeiffer, "Scientists Combine to Combat Pseudoscience," *Pyschology Today* 11 (November 1977): 38; John Derbyshire, "Teaching Science," *National Review* 30 (August 2005), http://old.nationalreview.com/derbyshire; Steven Dutch, "The Great Silly Season: 1965–1981," www.uwgb.edu/dutchs.

68. Martin Gardner, *In the Name of Science* (New York: G. P. Putnam's Sons, 1952).

69. "Proposed System of Classification for *Isis* Critical Bibliograpy," *Isis* 44 (1953): 229–31; http://www.arts.unimelb.edu.au/amu/ucr/student/1996. Major works

included Roy Wallis, ed., *On the Margins of Science: The Social Construction of Rejected Knowledge* (University of Keele, 1979); Marsha P. Hanen, Margaret J. Osler, and Robert G. Weyant, eds., *Science, Pseudo-Science and Society* ([Waterloo,ON]: Wilfrid Laurier University Press, 1980); Daisie Radner and Michael Radner, *Science and Unreason* (Belmont, CA: Wadsworth, 1982); Harry Collins and Trevor Pinch, *Frames of Meaning: The Social Construction of Extraordinary Science* (London: Routledge, 1982); Thomas Leahey and Grace Leahey, *Psychology's Occult Doubles: Psychology and the Problem of Pseudoscience* (Chicago: Nelson-Hall, 1983); Rachel Laudan, ed., *The Demarcation between Science and Pseudo-Science* (Blacksburg, VA: Center for the Study of Science in Society, Virginia Polytechnic Institute and State University, 1983); Nachman Ben-Yehuda, *Deviance and Moral Boundaries: Witchcraft, the Occult, Science Fiction, Deviant Sciences, and Scientists* (Chicago: University of Chicago Press, 1985); Terence Hines, *Pseudoscience and the Paranormal* (Amherst, NY: Prometheus Books, 1988); Nathan Aaseng, *Science versus Pseudoscience* (New York: Franklin Watts, 1994); Michael Zimmerman, *Science, Nonscience, and Nonsense: Approaching Environmental Liberacy* (Baltimore, MD: The Johns Hopkins University Press, 1995); Michael W. Friedlander, *At the Fringes of Science* (Boulder, CO: Westview Press, 1995); Paul R. Gross, Norman Levitt, and Martin W. Lewis, eds., *The Flight from Science and Reason* (New York: New York Academy of Sciences, 1996); Henry H. Bauer, *Science or Pseudoscience: Magnetic Healing, Psychic Phenomena, and Other Heterodoxies* (Urbana: University of Illinois Press, 2001); Robert L. Park, *Voodoo Science: The Road from Foolishness to Fraud* (New York: Oxford University Press, 2000). On pseudoscience, see Seymour H. Mauskopf, "Marginal Science," in *Companion to the History of Modern Science*, ed. R. C. Olby et al. (London: Routledge, 1990), 869–85.

70. David Bloor, *Knowledge and Social Imagery*, 2nd ed. (1976; Chicago: University of Chicago Press, 1991), 7 (tenets); Seymour H. Mauskopf and Michael R. McVaugh, *The Elusive Science: Origins of Experimental Psychical Research* (Baltimore, MD: The Johns Hopkins University Press, 1980); Roger Cooter, *The Cultural Meaning of Popular Science: Phrenology and the Organization of Consent in Nineteenth-Century Britain* (Cambridge: Cambridge University Press, 1984); Numbers, *The Creationists*; Winter, *Mesmerized*; Michael Shermer, ed., *The* Skeptic *Encyclopedia of Pseudoscience*, 2 vols. (Santa Barbara, CA: ABC Clio, 2002). For a scathing indictment of the "strong programme," see Larry Laudan, "The Pseudo-Science of Science?" *Philosophy of the Social Sciences* 11 (1981): 173–98.

71. T. Harry Leith, ed., *The Contrasts and Similarities among Science, Pseudoscience, the Occult, and Religion*, 4th ed. (Toronto, 1986); Charles Fort, *The Book of the Damned* (New York: Boni & Liveright, 1919).

72. Andrew Ross, *Strange Weather* (New York: Verso, 1991), 15–74.

73. In a lecture given in 1974, the physicist Richard P. Feynman packaged pseudoscience as "Cargo Cult Science"; see Feynman, *"Surely You're Joking, Mr. Feynman!": Adventures of a Curious Character* (1985; New York: W. W. Norton, 1997), 338–46. See also Gerald Holton, *Science and Anti-Science* (Cambridge, MA: Harvard University Press, 1993); and P. W. Huber, *Galileo's Revenge: Junk Science in the Courtroom* (New York: Basic Books, 1991).

CHAPTER 12

Scientific Methods

Daniel P. Thurs

In 1940, Howard Roelofs, then Wilson Professor of Ethics and chair of the Philosophy Department at the University of Cincinnati, recalled his upbringing as a "Trinitarian." "The members of the Trinity," he wrote, "were God, my Father, and Scientific Method." Understanding God presented no particular problem. He "went to church regularly and received peppermints during the sermons." He found the Bible and occasional "theological expositions" from his father and the minister "fascinating and intelligible." His father offered a harder puzzle. "God was good, my Father was intelligent; and the latter I found more perplexing than the former." Yet, here too, Roelofs was able to achieve an understanding sufficient for his needs. Scientific method proved a much more difficult challenge. When Roelofs put the matter to his father, who loved to invoke the scientific method as the solution to all kinds of problems from inefficient government to deficient education, the paternal reply would be, "think about it." Presumably the answer would come. Unfortunately, it did not. As he emerged from childhood, Roelofs realized with some horror that "I not only did not know what Scientific Method was but that to my knowledge I had never used it." This epiphany started him on a frequently frustrating quest for the true nature of the scientific method, a search that lasted well into his college years.[1]

We can take several lessons from Roelofs's reflections. The prominent place of "Scientific Method" in his upbringing was perhaps far from average—he recalled mentioning the third member of his trinity on the playground, only to discover that his schoolmates had no idea what he was talking about. But "scientific method" as a standardized slogan grew to prominence during the early years of the twentieth century, just those years Roelofs was recalling (see figures 12.1, 12.2, and 12.3).[2] Nor was its formation limited to specialized and technical discussions of scientific practice. Methodological rhetoric has typically served as a resource in debates, whether over details of scientific knowledge, across the borders of institutional or disciplinary organization, or in public. Partisans have cited very different images on each of these levels. The more "internal" a debate, the more likely methodological tools involved particular procedures, protocols, or techniques of interest to close colleagues instead of some global and often rather abstract characteristic of science as a whole. Scientific method was, and remains, a creature of the popular realm and, unlike many scientific terms, a shared element in the vocabularies of scientists and laymen alike. By the mid-twentieth century, "scientific method" had often found a place in that arbiter of standard English vocabulary, the dictionary. Likewise, many modern high school science textbooks have come to include a chapter or section entitled "Scientific Method." The 1981 Macmillan *Biology* even distilled its steps into a handy, easy-to-remember flow chart.[3]

Roelofs's recollections also rightly suggest that, though many people have agreed on the importance of scientific method, it has often eluded any precise definition, dictionaries aside. While a third of Americans surveyed in a 1958 poll conducted for the National Association of Science Writers identified the scientific method vaguely with being "thorough" and "getting to the bottom" of things, a quarter could not answer the question at all.[4] Nor was there any more clarity among practicing scientists. In 1874, British economist Stanley Jevons complained in his widely noted *Principles of Science* that "physicists speak familiarly of scientific method, but they could not readily describe what they mean by that expression."[5] Near the apparent height of its popularity in the 1920s, sociologist Stuart Rice attempted an "inductive examination" of the definitions of scientific method offered in social scientific literature. Ultimately, he lamented the "futility" of his mission. "The number of items in such an enumeration," he wrote, "would be infinitely large."[6] Unlike discussions about scientific knowledge, in which modern scientists have managed to convince many people, at least those with political and financial power, of their right to speak authoritatively, no single group has successfully claimed the right to determine or force consensus about the meaning of scientific method.

Identifying such variation in meaning is not to question the existence of reliable scientific knowledge or to undermine the conclusions of scientists. Rather, it is to accept the nature of language. Significant cultural

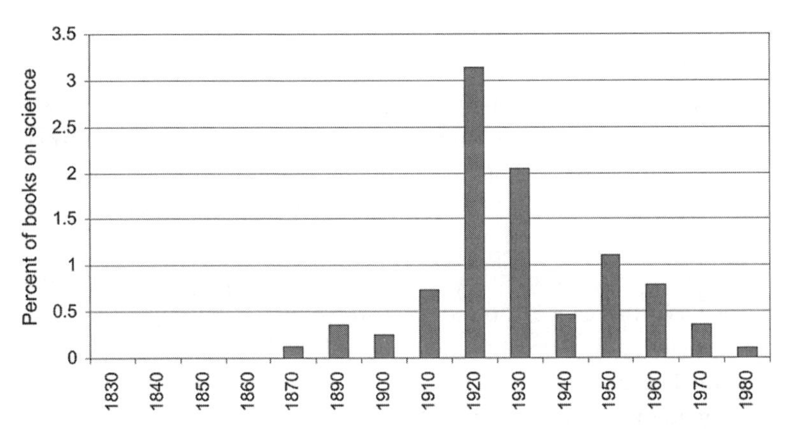

Figure 12.1. Books with the phrase "scientific method" in the title. (Source: Library of Congress.)

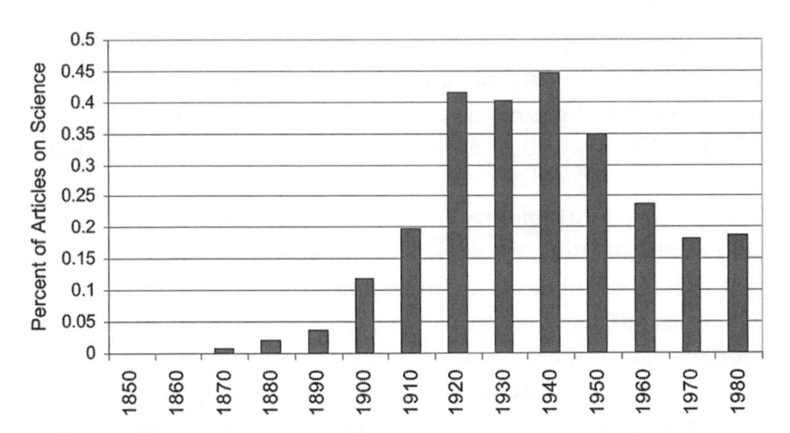

Figure 12.2. Magazine articles with phrase "scientific method" in the title. (Source: Periodicals Contents Index.)

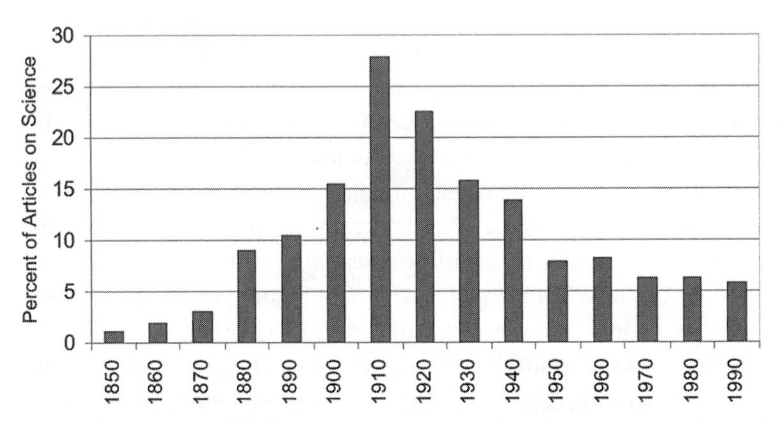

Figure 12.3. New York Times articles that include the phrase "scientific method." (Source: Proquest.)

keywords of the kind that scientific method became have often contained "radically different or radically variable, yet sometimes hardly noticed, meanings and implications of meaning."[7] In the absence of explicit questioning, the flexibility of such important terms as "family," "race," "freedom," or "America" has enabled very different groups and individuals to unite around them; in the midst of struggle, it has been a valuable source of rhetorical weaponry. While its slippery definition has limited its utility in any precise, technical discussion, it has made the term ideal for what Thomas Gieryn has called "boundary-work"—that is, attempts to fix the borders of science in ways that met the goals of individuals and groups and provided them with access to valuable resources. Multiple definitions of scientific method allowed multiple visions of science, and thus helped to clear out cultural space for boundary-work. UFO researchers with a desire to be included within the fold of science, for instance, have often emphasized that scientific method calls for open-mindedness, while those more orthodox scientists inclined to dispute the scientific pretensions of UFOlogy have stressed the requirement of a skeptical attitude. Scientific method, like science, was thus never one thing. It was many potentially useful things.[8]

Focusing on scientific method in this way might seem a little narrow. After all, prominent thinkers have offered advice about the proper ways to approach the study of nature for millennia. We might also focus on the full range of actual activities, insofar as they are accessible to the modern historian, that people around the globe have engaged in when they have tried to understand natural phenomena. Determining which of these prescriptions and behaviors we should examine here, of course, requires that we decide which are truly scientific. But if we accept the goal of this volume—that is, problematizing the nature of science instead of taking it for granted and opening it up to historical study—we cannot make such decisions without short-circuiting our overall project. Rather than imposing any particular notion of science onto the past, why not let historical figures determine what counted as "scientific" in their own words? This leads us to what may seem like a second symptom of myopia, namely, restricting our examination to English-language (and particularly American) discussion of "scientific method." Arguably, given the prominence of the United States as a center of scientific research and training for international students after World War II, there may be some reason to privilege American discussion. But even if no such rationale exists, focusing on English speakers is less a limitation than a beginning. In particular, it invites comparison with non-English traditions of talking about proper means of studying nature, from the methods of German *wissenshaft* to those of

Japanese *kagaku*. Differences and similarities can ultimately tell us much about alternative visions of science, nature, and their role in culture.

We therefore should not take seriously the claim offered by a lengthy Wikipedia entry on the scientific method that its development "is inseparable from the history of science itself."[9] Or rather, we should take it more seriously than the author perhaps intended. That is, talk about "scientific method" has been intimately related to talk about "science." The spread of scientific method as a widely used term reflects changing ideas about science and its role in the world; in particular, its growing prominence paralleled the emergence of science as a powerful category in the United States and elsewhere over the last two centuries. Such high-profile elements of modern science talk as "science and religion," "scientist," and "pseudoscience" grew in prominence over the last decades of the nineteenth century, the same period of time that witnessed "the scientific method" becoming a widely familiar phrase.[10] All of these expressions highlighted the growing distinctness of science. They eclipsed habits of talk about the scientific that emphasized its continuities with other intellectual, social, and cultural realms and shepherded its transformation from a category sometimes still synonymous with general knowledge during the early 1800s to an unquestionably special and distinct brand of information by the early 1900s. Ultimately, those changes opened the door to attestations of the cultural authority of science in contrast with other sorts of human activities.

THE REIGN OF BACON

We can set the stage for the emergence of scientific method and its role in the creation of ideas about modern science by glancing at methodological talk before the twentieth century. As figures 1–3 show, the term appeared only rarely in the vocabularies of English speakers prior to the last quarter of the nineteenth century. In several popular American periodicals before 1840, the number of articles that contained the phrase amounted to only 0.3 percent of those articles that used the word "science" somewhere in their texts. During the 1840s and 1850s, this percentage increased to only about 1 percent.[11] Conversely, discussion of the proper method for gaining knowledge has been traced back to the ancient Greeks, with a major flurry of activity during the seventeenth century.[12] Methodological discussion also blossomed in Britain and the United States after 1829, the year in which John Herschel published his *Preliminary Discourse on the Study of Natural Philosophy*. The work of Herschel, one of the most prominent English men of science during that period, in turn influenced the methodological meditations of William Whewell, whose 1840 *Philosophy of the Inductive Sciences* headed a long list of writings on method, and John Stuart Mill, whose 1843 *System of Logic* engaged Whewell in a prolonged battle over the proper means of making knowledge.[13] Instead of revolving around something called scientific method, however, such talk centered on a different series of keywords, especially "fact" and "induction."

Invocation of these two terms frequently culminated in what historians have often, and contemporaries sometimes, called Baconianism. The term derived from the name of the late-sixteenth- and early-seventeenth-century English essayist, courtier, and natural philosopher Francis Bacon. In his *New Organon*, he claimed that previous practitioners of natural philosophy had tended to fly upward to generalities too quickly. Bacon argued that the proper method of finding truth proceeded inductively from "senses and particulars, rising by a gradual and unbroken ascent, so that it arrives at the most general axioms of all." As the seventeenth century ended and the eighteenth progressed, Anglo-American authors increasingly cast the study of nature in Baconian terms. By the early 1800s, Bacon had achieved a kind of iconic status. In 1845, a columnist in the *Southern Quarterly Review* announced that "Lord Bacon, of all philosophers, was the only true one," who divined the only "natural and equitable mode of acquiring knowledge"—that is, proceeding inductively from particular facts to generalities.[14] Bacon was mentioned as a methodological role model far more often than his nearest competitor, Isaac Newton, who was typically associated with his work on gravity rather than with setting any ground rules for science. In 1831, Newton's biographer, David Brewster, complained that Newton "is represented as having owed all his discoveries to the application of the principles of" Bacon.[15]

From the modern point of view, Baconianism often implies a certain degree of parochialism. As Brewster's complaint suggests, some contemporaries seemed to agree. But in practice, like the scientific method of a later age, early-nineteenth-century methodological rhetoric was elastic enough to accommodate a wide range of interpretations. While many thinkers, such as John Stuart Mill, championed a strongly fact-centered and inductive method, others, including John Herschel and William Whewell, suggested the value of hypothesis in the study of nature. Whewell in particular rejected a single-minded focus on facts and defended not only the use of hypothesis but also the necessity of certain inherent intuitions, such as space and time, to make sense of basic observation. Yet, neither Herschel, who included an engraving of Bacon on the title page of his *Preliminary Discourse*, nor Whewell completely discarded Bacon's ideas and both packaged their proposals in terms of induction.[16] Many practicing American men of science also expressed some dissatisfaction with Baconian-style rhetoric, sometimes adding analogical reasoning to induction beginning in the 1820s. But, as historian of science George Daniels has written, their concerns "brought only 'reinterpretations' or 'natural extensions'; it never brought acknowledged abandonment or even serious questioning."[17] Only a very few contemporaries, such as the British poet and artist William Blake, rejected Bacon and his inductive method outright.

Invocations of facts, induction, and Bacon also circulated widely, becoming common features of public culture. Such catchwords particularly saturated popular sciences, including phrenology—in its most typical form, the belief that the relative sizes of organs in the brain, visible in the shape of the skull, determined personality and character. Itinerant phrenological lecturers and inexpensive phrenological literature did a great deal to spread scientific rhetoric, including talk about method. In his later years Mark Twain recalled the frequent visits of Orson Fowler, one of the most active of American phrenologists, to his boyhood home of Hannibal, Missouri, where Fowler's lectures and head examinations were "popular and always welcomed" and where people began to use phrenological terms themselves, batting them "back and forth in conversation with deep satisfaction."[18]

Phrenology also encountered resistance, often from critics invested in the status quo. In defense of their favorite science, advocates routinely declared that phrenology was "demonstrated CHIEFLY BY A WORLD OF PHYSICAL FACTS." Like all the "exact sciences" it was "discovered, and brought to its present state of perfection *entirely by induction*."[19] Elsewhere, its defenders relied on the name of Lord Bacon, noting that phrenology had been "perfected, by the true Baconian method" or rested securely "on the canons of the Baconial philosophy."[20] Detractors did not dispute the phrenologists' vision of method but rather fought them on methodological common ground. Paul M. Roget, professor of physiology at the Royal Institution, asked, "Who will dare to set up his opinion [against] ascertained facts?" The real issue for Roget and other skeptics regarded the "reality of these facts on which so much is made to depend," and the legitimacy of phrenologists' inductions. Critics accused phrenologists of "hasty generalization" when, in other areas of science, "facts must be accumulated for ages" prior to any conclusions, claiming for themselves the practice of "masterly induction."[21]

Such similar-sounding claims from enemy camps stressed the flexibility of antebellum American methodological rhetoric. A focus on facts and induction did sometimes come at the expense of what contemporaries called "theory" or "hypothesis," a limitation many modern observers have viewed as highly restrictive. Phrenological advocates widely claimed that their favorite science regarded a fact "as worth a million" theories, and its practitioners took "great care . . . to avoid everything hypothetical." Once the facts were in hand, they often guided the process of discovery. Some claimed to let "the facts *classify themselves*"—that is, to proceed "*wholly by induction.*"[22] However, the facts valued by phrenologists varied widely from the results of brain dissections to anecdotes about famous historical figures to the depictions of skulls in art, particularly ancient Greek and Roman sculpture. Popular invocation of induction and the influence of Bacon were also open enough to include such additional approaches as analogical reasoning. Orson Fowler "boldly" asserted "that all real 'analogy' is an unerring guide to truth."[23] In fact, he claimed elsewhere, phrenology deserved the label Baconian *because* it depended on analogy. When one had learned a general principle inductively, the next step involved applying it "to all new but analogous facts."[24]

Meanwhile, even the most Bacon-happy phrenologists made room within inductive method—and within the human skull—for more than just simple observation. Overdevelopment of the cerebral organs responsible for reflection, as we might expect, led to "metaphysical *theorizing*, which is valueless." But if the observing organs dominated, then one would be unable to ascend from facts to principles, which was "the only possible means of arriving at *truth*." Lydia Fowler, sister of Orson Fowler, echoed these sentiments in a book aimed at children, counseling her young readers to "think, inquire, and be not satisfied with simple *facts*, but search for the *principle*, and endeavor to understand it."[25] It was the reflective organ of comparison, and not those engaged by basic observation, on which inductive method was founded.

Such an amorphous and weakly bounded vision of science was also perfectly consistent with the broad strokes of contemporary American culture. The desire of many Americans to assert their rights in various fields of activity without regard to distinctions or divisions was in part related to celebration of the spirit of democracy and the rejection of aristocracy in all its forms. Information generally became less confined to the hierarchical systems of distribution that had prevailed during the 1700s. Newspapers and popular lectures flourished to give citizens access to knowledge of all kinds that was, according to some democratic enthusiasts, "every man's birth-right."[26] At the same time, the conceptual tools available for dissecting the world into well-bounded categories remained rare. The resulting cultural landscape was, on the whole, far more flexible, permeable, and indeterminate than the one that emerged over the next two hundred years. George Daniels observed of the second quarter of the nineteenth century that if "one expression were to be used to characterize the period as a whole, it would be 'intellectual ambiguity.'"[27] Likewise, while writing about the formation of identity in early America, historian Greg Dening claimed that the context was one "of ambivalence and unset definition" and that the search for identity was "multivalent and unending."[28]

In the absence of clear boundaries, few Americans lauded science because it provided a totally unique view of the world or because it was a single beacon of reliable knowledge in a haze of subjectivity, blind faith, and make-believe. Rather, science was important because its revelations fit neatly into a larger system of truth that ultimately transcended any single source of information. One scholar has called it the nineteenth-century "truth complex."[29] In the highly religious popular culture of the day, science often entered general discussion through the natural theological hybridization of mutually supportive physical phenomena and divine intention rather than in any purely or self consciously scientific guise. Orson Fowler proclaimed in a treatise on memory that instructors should not teach "children ANY thing in science or nature, without teaching them GOD in it all."[30] Asserting the value of science because of its reinforcement of other areas of knowledge was much closer to the model of the 1700s than the one that would dominate the 1900s. Historian Thomas Broman has pointed out that eighteenth-century scientific pronouncements gained their authority from the appearance of accessibility and openness. Those who made them spoke "for anyone sufficiently apprised of the facts to formulate a scientific comprehension of the matter."[31] As the 1800s progressed, however, the greater boundedness of science opened new opportunities for boundary-work and for assertion that scientific knowledge was valuable precisely because it was so unlike other, less reliable sorts of information about the world.

MAKING SCIENTIFIC METHOD

Indeed, during the last quarter of the nineteenth century, as we have seen, a larger number of English speakers began turning to a new rhetorical tool, something one contemporary called "the strictly scientific method, as it labels itself at present."[32] While the various meanings associated with it varied every bit as much as those attached to Baconian-style rhetoric—and sometimes overlapped with talk of facts and induction—it differed insofar as it carried with it a special link with something called "science" and implied that the method used in the scientific world differed somehow from those employed in other realms. Its modest but noticeable proliferation coincided with a general trend in Anglo-American culture toward stronger boundaries, whether between regions of knowledge, organizations, or the tangible products of industry. Such boundaries helped to organize intellectual, social, and cultural worlds whose expanding scale challenged attempts at management.

Meanwhile, and by contrast, the public face of science took on a more cohesive and unified appearance. The late nineteenth century witnessed the creation of additional terms that people used to make finer distinctions between science and nonscience, including such chestnuts as "popular science" or "pseudo-science," terms that spread, as I have already noted, in much the same way as scientific method. The greater separateness of science was also often reflected in the tendency to ascribe scientific knowledge to a special group of people called "scientists."[33] In other areas, too, from medicine to journalism, aspiring "professionals" sought to claim areas of knowledge and practice for themselves. Firmer boundaries between kinds of knowledge affected even the most mundane cultural maps. Readers of American newspapers encountered more diverse and segregated content as the end of the century neared, including special reporting for women, sports fans, and others.[34] Likewise, in the emerging consumer culture, the advent and growth of brand names helped to provide some order to the growing number of commercial goods.

By the closing decades of the 1800s, invocations of scientific method had not yet displaced the rhetoric of fact, induction, and Bacon, which continued to provide a widely shared resource. Stories about the history of evolutionary ideas have sometimes depicted their poor reception by the conservative, confining Baconianism that supposedly saturated Anglo-American culture. Critics of Charles Darwin, often those writing in religious periodicals, did frequently rely on a fact-based, inductive view of science, presenting him and his supporters as "investigating nature, not in the interests of science properly so-called, but, consciously or unconsciously, to find facts to fit a hypothesis." An author writing in the *Princeton Review* in 1878 claimed that Darwin's theory "does not rest completely on observation, and induction from such observation." If Darwin's ideas found general acceptance, another columnist worried, "our own Baconian mode of viewing nature will be quite reversed."[35] But supporters of evolution were not above the use of traditional rhetoric. During the 1870s and 1880s, Darwin was often presented, especially in periodicals such as *Popular Science Monthly*, as the cautious and unprejudiced searcher for facts, guided by "a system of the most severe inductive philosophy," who accepted evolution only after it had become "an obvious inference from facts." Even opponents of those "ultra-evolutionists" who had supposedly taken evolution too far in explaining the history of humanity or the origin of life, sometimes acknowledged Darwin as the epitome of Bacon's ideal scientists or at least admitted that, though he was "over-fond of theorizing," science owed "him something for his collation of facts."[36]

But traditional rhetorical tools, though still widespread, were plagued by increasing controversy. The variation in their meaning was probably never more than earlier in the century, but the dramatic battles over Darwin's evolution in particular exposed methodological disagreements to the harsh light of publicity and deprived catchwords such as fact, induction, and Bacon of some of their inherent ambiguity, and hence flexibility. This difference of opinion became particularly visible in discussion of religion and the supernatural. In his famous (or infamous) address before the Belfast meeting of the British Association for the Advancement of Science in 1874, John Tyndall declared that modern men of science "shall wrest from theology the entire domain of cosmological theory" and that "schemes and systems which thus infringe upon the domain of science must, in so far as they do this, submit to its control, and relinquish all thought of controlling it." An unyielding commitment to naturalism provided one means of establishing such a fiercely independent and well-bounded science. In 1867, geologist William North Rice commented that the "great strength of Darwinian theory lies in its coincidence with the general spirit and tendency of science," which aimed to "narrow the domain of the supernatural."[37]

A naturalistic science admitted only physical, sensible facts, a method often associated with the "positivism" of French philosopher Auguste Comte. Such assertions were hardly popular with Christians, who preferred strong consensus on matters religious and scientific and carried with them a broader view of the nature of science. The geologist-cleric George Frederick Wright claimed that the supernatural inspiration of Scripture taxed the imagination no more than Darwin's pangenesis. Others objected that any "mechanical, external, superficial, false" method threatened to "exalt the senses, which are the servitors of the mind, into the mind's masters." True science considered *all* the facts, including those of divine Revelation; the true method of studying nature, which included the method of faith, aimed to read God's handiwork in these facts.[38]

Not all of those concerned about defending Christianity in the perceived face of materialistic assault championed a broad view of scientific boundaries. Less permeable distinctions around science, especially in light of Tyndall's aggressive comments, were often in the interests of theologians, too, many of whom also sought greater autonomy in their own work. As one forward-looking observer noted in 1839, "If phrenologists and divines neglect to settle where the boundary lines are in the disputed territory," which in this case was the explanation of religious conversion, "infidel trespassers will commit depredations."[39] However, an increasing number of attempts to confine the study of nature with traditional methodological catchwords involved abandoning the universal application of inductive and Baconian method. The Reverend A. F. Hewit, a prominent Catholic apologist, sounded a common refrain when he reminded his readers that induction could not lead one to God. The astronomer could not see Heaven. Yet, lamented another author, the scientist "used to the precise terminology of science" in the material realm and "unused to any sharp thinking or precise terminology in intellectual or spiritual phenomena" seemed to believe he could solve the mysteries of the life and the universe from "within the narrow limits of his lab, among instruments of death." In such cases, people had a "right to demand that Science shall confine herself to facts." This view often implied a methodological gap between natural science, which relied on induction, and theology, which "as a science, is deductive."[40]

It was a short step from limiting induction to the realm of natural science to discussion of some method particularly scientific, whether inductive or not. In such cases, a stricter image of science, reflected in references to a peculiarly scientific method, operated to keep scientists confined to their chosen field of expertise rather that to protect it from nonscientists. In his *Natural and the Supernatural*, Horace Bushnell warned readers of too much confidence in the methods of science, "as if nothing could be true, save as it is proved by the scientific method." Instead, he noted that "the method of all the higher truths of religion is different, being the method of faith; a verification by the heart, and not by the notions of the head." Similarly, an author speculating on the future of human character in the *Ladies' Repository* reflected that "every generation, as it accumulated fresh illustrations of the scientific method, is more and more embarrassed at how to piece them in with that far grander and nobler personal discipline of the soul which hears in every circumstance of life some new word of command from the living God."[41]

A narrow and exclusive view of the scientific method appeared in other contexts, too, far removed from religious controversy. Using a scientific method often meant simply being thorough and careful, without any particular connection to the study of nature. In 1860, an author of a serialized work of fiction described a bank customer's impatience with the teller's "scientific method" of counting money.[42] Increasingly, however, in some usage that caution and concern for detail came to be contrasted with being lively and engaging in a way that more firmly inscribed a line between the scientific and popular. An 1878 review of a work on birds complimented the author for "avoiding the technicalities of the ordinary scientific method" and thus exhibiting "the study of ornithology in its more agreeable and fascinating aspect." Numerous reviews of books on scientific subjects meant for wide audiences—all a part of the emerging genre of popular science—echoed such sentiments, praising the "art of combining a 'scientific method' with an animated exposition," so that such works were "neither too abstract nor too popular."[43] The potentially rather dull and strict requirements of scientific methodology also sometimes heightened distinctions between science and art. In 1872, actor and playwright James MacKaye tried to apply scientific method to a stage performance. His experiment was not entirely successful and one observer noted that "science had destroyed the artist. Rule and method underlie the efforts of every accomplished actor; but the moment they become apparent, the passion loses every touch and quality of genuineness."[44]

Scientific method thus did help lend science a more substantial and independent existence, but in practice offered little more flexibility than the increasingly narrow inductive rhetoric. A relative lack of flexibility, in turn, discouraged any significant spread or formalization through repetition. Prominent advocates of a powerful science, such as William James and Thomas Henry Huxley, both rejected the value of Baconian induction, though in a highly naive form that did little justice to its previous adaptability. Huxley claimed that anyone familiar with the actual operation of science was "aware that those who refuse to go beyond fact, rarely get as far as fact." James suggested that the Baconian believer might as well expect "a weather table to sum itself up into a prediction of probabilities, of its own accord, as to hope that the mere fact of mental confrontation with a certain series of facts will be sufficient to make *any* brain conceive of their law."[45] But neither replaced induction with anything called scientific method. Ultimately, neither set of methodological boundary-making tools found much use among those who sought a robustly bounded science to keep dabblers and outsiders away while still allowing a powerful and far-reaching scientific enterprise.

Instead, a more active and imaginative view of science often proved more useful in depicting a science that was capable of wresting from theology, or any other competitor for that matter, "the entire domain of cosmological theory." In his address "On the Use of the Scientific Imagination," John Tyndall argued that despite occasional excesses, society should be tolerant of the imaginative flights of geniuses like Darwin rather than trying to limit them.[46] The rhetoric of an active, imaginative science appeared increasingly in general discourse too. During the 1880s especially, the term "working hypothesis" became common fare. And the pantheon of great scientists also began to change. In his *Principles of Science*, Stanley Jevons championed the ascendancy of a hypothesis-friendly Newton over Bacon, whose method he dismissed as a "kind of scientific bookkeeping." A more active view of methodology often resulted in a division of labor between the observer, "who contents himself with merely ascertaining facts," and the thinker, "who gives shape to science." This distinction was often projected onto history, though such depictions distorted the true methodological beliefs of an earlier generation. In 1890, Joseph LeConte, a well-known geologist and science popularizer, claimed that natural history before Darwin gathered a "dead mass of facts" without inquiry into their meaning. Evolution, for LeConte, led to a more advanced and synthetic method.[47] The cost of such rhetoric, however, was a potential return to more permeable boundaries. Karl Pearson, one of the most notable advocates of such a view, wrote in 1892 that "all great scientists have, in a certain sense, been great artists; the man with no great imagination may collect facts, but he cannot make great discoveries."[48]

Its strong links with very narrow and confining depictions perhaps made scientific method less flexible and multivalent than it might have been, and thus limited its spread as a prominent keyword. The term was only beginning to become a popular slogan during the late 1800s. As a fraction of magazine articles catalogued by the American Periodicals Series Online that were about science generally, those that used the phrase "scientific method" grew from 1.5 percent to 2.5 percent during the 1860s. By the 1870s, that number had increased to 6.5 percent. These were noticeable but not massive shifts in public rhetoric. The contemporary nature of scientific knowledge itself also hampered too ready an appeal to scientific method. Instead, most authors of popular literature offered ideas—such as evolution or the conservation of energy or new devices such as the electric light or the telegraph—as the primary products of nineteenth-century science. Development by evolution, for instance, provided a grand metaphor that people could and did use to describe progressive change, whether discussing the history of human society or tracing the path of the soul during life and after death. In doing so, they could make an area of thought scientific in a much more obvious way without any overt appeal to methodology.

THE GREATEST GIFT OF SCIENCE

By the twentieth century, the relative value of methodology in making areas scientific increased. During the early 1900s, much of the latest science seemed too abstract to provide even metaphorical guidance in everyday life. As early as 1906, a columnist in *The Nation* lamented the greater complexity and specialization of scientific knowledge. "One may say," the author observed, "not that the average cultivated man has given up on science, but that science has given up on him."[49] The perception that modern scientific knowledge was receding from the world of ordinary experience became more intense in light of the revelations of modern theoretical physics during the second quarter of century. A *New York Times* article noted the possibility, with tongue firmly in cheek, of "electronic marriage morality" or a "wave code for fathers and children." The growing specialization of science was bolstered by its movement into particular social locations, such as industrial and government laboratories, institutions of higher learning, and professional organizations. More and more, method rather than the particular details of scientific knowledge captured what was accessible and portable about such a science. In his 1932 address to journalists in Washington, D.C., physicist Robert Millikan asserted that the "main thing that the popularization of science can contribute to the progress of the world consists in the spreading of a knowledge of the method of science to the man in the street" and showing how it could be used to solve the problems of life.[50] Howard Roelofs's father was in perfect tune with Millikan's sentiment.

At the same time, science was emerging as an increasingly powerful notion in American culture. Practitioners found that their science had new abilities to leverage an expanded variety of material resources from new patrons, including philanthropic organizations and large corporations. After World War I, many scientists pushed to protect and to extend the benefits of their participation in the war effort.[51] Science also became a more significant source of cognitive prestige both sought after and occasionally resisted. In his memoirs of the 1920s, journalist Frederick Lewis Allen claimed that "the prestige of science was colossal" and that "the man in the street and the woman in the kitchen" were "ready to believe that science could accomplish almost anything."[52] There is good reason to question the universality of Allen's impression—his somewhat biased assertions about the aftermath of the famous 1927 Scopes Trial have misled subsequent generations of Americans—but it likely reflected his own view and that of the professionals and intellectuals with whom he worked and socialized. The ability of science to sweep away examples of outmoded thinking made it a particularly powerful agent of modernization. That same power could also be threatening. By the early 1930s, in light of the Great Depression and the presumed erosion of traditional values, some members of the American intelligentsia questioned the uncritical celebration of the modern scientific and technological juggernaut.[53]

Despite such increased prominence, initial success at professionalizing scientific work mitigated the desire of many early-twentieth-century scientific enthusiasts for any far-reaching "cosmological theory." Growing specialization and the practical need to attend to specific patrons rather than to inspire the imagination of the public at large made a narrower view of science's boundaries easier to bear. Indeed, scientific method largely assimilated more traditional methodological notions. Millikan described the scientific method as an approach that "starts by trying to get to definite facts, and then allows these facts themselves, with their inevitable consequences, to determine the direction in which conclusions are formed." It was a statement worthy of the Fowlers. Inductivist rhetoric survived even among educators. In 1929, M. Louise Nichols, from the science department at South Philadelphia High School, declared that the scientific method, "as all educated persons are nowadays aware, consists in the accurate observation and comparison of a number of particulars followed by the drawing of a conclusion."[54]

The growing prominence and power of science as a category of thought also corresponded with attempts to expand its methodological boundaries. After World War I, some advocates of Albert Einstein's relativity theory announced its inevitable and revolutionary impact on religion, philosophy, even ethics. The method of the German physicist himself was "surprisingly analogous to that of the artist" who created "a clear, harmonious world of thought." Others were less sanguine about such claims. Critics charged that Einstein had turned physics into geometry; even sympathizers admitted that his work appeared more characteristic of the age of the Greeks when the universe "could only be understood by . . . reason" rather than the modern world, where "physics rather than mathematics is the dominant science." An expansive methodology not only threatened the distinction between science and mathematics but also the one between science and religion. Einstein's image in particular seemed to waver between scientist and prophet, and his theories between mere tools and divine truth. "If his studies and intuitions turn out to be an important new revelation," a columnist in *Harper's* wrote, "it will be interesting to observe that they have come through the Jewish mind," the same mind "through which have proceeded in times past revelations of the highest importance to humanity."[55]

While such expansive methodological depictions often threatened to blur the borders of science and frequently conflicted with narrower visions of scientific method, the term had achieved enough flexibility by the early years of the twentieth century to encompass even broad visions of scientific practice. Howard Roelofs may have initially been confused about its precise nature, but he knew that it worked "miracles," something mere calculation could never do. It involved intelligence and imagination, elements that could not be taught "by a year's course in a laboratory science," where "all too often only measuring and counting are learned." For others, scientific method relied on "insight" and on thinking about events "until they become luminous."[56]

Such newfound flexibility, in turn, encouraged its spread as a convenient slogan, one that could mean many things to many people and thus served as a point around which numerous groups and individuals could find the appearance of consensus. An expanded range of meaning also meant a widened scope of use as a means of inclusion within the scientific world. The diffusion of scientific method as a mindset rather than a technique, and sometimes associated with virtues such as honesty and healthy skepticism, occasionally became vital to the continued functioning of democracy. This was true especially during the spread of totalitarianism in the late 1930s and 1940s. Scientific method provided a balance between an open and a critical mind, which allowed citizens to evaluate claims made by political figures. Its application by the American public foreshadowed a true "science of democracy."[57] References to scientific method migrated further to such diverse realms as law, parenting, education, and journalism, and beyond. During the 1920s and 1930s, consumers in the new advertising-driven marketplace encountered books such as *Eby's Complete Scientific Method for Saxophone* (1922), Martin Henry Fenton's *Scientific Method of Raising Jumbo Bullfrogs* (1932), and Arnold Ehret's *A Scientific Method of Eating Your Way to Health* (1922). Authors of such works used scientific method to sell new ways of solving old problems. Sometimes this method involved the application of scientific knowledge, such as biological information about bullfrogs. At other times, it meant something more like paying attention to the facts. Often, the ambiguity of the term was left intact. Eby never spelled out his complete scientific method for saxophone, nor even mentioned it once in the small amount of text included in his book. Nevertheless, its invocation was in perfect harmony with the growing sense of many Americans during the first third of the twentieth century that the "greatest gift of science is the scientific method."[58]

Educators especially sought to spread this newly important method, the meaning of which they tailored to fit the environment of the classroom. Speaking to the educational section of the American Association for the Advancement of Science in 1910, John Dewey charged that "science has been taught too much as an accumulation of ready-made material with which students are to be made familiar and not enough as a method of thinking." By 1947, the *47th Yearbook of the National Society for the Study of Education* was declaring that there "have been few points in educational discussions on which there has been greater agreement than that of the desirability of teaching the scientific method." Despite the occasional inductive-sounding descriptions, educational reformers increasingly identified this method with the collection of observations and the formulation of a hypothesis, followed by its testing and rejection or modification if it proved unfit—the series of easy-to-teach-and-memorize steps, a sequence familiar to many students today. As early as the 1880s, the geologist Grove Karl Gilbert described the scientific method that should be taught in schools as generating a working hypothesis (or rather a large number of working hypotheses to prevent attachment to any one theory), testing, then discarding as necessary. By the second quarter of the twentieth century, this sequence was often formalized into a numbered list. In 1933, John E. Almack, professor of education at Stanford, offered a sixfold division, which included finding a problem, planning an approach, collecting data, generalizing, verifying, and evaluating.[59]

Scientific method thus provided an important means of identifying and attesting to the presence of science. Americans also continued to put it to use exorcising error and misconception from the realm of the truly scientific. There were numerous opportunities for such exorcism after the turn of the twentieth century. The antievolution crusade of the 1920s, which led to the Scopes Trial, pit those, such as William Jennings Bryan, who decried the perversion of true God-friendly science by the forces of atheism and radicalism, against self-described defenders of independent and agnostic scientific truth.[60] The interwar years also witnessed growing interest in occultism and spiritualism, which sometimes claimed the sanction of science. Yet, objected one critic of investigation of psychic mediums (who often worked in informal circumstances rather than in laboratory conditions), the layman "does not realize how rigid is the control required by scientific methods."[61]

Self-perceived outsiders could also use the same tactics against those in the scientific mainstream, sometimes in ways that occasionally limited the potential scope and reach of science. Charles Lane Poor, an astronomer at Columbia University and one of the few outspoken American critics of relativistic physics, suggested that the rapidity with which advocates declared relativity proven contrasted with the caution of the "accepted scientific method" as demonstrated by Darwin.[62] Even nonscientists turned to scientific method in this way. The writings of some prominent advocates of the new physics and relativity during the 1920s and 1930s, including the British astronomers Arthur Stanley Eddington and James Jeans, on topics that verged on the mystical prompted one Hollywood resident to complain that scientists seemed "hopeless" outside their specialties. When he heard, falsely as it turned out, that Einstein himself had endorsed a psychic named Jean Davis, he lamentably exclaimed "and so the scientific method goes crashing to the ground."[63]

Debate about the scientific nature of the social sciences demonstrated both the inclusionary and exclusionary uses of scientific method, and the many different meanings that it contained. In the interwar years, many authors in magazines accused social scientists of using less-than-rigorous methods in their work with intangible, subjective qualities, such as emotion. In such views, true scientific method belonged to the physical sciences, particularly physics. John B. Watson, the central figure in the behaviorist program, agreed in 1926 that psychology's methods "must be the methods of science in general." That same year, the Social Science Research Council retooled one of its subgroups into the "Committee on Scientific Method." A conference held under its auspices eventually generated the massive *Methods in Social Science*.[64] To counter concern over the human and emotional elements in their work, both on the parts of subjects and observers, many social scientists stressed the objectivity provided by legitimate scientific method.

Thus, for example, behaviorists insisted on sticking to measurable quantities, such as actual behaviors rather than internal sensations and feelings. Other social scientists turned to statistics in search of an objective, accurate, and repeatable method, or to the rhetoric of the laboratory. In her *Coming of Age in Samoa* (1928), an important work in the budding field of cultural anthropology, Margaret Mead compared other cultures to laboratories where anthropological ideas could be tested.[65] Many journalists during the 1920s and 1930s, who looked increasingly to social science as a guide for their own work, also saw the essence of scientific method in its supposed objectivity. In 1928, George Gallup, the founder of the Gallup poll, completed a dissertation at the University of Iowa titled "An Objective Method for Determining Reader-Interest." Two years later, he presented a shortened version in an article called "A Scientific Method for Determining Reader-Interest." In both cases, he advocated examining newspapers along with readers, noting their reactions as they read.[66]

THE DECLINE OF SCIENTIFIC METHOD

After the end of World War II, the oft-noted "age of science" began and scientific method continued to play a role in determining scientific boundaries. Whether involved in the controversy around the cosmic theories of Immanuel Velikovsky in the 1950s, UFOs in the 1960s, or creation science in the 1980s, the slogan became an important weapon in the arsenals of those seeking the include ideas in or exclude them from the realm of real science. J. Allen Hynek, celebrated by later generations of UFO researchers as the founder of UFOlogy, used the rhetoric of scientific method to defend his chosen field of study from negative attacks by other scientists. Still hopeful that a University of Colorado study of UFOs led by the prominent physicist Edward Condon might approach the problem of strange aerial phenomenon more seriously than had the Air Force during the 1950s and mid-1960s, Hynek celebrated the fact that at last the public would be able "to see the scientific method applied thoroughly."[67] Sadly, such optimism quickly turned sour when Condon's report dismissed UFOs as illusions and not worth further study and Condon himself publicly assaulted those who continued to believe in them. In answer, numerous UFO advocates repeated Hynek's later assertion that "ridicule is not part of the scientific method."[68] For skeptics, meanwhile, belief in strange ideas like flying saucers "involved rejecting the scientific method and standards of evidence and credibility."[69]

Aside from battles over what orthodox opponents increasingly labeled "pseudoscience," however, the invocation of scientific method seemed to have declined somewhat during the latter half of the twentieth century. The notion of scientific method assimilated new kinds of ideas. The work of Karl Popper, who depicted scientific methodology as a logical technique in which scientists tested theories by seeking to falsify their claims, provided a favorite definition of scientific method on all sides, particularly among working scientists.[70] But it also endured more scrutiny. A small spate of magazine articles appeared with titles along the lines of "Is There a Scientific Method?"[71] The answer was generally "yes," though it was sometimes an anguished affirmative. An author of one such article in a 1964 issue of the *University of Toronto Quarterly* suggested that "the scientific method has become a sort of demi-god whose worshippers believe that it is only by this method that truth may be known." Nevertheless, she could not reject either science or the scientific method entirely "for I love and admire both of them and cannot help doing so since I am a child of my own time."[72]

Such examination seemed to take a toll on the flexibility of the term, however. The 1968 edition of *Teaching Science in Today's Secondary Schools* lamented that "thousands of young people have memorized the steps" of the scientific method as they commonly appeared in textbooks "and chanted them back to their teachers while probably doubting intuitively their appropriateness."[73] The authors of this critique were not alone. Historian John Rudolph has charted the rise of a movement among education reformers, scientists, and university educators who, amidst cold-war concerns that the United States might fall behind the Soviet Union in science, lobbied against elementary and high school science teaching that targeted "the scientific method" as a stifling and rigid picture of how science worked. Instead, they promoted a far more complex, subtle, and informal vision, often depicted simply as exploring the world, that explicitly rejected the term and that thus seemed more likely to attract the brightest students.[74]

New ideas about the methodology that seemed at odds with any single term also appeared after World War II in the emerging disciplines of history and philosophy of science, furthering the brittle and naive appearance of scientific method. Thomas Kuhn's writings on scientific paradigms stressed the place of worldview and state of mind in the process of science. To one reader, the work of Kuhn suggested that in order to truly understand scientific method, one should look at what scientists did, not at what they said.[75] By the end of the twentieth century, such a focus had come to dominate the scholarly study of science, largely replacing the notion of science as a homogeneous and universal method with that of science as a heterogeneous and local practice. From this perspective followed perhaps the ultimate development in methodological rhetoric: the denial of scientific method itself. In *Against Method*, which appeared in 1975, philosopher Paul Feyerabend rejected the very notion of a singular and definable scientific method, suggesting instead that scientists did whatever worked. In a milder form, one in which methodological assertions continued to play a role as a rhetorical device for gaining authority, such a critical frame of mind has influenced much of the recent work in the history and sociology of science, helping to construct a program bitterly opposed by some late-twentieth-century scientists, and at the heart of the so-called science wars.

Broadly public invocations of the scientific method have also contained increased tones of brittleness, fragility, and narrowness. Vivienne Simon, executive director of the Center for Psychological and Social Change and a supporter of the reality of alien abduction, told a *Time* magazine reporter that the goal of her organization was to "challenge the current scientific method, which is to deny all things you cannot reduce to statistics."[76] Even among defenders of more orthodox science, scientific method has sometimes appeared restricted to the limited confines of safely scientific subjects or contexts. A report on Condon's findings about UFOs in a 1968 issue of *Science* pointed to its widespread rejection among advocates as a "reminder that scientific methods are not always able to resolve problems in fields where emotions run high and data are scarce."[77] Amidst recent debates over intelligent design, which claims to detect clear evidence of conscious engineering in living things, scientific method has rarely served as a bridge to involve ordinary people in science or to export the intellectual power of science beyond the limits of the purely scientific world. Fred Spilhaus, executive director of the American Geophysical Union, claimed in 2005 that scientists and students were "bound by the scientific method" inside "their laboratories and science classrooms," but once outside of these special locations they could "believe what they choose about the origins of life."[78] Alternatively, high-profile advocates of intelligent design, such as Michael Behe and William Dembski, undermined a formal notion of scientific method, quoting physicist Percy Bridgman's quip that "the scientific method, as far as it is a method, is nothing more than doing one's damndest with one's mind, no holds barred" and citing Feyerabend's *Against Method*.[79]

Similar kinds of questioning and controversy had weakened inductive, Baconian rhetoric a century earlier. But factors more particular to the late twentieth century were at work too. In particular, scientific method was increasingly obscured by the same scientific boundaries it helped to create and exploit. In one sense it is a victim of its own success. Insofar as widespread invocation of scientific method resembled an advertising campaign on behalf of a well-bounded concept of science, then by the mid-1900s, the product had already been largely sold. Scientific method had already done its job in helping to construct, maintain, and make sense of the boundaries of science during a crucial transition from one kind of enterprise to another—that is, from an individual, accessible, and often self-supported practice to a more professional, technical, and well-bounded one supported by institutions such as universities, government and industry. The increasing security of professional, especially natural, scientists during the twentieth century left scientific method primarily in the hands of those wrangling about the status of fields on the margins, from social science in the decades after 1920 to parapsychology and UFOlogy after World War II and intelligent design in the present day. The larger cultural context of postwar American science-talk also played an important role in the fate of scientific method. Its continued invocation has taken place in a public culture largely structured by the mechanisms and assumptions of advertising. In such a framework, words and images gain power through endless repetition but for that reason seem to have limited lifetimes. One can only ask where the beef is for so long before people become annoyed.

Over the last few decades, scientific method has also been displaced by the tangible and powerful postwar products of science, from nuclear weapons to spacecraft to computers as means of making science appear culturally significant and relevant. Those public figures defending evolution against the inroads of intelligent design rarely cited the broad intellectual or ethical relevance of scientific ideas to justify their labors. Rather, many held up a presumed link between Darwinian evolution and the technological fruits of scientific research. Zoologist and educator Robert George Sprackland asserted that "America will continue to fall behind in medicine, technology, and other fields of science as long as our children are denied a good science education—one based on an understanding of what is, and is not, science."[80] Opponents of evolution also sometimes presented technology as the least problematic symbol of science and its relevance. An early draft of revised Kansas science standards that sought to de-emphasize the teaching of evolutionary ideas in public schools, offered in collaboration with a local creation science organization, focused on "technological science" and rejected "theoretical science," including Darwinism, as speculative and unreliable.[81] Intelligent design was itself, according to some of its foremost theorists, profoundly technological, borrowing mechanical metaphors with gusto and depicting living things as finely tuned machines. One prominent advocate noted that "the reason evolutionary biology has lost all sense of proportion about how much evolution is possible as a result of blind material mechanisms (like random variation and natural selection) is that it floats free of the science of engineering."[82]

Talk about technology has become such an attractive rhetorical strategy for making science culturally relevant because it has provided a far more robust and immediate bridge between science and the world of ordinary human experience. Yet, when the opposite shore is so far away, sometimes the bridge itself becomes the focus of attention. In fact, just as the scientific once began to obscure the harmonious world of truth that formerly stood behind all specific forms of human knowledge, technology has come to eclipse science in much of public parlance. In a way that resembles the later hyphenation of "American," numerous Americans began to engage in a kind of linguistic expansion of science, rhetorically at least, during the decades straddling the turn of the last century. References to scientific medicine, scientific engineering, scientific management, scientific advertising, and even scientific motherhood all spread, often justified by adoption of scientific method.

Few of these scientific fields have survived into the present day as widely used categories. Rather, we find biotechnology, information technology, and nanotechnology increasingly in the headlines, on television, in films, or on the Internet. Appeal to the new technologies available in everything from electronic devices to hair products has also become a staple of advertising. Likewise, modern intellectuals routinely make use of technological metaphors in analyzing the everyday, including allusions to systems, platforms, construction, or "technologies" as general methods of working. "Technoscience" has achieved widespread popularity among sociologists of science to refer to the intertwined production of abstract knowledge and material devices. We will see whether a century or two from now, coming generations will replace reference to science entirely with a term that strikes them as more relevant, powerful, and meaningful, whether technoscience or something more fanciful and exotic such as "techknowledgy."

Thus, the progress of boundedness has been a mixed blessing for science as well as scientific method. The same polls that have revealed Americans' respect for science have found that the average citizen's grasp of scientific knowledge is often fairly feeble.[83] Science possesses, according to some observers, "awesome authority" and is respected "as a kind of religion."[84] And yet, a large segment of the American public appears "merely uninterested in, or perhaps bored by" it.[85] Unable to explain similarly divergent indications, zoologist and popularizer Richard Dawkins exclaimed that it was "bafflingly paradoxical that the United States is by far the world's leading scientific nation while *simultaneously* housing the most scientifically illiterate populace outside the Third World."[86] A wide array of scientists, educators, and popularizers have explained this paradox in terms of the influence of extra- or anti-scientific factors, such as religion, political extremism, or superstition; the barrier of attention spans ruthlessly shortened by exposure to entertainment-driven mass media; the failure of those responsible for popularizing scientific knowledge to communicate the essence of "real" science; and the spread of fear or even resentment of scientific power. Yet the subtle place of science in America has even more directly followed from the ways Americans learned to talk about it and from its fundamental and inevitable multivalence as a significant cultural keyword.

NOTES

1. Howard D. Roelofs, "In Search of Scientific Method," *American Scholar* (Summer 1940): 296–304.

2. The three databases used for the figures in this chapter include the catalog of

the Library of Congress (http://catalog.loc.gov/), Periodical Contents Index (recently renamed Periodicals Index Online) (http://pio.chadwyck.co.uk/), and the catalog of the contents of the *New York Times*, offered by Proquest Historical Newspapers (http://www.proquest.com/products_pq/descriptions/pq-hist-news.shtml). I performed all these searches between May and June 2004. The first two were on "scientific method" in titles. The third was on "scientific method" in the full text of articles. In all cases, I divided each result by the total number of items about science indexed in the given database to gain a better image of the relative prominence of scientific method as an aspect of science talk.

3. John Schuster and Richard Yeo, introduction to *Politics and Rhetoric of Scientific Method* (Dordrecht: D. Reidel, 1986), xi; Joan G. Creager, Paul G. Jantzen, and James L. Mariner, *Biology* (New York: Macmillan, 1981), 40.

4. Hillier Kreighbaum, *Science; Who Gets What Science News—the News, Where They Get It, What They Think About It—and the Public* (New York: New York University Press, 1958), 183.

5. Stanley Jevons, *Principles of Science* (London: Macmillan and Co., 1874), vii.

6. Stuart Rice, introduction to *Methods in Social Science* (Chicago: University of Chicago Press, 1931), 5.

7. Raymond Williams, *Keywords: A Vocabulary of Culture and Society* (New York, 1976), 17.

8. Thomas F. Gieryn, *Cultural Boundaries of Science* (Chicago: University of Chicago Press, 1999); Thomas F. Gieryn, "Boundary-Work and the Demarcation of Science from Non-Science," *American Sociological Review* 48 (1983): 781–95; Thomas F. Gieryn, "Boundaries of Science," in *Handbook of Science and Technology Studies*, ed. Shiela Jasonoff et al. (Thousand Oaks, CA: Sage Publications, 1994), 393–443. In many ways, the phrase "scientific method" resembles the "boundary objects" of Susan Leigh Star and James R. Griesemer in their ability to present different meanings and aspects for several people or groups at one time; see Star and Greisemer, "Institutional Economy, 'Translations' and Boundary Objects: Amateurs and Professionals in Berkeley's Museum of Vertebrate Zoology, 1907–39," *Social Studies of Science* 19 (1989): 387–420.

9. "Scientific Method," Wikipedia, http://en.wikipedia.org/wiki/Scientific_method (accessed September 21, 2006).

10. See Daniel Patrick Thurs, *Science Talk: Changing Notions of Science in American Culture* (New Brunswick, NJ: Rutgers University Press, 2007).

11. This data was gathered from an online database called Making of America in August 2000. The periodicals indexed include *Ladies Repository*, *Debow's Agricultural Review*, *Princeton Review*, *Southern Quarterly Review*, *Southern Literary Messenger*, *Overland Monthly*, *Catholic World*, and *Appleton's Scientific Monthly* (see http://www.moa.umdl.umich.edu).

12. For a general methodological overview see Laurens Laudan, "Theories of Scientific Method from Plato to Mach," *History of Science* 7 (1968): 1–63; and Barry Gower, *Scientific Method* (London: Routledge, 1997). On the scientific revolution, see Steven Shapin, *The Scientific Revolution* (Chicago: University of Chicago Press, 1996).

13. Laurens Laudan, "Theories of Scientific Method," 30–32; see also Larry Laudan, *Science and Hypothesis* (Dordrecht: D. Reidel, 1981); Gower, *Scientific Method*, 109–29; Richard Yeo, *Defining Science: William Whewell, Natural Knowledge, and Public Debate in*

Victorian Britain (Cambridge: University of Cambridge Press, 1993); Richard Yeo, "Scientific Method and the Rhetoric of Science in Britain, 1830–1917," in Schuster and Yeo, *Politics and Rhetoric*, 259–97.

14. Francis Bacon, *New Organon*, quoted in Ernan McMullin, "Conceptions of Science in the Scientific Revolution," in *Reappraisals of the Scientific Revolution*, ed. David C. Lindberg and Robert S. Westman (Cambridge: Cambridge University Press, 1990), 45; Gower, *Scientific Method*, 40–62; Larry Laudan, *Science and Hypothesis*, 9–10; "Education in Europe," *Southern Literary Messenger*, January 1845, 3.

15. David Brewster, *The Life of Sir Isaac Newton* (New York: Harper and Brothers, 1831), 294.

16. Northrop Frye, *Fearful Symmetry: A Study of William Blake* (Princeton, NJ: Princeton University Press, 1969), 22–3; Larry Laudan, *Science and Hypothesis*, 114–127.

17. George Daniels, *American Science in the Age of Jackson* (New York: Columbia University Press, 1968), 198.

18. Mark Twain, *The Autobiography of Mark Twain*, ed. Charles Neider (New York, 1959), 64. On phrenology in America, see Madeleine B. Stern, *Heads and Headlines* (Norman: University of Oklahoma Press, 1971); John D. Davies, *Phrenology, Fad and Science* (New Haven, CT: Yale University Press, 1955).

19. Nathaniel Bradstreet Shurtleff, *An Epitome of Phrenology* (Boston, 1835), 8; Orson Fowler and Lorenzo Fowler, *Phrenology Proved, Illustrated and Applied* (New York, 1837), 44 (original emphasis).

20. Orson Fowler, *Fowler's Practical Phrenology* (New York, 1848), 8; Andrew Boardman, *Defence of Phrenology* (New York, 1850), 10–11.

21. Paul M. Roget, *Outlines of Physiology* (Philadelphia, 1839), 35, 487; Dr. Thomson, "Phrenology," *Ladies Repository*, December 1841, 366; "Examination of Phrenology," *Southern Literary Messenger*, November 1839, 742.

22. George S. Weaver, *Lectures on Mental Science* (New York, 1852), 35; Shurtleff, *An Epitome of Phrenology*, 8; W. Byrd Powell, *Natural History of Human Temperaments* (Cincinnati, 1856), 203; Fowler and Fowler, *Phrenology Proved*, 44.

23. Orson Fowler and Lorenzo Fowler, *New Illustrated Self-Instructor in Phrenology* (New York, 1859), 65–66.

24. Orson Fowler, *Phrenology Defended* (New York, 1842), 14.

25. Ibid., v; Lydia N. Fowler, *Familiar Lessons on Phrenology* (New York, 1848), 2:176 (original emphasis).

26. "Introductory Statement," *American Phrenological Journal*, October 1, 1838, 3–4.

27. Daniels, *American Science in the Age of Jackson*, 197.

28. Greg Dening, "Introduction: In Search of a Metaphor," in *Through a Glass Darkly: Reflections on Personal Identity in Early America*, ed. Ronald Hoffman, Fredrika Teute, and Mechal Sobel (Chapel Hill: University of North Carolina Press, 1997), 2.

29. Susan Faye Cannon, *Science in Culture* (New York: Science History Publications, 1978).

30. Fowler and Fowler, *New Illustrated Self-Instructor in Phrenology*, 65–66; Orson Fowler, *Fowler on Memory* (New York, 1842), 80.

31. Thomas Broman, "The Habermasian Public Sphere and 'Science *in* the Enlightenment,'" *History of Science* 36 (1998): 142.

32. "Angela, Chapter V," *Catholic World*, November 1869, 167.

33. Thurs, *Science Talk*, 69–84.

34. Gerald J. Baldasty, *The Commercialization of the News in the Nineteenth Century* (Madison: University of Wisconsin Press, 1992), 113–38.

35. "Origin of Civilization," *Catholic World*, July 1871, 493; J. W. Dawson, "The Present Aspect of Inquiry as to the Introduction of Genera and Species in Geological Time," *Canadian Monthly*, August 1872, 154; Henry Calderwood, "The Problems Concerning Human Will," *Princeton Review*, July–December 1878, 334.

36. "Charles Robert Darwin," *Popular Science Monthly*, February 1873, 497; David Starr Jordan, "Darwin," *Dial*, May 1882, 3; "Charles Darwin and Evolution," *Living Age*, September 16, 1882, 643, 646; E. Lawrence, "Modern Forms of Theistic Naturalism," *Ladies' Repository*, May 1870, 334; "New Publications," *Catholic World*, May 1884, 284.

37. William North Rice, "The Darwinian Theory on the Origin of Species," *New Englander* 26 (1867): 608, quoted in Ronald Numbers, *Darwinism Comes to America* (Cambridge, MA: Harvard University Press, 1998), 48.

38. George F. Wright, "Recent Works Bearing on the Relation of Science to Religion," *Bibliotheca Sacra*, January 1880, 70–72; "Science and Atheism," *Ladies' Repository*, March 1868, 211; "Origin of Civilization," 503; "On the Higher Education," *Catholic World*, March 1871, 729.

39. Frank M. Turner, "The Victorian Conflict Between Science and Religion: A Professional Dimension," in *Contesting Cultural Authority* (Cambridge: Cambridge University Press, 1993), 171–200; John Tyndall, "The Belfast Address," in *Fragments of Science* (New York: A. L. Burt Company, 1919), 491; "Remarks upon the Scriptural Doctrine of Regeneration," *American Phrenological Journal*, May 1, 1839, 253–54.

40. A. F. Hewit, "Scriptural Questions, Part III," *Catholic World*, February 1887, 656; "Christianity and Modern Science," *Ladies Repository*, May 1868, 361, 364; Noah Porter, "The New Atheism," *Princeton Review*, January–June 1880, 370; "The Evolution of Life," *Catholic World*, May 1873, 155; "The Skepticism of Science," *Princeton Review*, January 1863, 52–3.

41. Horace Bushnell, *Nature and the Supernatural* (New York: Charles Scribner, 1858), 20; "The Future of Human Character," *Ladies' Repository*, January 1868, 43.

42. "Blue-Eyes and Battlewick, Chapters V–XI," *Southern Literary Messenger*, February 1860, 105.

43. "Books of the Day," *Appleton's Scientific Monthly*, December 1878, 576; "Contemporary Literature," *Princeton Review*, April 1876, 370.

44. "Dramatic Notes," *Appleton's Scientific Monthly*, Febuary 3, 1872, 135

45. Thomas Henry Huxley, "The Progress of Science," in *Methods and Results* (New York: Appleton, 1899), 61–2; William James, "Great Men, Great Thoughts, and the Environment," *American Magazine*, October 1880, 457.

46. John Tyndall, "On the Scientific Use of the Imagination," *Appleton's Scientific Monthly*, October 29, 1870, 527.

47. Jevons, *Principles of Science*, 576; F. W. Clarke, "Evolution and the Spectroscope," *Popular Science Monthly*, January 1873, 320; Joseph LeConte, "The Effect of the Theory of Evolution on Education," *Educational Review*, September 1895, 123–4. For more on Jevons's methodology, see Margaret Schabas, *A World Ruled by Number: William Stanley Jevons and the Rise of Mathematical Economics* (Princeton, NJ: University of Princeton Press, 1990), 54–79. On Thomas C. Chamberlin, a geologist who took a lead-

ing role in popularizing the notion of multiple working hypotheses, see Susan Schultz, "Thomas C. Chamberlin: An Intellectual Biography of a Geologist and Educator" (PhD diss., University of Wisconsin–Madison, 1976).

48. Karl Pearson, *The Grammar of Science* (London, 1892), 31.

49. Quoted in Daniel Kevles, *The Physicists* (Cambridge, MA: Harvard University Press, 1995), 98.

50. "A Mystic Universe," *New York Times*, January 28, 1928; Robert Millikan, "The Diffusion of Science—The Natural Sciences," *Science Monthly*, September 1932, 205.

51. Howard S. Miller, *Dollars for Research* (Seattle: University of Washington Press, 1970), 166–81; Kevles, *The Physicists*, 91–154; Ronald J. Tobey, *The American Ideology of National Science, 1919–1930* (Pittsburgh: University of Pittsburgh Press, 1971), 3–61.

52. Frederick Lewis Allen, *Only Yesterday: An Informal History of the Nineteen-Twenties* (New York: n.p., 1931), 197.

53. Kevles, *The Physicists*, 236–51.

54. Millikan, "The Diffusion of Science," 204; M. Louise Nichols, "The High School Student and Scientific Method," *Journal of Educational Psychology* (March 1929): 196–97.

55. Anton Reiser, quoted in Henry Hazlitt, "Einstein," *Nation*, November 19, 1930, 553; "Einstein's Reality," *Time*, March 16, 1936, 74; F. S. C. Northrop, "The Theory of Relativity and the First Principles of Science," *Journal of Philosophy* (August 2, 1928): 422; Edward S. Martin, "Einstein Gets Us Guessing," *Harper's*, April 1929, 654.

56. Roelofs, "In Search of Scientific Method," 296, 301, 304; Willard Waller, "Insight and Scientific Method," *American Journal of Sociology* (November 1934): 288.

57. David Hollinger, "Justification by Verification," in *Religion and Twentieth-Century American Intellectual Life*, ed. Michael H. Lacey (Cambridge; Cambridge University Press, 1989), 116–35; William Ritter, quoted in Watson Davis, "Science, Philosophy, Religion Find Ground for Common Front," *Science News Letter*, September 21, 1940, 190.

58. W. C. Croxton, *Science in the Elementary School* (New York: McGraw Hill, 1937), 337.

59. John Dewey, quoted in Nichols, "The High School Student," 196; Nelson B. Henry, ed., *46th Yearbook of the National Society for the Study of Education* (Chicago: University of Chicago Press, 1947), 62; G. K. Gilbert, "The Inculcation of the Scientific Method by Example," *American Journal of Science* (April 1886): 285; John C. Almack, "Scientific Method in Teaching," *Educational Method* (March 1933): 323–24.

60. James Gilbert, *Redeeming Culture: American Religion in an Age of Science* (Chicago: University of Chicago Press, 1997), 23–35.

61. "Another Chance for Mediums," *Literary Digest*, July 4, 1925, 27; Marcel LaFollette, *Making Science Our Own* (Chicago: University of Chicago Press, 1990), 112–18.

62. Charles Lane Poor, *Gravitation versus Relativity* (New York: G. P. Putnam, 1922), iv.

63. C. Hartley Graham, "Why, Dr. Einstein!" *New Republic*, March 9, 1932, 94–95.

64. LaFollette, *Making Science*, 122; John B. Watson, "What is Behaviorism?" *Harper's*, May 1926, 724; Dorothy Ross, *The Origins of American Social Science* (Cambridge: Cambridge University Press, 1991), 401–2.

65. Margaret Mead, *Coming of Age in Samoa* (New York: William Morrow, 1928), 5–7.

66. Michael Schudson, *Discovering the News* (New York: Basic Books, 1978), 7–8;

George Gallup, "A Scientific Method for Determining Reader-Interest," *Journalism Quarterly*, March 1930, 1–13.

67. "A Hard Look at 'Flying Saucers,'" *U.S. News & World Report*, April 11, 1966, 15.

68. J. Allen Hynek, *The UFO Experience: A Scientific Inquiry* (Chicago: Henry Regnery, 1972), 237; "A Hard Look at 'Flying Saucers,'" 15; Patrick Huyghe, "Scientists Who Have Seen UFOs," *Science Digest*, November 1981, 119.

69. Robert Cowan, "Explanations of the First Kind," *Technology Review*, March–April 1979, 83.

70. On falsification, see Malachi Haim Hacohen, *Karl Popper: The Formative Years, 1902–1945* (Cambridge: Cambridge University Press, 2000).

71. See for instance, Stephenie G. Edgerton, "Is There a Scientific Method?" *History of Education Quarterly*, Winter 1969, 492–95; A. Cornelius Benjamin, "Is There a Scientific Method?" *Journal of Higher of Education* (May 1956): 233–38; Joseph Turner, "Is There a Scientific Method?" *Science*, September 6, 1957, 431.

72. Helen P. Libel, "History and the Limitations of Scientific Method," *University of Toronto Quarterly*, October 1964, 15–16.

73. Walter A. Thurber and Alfred T. Collette, *Teaching Science in Today's Secondary Schools* (Boston: Allyn and Bacon, 1964), 7.

74. John L. Rudolph, *Scientists in the Classroom* (New York: Palgrave, 2002).

75. Edgerton, "Is There a Scientific Method?" 493.

76. James Willwerth, "The Man from Outer Space," *Time*, April 25, 1994, 75.

77. Philip M Buffey, "UFO Study: Condon Group Finds No Evidence of Visits from Outer Space," *Science*, January 17, 1969, 262.

78. "AGU: President Confuses Science and Belief, Puts Schoolchildren at Risk," *Skeptical Inquirer*, November–December 2005, 45.

79. Michael J. Behe, "The God of Science: The Case for Intelligent Design," *Weekly Standard*, June 7, 1999, 35; William Dembski, *Intelligent Design: The Bridge Between Science & Theology* (Downers Grove: InterVarsity Press, 1999), 258.

80. Robert George Sprackland, "A Scientist Tells Why 'Intelligent Design' Is NOT Science," *Educational Digest*, January, 2006, 30.

81. Keith B. Miller, "The Controversy over the Kansas Science Standards," http://www.wheaton.edu/ACG/essays/miller1.html.

82. William Dembski, *The Design Revolution* (Downers Grove: InterVarsity Press, 2004), 312; Dembski, *Intelligent Design*, 108.

83. By one estimate, between 10 and 17 percent of U.S. adults qualified as scientifically literate during the 1980s and 1990s. See Jon D. Miller, "Public Understanding of, and Attitudes toward, Scientific Research: What We Know and What We Need to Know," *Public Understanding of Science* 13 (2004): 288.

84. Broman, "The Habermasian Public Sphere," 143; Christopher Toumey, *Conjuring Science* (New Brunswick: Rutgers University Press, 1996), 153.

85. Jim Holt, "Madness About a Method," *New York Times Magazine*, December 11, 2005, 25.

86. Richard Dawkins, *The Best American Science and Nature Writing 2003* (Boston, 2004), xvii (original emphasis).

CHAPTER 13

Science and the Public

Bernard Lightman

Science, William Whewell warned in 1834, is disintegrating like "a great empire falling to pieces." The Master of Trinity College at Anglican Cambridge, Whewell was particularly concerned about the endless subdivision of the physical sciences. One striking result of the loss of "all traces of unity" could be seen "in the want of any name by which we can designate the students of the knowledge of the material world collectively." Whewell reported that "this difficulty was felt very oppressively" by the members of the recently founded British Association for the Advancement of Science, who could find no "general term" to "describe themselves with reference to their pursuits." A series of terms was considered and discarded by the leading figures in the association. "Philosophers" was "too wide and too lofty"; "*savans*" was "rather assuming," besides being French; and undignified compounds based on the experimental and observational methods of science, such as "nature-poker" or "nature-peeper," were "indignantly rejected." Whewell mentioned one more suggestion, which came from "some ingenious gentlemen," who proposed, "by analogy with 'artist,' they might form 'scientist,'" but this too "was not generally palatable."[1] The unnamed "ingenious" gentleman was Whewell himself, and this was the first appearance of the word "scientist" in print.

Whewell's new coinage "scientist" made its public debut in a review of Mary Somerville's *On the Connexion of the Physical Sciences* (1834). Whewell had nothing but praise for the book, despite the gender of the author and her intention that it be "a popular view of the present state of science." Feelings of admiration would arise in his readers, he believed, when they discovered that "the work of which we have thus to speak is that of a woman." Though there was a "sex in minds," Whewell believed that women could acquire learning in science. In the rare case of women like Somerville who became philosophers, they actually had an advantage since "one of the characteristics of the female intellect is a clearness of perception." Strikingly, Whewell looked to Somerville's book to help counteract the growing fragmentation of the physical sciences. "If we apprehend her purpose rightly," Whewell declared, "this is to be done by showing how detached branches have, in the history of science, united by the discovery of general principles." Whewell therefore recommended the book to his colleagues in the British Association, notwithstanding Somerville's claim that the book was addressed to her countrywomen. "We believe," Whewell remarked, "that there are few individuals of that gender which plumes itself upon the exclusive possession of exact science, who may not learn much that is both novel and curious in the recent progress of physics from this little volume."[2]

As a commentary on the state of science, Whewell's review is telling. The failure of Whewell and his British Association colleagues to reach a consensus on a fixed term for those who sought knowledge of the natural world is some indication of the instability of the meaning of science in the shifting geography of early nineteenth-century culture. But Whewell's new term eventually "stuck," which tells us how important the nineteenth century has been in shaping the modern conception of science. The coining of the term "scientist," with its overtones of specialism and professionalism, has been interpreted as marking the transition of the cultivation of science from the hands of the amateur to those of the professional.[3] However, it is remarkable that the first use of the term is in an article singing the praises of a female popularizer of science whose book is presented as the solution to the most pressing problem confronting scientists. It would be unthinkable for a professional scientist in the latter half of the century to look to a work of popular science, let alone a book by a woman, for answers to any of the big questions in science. The meaning of the term "scientist," at least in the mind of the individual who coined it, was not identical to "professional scientist" as we understand it today.

Though Whewell worried about the increasing fragmentation of science, at least during this period he could take some comfort in knowing that his powerful brethren in the Anglican clergy guaranteed that a strong religious framework unified science. But by the middle of the nineteenth century, when the scientific naturalists led by Thomas Henry Huxley arrived on the scene to challenge the cultural authority of the Anglican clergy, even the unifying thread of religion was threatened. Scientific naturalists repudiated supernaturalism and offered new interpretations of humanity, nature, and society from the theories, methods, and categories of empirical science rather then from rational analysis or Christian faith. In effect, they defended a far more secular perspective on both science and the worldview, which they sought to base upon it. Victorian scientific naturalism represented the English version of the cult of science that dominated Europe during the second half of the nineteenth century connected with such intellectual currents as scientific materialism and scientific socialism and with influential intellectuals such as Renan, Taine, Bernard, Büchner, and Haeckel.[4] During a period when science was seen as providing the key to all knowledge, it became even more important to fix a stable meaning to an agreed-upon term. Whoever determined the boundaries between legitimate and illegitimate scientific knowledge, between valid and invalid ways of doing science, and between the professional scientist and the mere amateur controlled the definition of science and could dominate the intellectual, social, and political agenda of the day. If Huxley and the other scientific naturalists had their way, all those without proper scientific training would be excluded from the serious scientific societies and hence the scientific community. This included women. Huxley had no use for popularizers of science like Mary Somerville. He also wanted to exclude parson naturalists and upholders of natural theology. During the latter half of the nineteenth century, Huxley's generation subtly altered the meaning of the term "scientist" so that it began to take on some of the modern connotations with which we are familiar—but the change did not take place without a fight.

Members of the intellectual elite debated the key issues, as defenders of the Anglican establishment, allied with physicists such as William Thomson and James Clerk Maxwell, fought with scientific naturalists over the meaning of science and both groups attempted to enroll the support of the British public. The physicist John Tyndall, one of Huxley's closest allies, boldly threw out a challenge to those who resisted the cultural authority of scientific naturalism in his infamous "Belfast Address" (1874). In his presidential address to the British Association for the Advancement of Science, traditionally delivered before the public, he aggressively declared, "we shall wrest from theology, the entire domain of cosmological theory."[5] For his efforts, Tyndall was savagely attacked in books, pamphlets, and periodicals for attempting to convert the public to materialism and for unsettling the very basis of British society. *Blackwood's Magazine* was comparatively mild in its condemnation of Tyndall as a symbol for all those modern scientists who insisted that "'religious theories' must be brought to their lecture-rooms and tested." No doubt, the journalist admitted, it is a great thing "to extend the boundaries of science, and to apply its verifying tests to the explanation of all phenomena; but it is also a serious thing to meddle rashly with the foundations of human belief and society."[6]

But this was also an age when a vast transformation in human communication, made possible by mechanized printing presses, railway distribution, improved education, and the penny post, played a crucial role in the creation of a large reading public not always easily controlled by professional scientists. Though Huxley and his allies attempted to dominate the world of scientific journalism, their aims were thwarted by large numbers of popularizers of science, many of them women, who read a different meaning into contemporary science and who proposed a more egalitarian relationship between scientist and layman. Moreover, many Victorian readers did not accept the role of passive consumers of knowledge conferred upon them by scientific naturalists stressing their expertise and professional status. Although this was the period when categories like "popular" science and "professional" science began to emerge, there was tremendous opposition to the attempt of scientific naturalists to separate the first from the second and then relegate it to a subordinate rank. The crucial story in the latter half of the nineteenth century revolves around processes of control and resistance. Huxley and his scientific naturalist comrades aimed at controlling the meaning of science; but they were resisted by defenders of the Anglican establishment, by physicists within the ranks of professional scientists, by popularizers of science, and by members of the public. The details of this story are distinctively British, but the larger outlines of the tale are relevant for understanding the fortunes of science in other parts of the west. During the nineteenth century the meaning of science was changing throughout Europe and the United States, and in most cases the dynamic involved would-be professionals bent on reforming institutions and ideas opposed by a diverse set of groups both "inside" and "outside" the scientific community, as we now conceive it. I use quotation marks around the words "inside" and "outside" because this was precisely what was stake in these contests: who would be counted as being part of the scientific community and who, therefore, had the authority to define the meaning of science. In this chapter I will provide an example of the transformation of science in the nineteenth century by focusing on Victorian Britain, for we can only understand the complexity of the process if examine a specific case and the social and cultural context particular to it.

THE GENTLEMEN OF SCIENCE AND NATURAL THEOLOGY

In the first half of the nineteenth century, the scientific scene was dominated by such figures as geologists Charles Lyell (1797–1875) and Adam Sedgwick (1785–1873), mathematicians Charles Babbage (1791–1871) and Augustus De Morgan (1806–1871), biologist Richard Owen (1804–1892), chemist Humphry Davy (1778–1829), and astronomer John Herschel (1792–1871), in addition to Whewell. These men were devoted to the serious pursuit of knowledge as a vocation, but not for pay. Though relatively few of them were noble by birth, they adhered to an ideal of gentlemanly science based on a conception of a hierarchical society, masculine authority, and government by an Anglican, aristocratic elite.[7] The British Association for the Advancement of Science symbolized a scientific alliance between the metropolitan gentlemen and the ancient Anglican universities (Cambridge and Oxford) and embedded the ideals of gentlemanly scientific practice in its presidential addresses, discussions of papers, and after-dinner toasts.[8] Founded in 1831, the British Association was the very first society devoted to natural science to use the term "science" in its title. The allegiance of the gentlemen of science to the Anglican-aristocratic establishment was reflected in the way they grounded their science on the theological and political principles formulated by Anglican divine William Paley (1743–1805), author of *Natural Theology* (1802). Herschel, Whewell, and other gentlemen of science made their contributions to the eleven volumes of the *Bridgewater Treatises* (1833–1836), which all set as a goal the analysis of the wisdom, goodness, and benevolence of God as manifested in the works of creation. The design in the natural world was paralleled by a design in the social world. Writing in the shadow of the French Revolution, Paley urged the British workers to be content with their lot in life. It was impious to complain about "the necessity to which human affairs are subjected" since it was God who had "contrived, that, whilst fortunes are only for a few, the rest of mankind may be happy without them."[9]

The gentlemen of science argued that natural theology could shore up the political and social status quo and they integrated it into their popular scientific lectures and writings. From his headquarters in the fashionable Royal Institution, Sir Humphry Davy introduced natural theology into chemistry in his spectacular public lectures. Davy pioneered a rhetoric that presented experimental science as an activity, which was genteel, theologically safe, and socially conservative. In his "Discourse Introductory to a Course of Lectures on Chemistry," first read in the Royal

Institution lecture theatre in 1802, Davy concluded with a comforting vision of the social function of chemistry. He suggested that a wider knowledge of experimental philosophy would lead men to prefer stability and harmony in the both the natural and the social worlds. "The man who has been accustomed to study natural objects philosophically," Davy declared, "perceiving in all the phenomena of the universe the designs of a perfect intelligence, will be averse to the turbulence and passion of hasty innovations, and will uniformly appear as the friend of tranquillity and order."[10] In a new move, which was adopted by more and more scientists in the nineteenth century who came into contact with the British public, Davy assigned his audience a passive role—to admire and support scientific work.[11]

However, the gentlemen of science by no means controlled the meaning of science in the first half of the nineteenth century. They did not form a well-defined group of authoritative leaders.[12] Though the natural sciences became an increasingly significant part of the establishment's response to unbelief, the gentlemen of science were not always successful in their attempts to persuade fellow members of the Anglican intellectual elite that science was more than a secondary adjunct to theology. The support for science at Oxford and Cambridge, where an education in languages and the classics was still considered the ideal, was woefully inadequate. Geology, chemistry, and botany were taught, but they were merely optional lectures in subjects that were not examinable until after midcentury. Science teaching at the ancient universities was not intended to institute a modern, professional education, but instead to educate Christian gentlemen.[13] In spite of the connection with natural theology, the intellectual status of "science" was a matter of contention in the first half of the century. This is precisely why Whewell attempted to coin a new term to highlight the common enterprise in which he and his British Association colleagues were engaged. It was part of the role that he had created for himself as critic, reviewer, adjudicator, and legislator of science at a time when the scale and character of the scientific enterprise were undergoing fundamental change.[14] Moreover, Whewell and his allies found that various groups with different political and social agendas contested their vision of science. Whether it was the middle-class Philosophical Radicals with their utilitarian ideology and sensationalist theory of knowledge, or more radical Nonconformists who attacked the Anglican oligarchy by drawing on egalitarian conceptions imported from continental anatomy and Lamarckian evolution, the gentlemen of science were hard-pressed to defend the legitimacy of their own vision of science and the aristocratic notion of society so intimately connected with it. Whereas Whewell and the gentlemen of science were convinced that true knowledge would lead to the recognition of a divine, static and hierarchical order behind nature, the science of their enemies was naturalistic and emphasized an egalitarian natural and social world characterized by progress and development.[15]

THE RISE OF SCIENTIFIC NATURALISM AND THE PROFESSIONAL IDEAL

By the 1840s, the circulation of knowledge was no longer limited predominantly to the wealthy and to the aristocracy. Science could be "exhibited, learnt, practised, witnessed and consumed by a much larger cross-section of society and in a multitude of new metropolitan settings."[16] As new social groups came into contact with science, they instilled it with new meanings. From 1832 to 1845, the Adelaide Gallery provided a venue for mechanics and London instrument makers to present spectacular shows of light and sparks, and to appear before the public as men of science.[17] The establishment of institutions such as the non-Anglican University College, London, in 1826, the Museum of Economic Geology in 1839, and the Royal College of Chemistry in 1845 offered the sons of middle-class dissenters the opportunity to learn practical skills and gain expertise in scientific and medical subjects. For some, these new institutions of science provided opportunities to earn a living by teaching science or engaging in scientific work of various sorts. Those who could not find jobs in these institutions eked out a precarious existence on the meager wages garnered from freelance lecturing or journalism. But it was more and more possible for those without genteel upbringing or connections to forge careers in science from metropolitan sites that were part of the bourgeois landscapes of new science colleges and medical schools, exhibition halls and engineering workshops.[18]

When Huxley returned from a four-year voyage as assistant surgeon on the HMS *Rattlesnake* in October 1850, this was the volatile situation that confronted him. Born to a humble lower-middle-class Anglican family and trained during the hungry forties at a medical academy in London where high-profile Nonconformists lectured, Huxley was deeply affected by the poverty he saw in the East End slums. It was a grim reminder of how the Anglican-aristocratic establishment had failed to provide for many members of English society. The second-class status of his own teachers, engaged in a dirty war with elite surgeons, opened his eyes to how Anglican privilege operated in the world of science. Without an education at Anglican Oxford or Cambridge, his prospects for obtaining a paying job in science were slim. Sending papers on marine biology back to England while on the *Rattlesnake* brought him to the attention of some of the gentlemen of science. When he returned, his scientific credentials now established, he endured five years of frustration searching for a suitable position. There were few paid posts in science, and when one of the few became open, he was passed over for lesser-qualified individuals with more respectable social and religious backgrounds. Cultivating powerful patrons among the gentlemen of science finally paid off in 1855 when he obtained a position at the Royal School of Mines, but in the meantime he had built up a nasty grudge against the Anglican establishment that had delayed the start of his scientific career. Huxley never forgot his early struggles to establish a scientific career for himself, and throughout his life he set as one of his main goals the redefinition of both the meaning and institutional infrastructure of British science, which had in the past been dominated by the Anglican-aristocratic establishment.

Huxley found like-minded friends, some with similar backgrounds and experiences. The German-trained physicist John Tyndall (1820–1893), professor of natural philosophy at the Royal Institution, and the philosopher of evolution Herbert Spencer (1820–1903), both from Nonconformist backgrounds, shared Huxley's lower-middle-class roots. Like Huxley, they had been educated outside the privileged Anglican institutions of Oxford and Cambridge. Some allies were men whose Anglicanism had been demolished by a tumultuous crisis of faith, such as William Kingdon Clifford (1845–1879), professor of applied mathematics at University College, who rebelled against his High Church upbringing while at Cambridge in the late sixties, and literary critic and editor of the *Dictionary of National Biography* Leslie Stephen (1832–1904), whose agony at Cambridge in the early sixties was so intense that he considered suicide. Other members of the group, who scholars refer to as the "scientific naturalists," included the anthropologist Edward Tylor (1832–1917), the biologist E. Ray Lankester (1847–1929), and the medical doctor Henry Maudsley (1835–1918).

During the latter half of the nineteenth century, intellectuals believed that science provided the most legitimate path to certain truth. Fact, objectivity, and practicality were catchwords of the day. In her autobiography, Beatrice Webb, a Fabian socialist, recalled the "cult of science" that pervaded the mid-Victorian world when she was a young disciple of Herbert Spencer. "Two outstanding tenets, some would say, two idols of the mind, were united in this mid-Victorian trend of thought and feeling," she remembered. "There was the current belief in the scientific method . . . by means of which alone all mundane problems were to be solved" and "the transference of the emotion of self-sacrificing service from God to man."[19] In this intellectual atmosphere whoever could lay claim to speak on behalf of science could present themselves as authoritative leaders who knew how to understand the larger significance of modern science. Scientific naturalists often asserted their authority beyond the domain of knowledge of the natural world. Clifford denied that science pertained merely to thought about scientific subjects with long names. "There are no scientific subjects," he insisted. "The subject of science is the human universe; that is to say, everything that is, or has been, or may be related to man."[20] Scientific naturalists put forward new interpretations of man, nature, and society derived from theories, methods, and categories of empirical science. This cluster of ideas and attitudes was naturalistic in the sense that it would permit no recourse to causes not empirically observable in nature, and scientific because it drew on three major mid-nineteenth-century theories: the atomic theory of matter, the conservation of energy, and evolution. In his essay "The Progress of Science" (1887), Huxley referred to these theories as "three great products of our time," which proved that "our epoch can produce achievements in physical science of greater moment than any other has to show." Huxley believed that the "peculiar merit of our epoch is that it has shown how these hypotheses connect a vast number of seemingly independent partial generalisations; that it has given them that precision of expression which is necessary for their exact verification; and that it has practically proved their value as guides to the discovery of new truth." Since these three theories were "intimately connected" and "applicable to the whole physical cosmos," they provided the basis for a comprehensive view of the world.[21] Scientific naturalists often presented themselves as the cultural elite best equipped to guide Britain as it was being transformed into a modern, industrialized nation. Scientific naturalism served the interests of sections of the new professional middle class and provided a rationale for their leaders to wrest cultural and social control from the Anglican-aristocratic establishment.[22]

Throughout the latter half of the nineteenth century, Huxley, Tyndall, and Spencer attempted to redefine the nature of British science from their base in London. In 1864 they helped found the X-Club, whose membership included men who were to become prominent in their respective fields of research, including mathematicians William Spottiswoode (1825–1883) and Thomas Archer Hirst (1830–1892), botanist Joseph Dalton Hooker (1817–1911), chemist Edward Frankland (1825–1899), archaeologist John Lubbock (1834–1913), and zoologist and paleontologist George Busk (1807–1886). The X-Club was a private, informal society where the members could exchange ideas on literature, politics, and science over dinner. For twenty years the members met once a month from

October to June. The X-Club wielded tremendous power in the scientific world.[23] Its formation allowed club members to pursue a number of common objectives, the foremost among them to turn science into a professional, meritocratic, publicly respected, and state-endowed activity. The model for reforming British science existed in the state-funded system of the German universities, with their emphasis on research and the laboratory. In order to succeed they had to be aggressive, opportunistic, and politically savvy. They served on various government commissions related to science, angled for and won high posts in scientific societies, participated in reforms of scientific education, and remade the scientific institutions in which they worked. They engaged supporters of the Anglican Church in the pages of the periodical press or in public debate. Taking advantage of the controversy surrounding *On the Origin of Species* (1859), they defended the right of scientists like Darwin to put forward naturalistic theories without fear of reprisal from scientifically unqualified Anglican intellectuals and to be judged by their scientific peers on the basis of the evidence. Professionalization could be used as a tool for forcing parson naturalists, wealthy amateurs, and women out of science.[24] Though Huxley and his friends were by no means consistent in their pursuit of the goals of professionalization as defined by modern standards, by the end of the century the scientific landscape had been profoundly transformed.[25] In the first half of the century, when the gentlemen of science were devoted to the serious pursuit of knowledge as a vocation, the sites of scientific work were the field museum, the lecture hall, and the hospital. To learn about nature everyone, gentlemen of science included, attended lectures and experienced nature directly through fieldwork. After Huxley and his allies had done their work, the typical figure of the later period was the paid professional "scientist" who conducted his experimental work in a government, industrial, or academic laboratory.[26]

A similar process of professionalization took place in other Western nations during the nineteenth century, though the pace and exact nature of change varied depending on the context. In the United States, for example, historians have identified 1820 to 1860 as a key period of transition. By the middle of the century, earlier patterns of gentlemanly scientific activity were becoming obsolete as a community of professional scientists emerged. The first professional scientific association in America was founded in 1840 by a small group of working geologists, becoming the American Association for the Advancement of Science in 1847. Also in the late forties, the astronomer Benjamin Gould (1824–1896) and other Americans were urging the use of Whewell's new term "scientist."[27] Trained at Göttingen University, where he received a PhD in astronomy,

Gould pursued a program of improving the state of American astronomy by disseminating German methodology and founding scientific institutions based on the German model. Though Gould failed to transform the Dudley Observatory into his vision of a research institution while he was director, he successfully developed the *Astronomical Journal* (established in 1849) into a well-respected scientific periodical with an international reputation. Like Huxley, he had a formidable group of allies who belonged to an informal group of research scientists, known as the Lazzaroni, led by Alexander Dallas Bache (1806–1867), a physicist and director of the U.S. Coast Survey. Other members included mathematician, physicist, and Harvard professor Benjamin Peirce (1854–1914); Harvard professor of natural history Louis Aggasiz (1807–1873); and Joseph Henry (1797–1878), physicist and first director of the Smithsonian Institution.[28]

Professionalization did not merely involve the transformation of scientific institutions and practices. The revolution in the meaning of science, and its larger cultural significance, can be detected in two of Huxley's essays, published almost twenty years apart. In his "On the Educational Value of the Natural History Sciences" (1854), Huxley stressed that the methods of all the sciences were identical, whether the scientist was dealing with the physical or the life sciences. He declared that "Science is, I believe, nothing but 'trained' and 'organized common sense,' differing from the latter only as a veteran may differ from a raw recruit: and its methods differ from those of common sense only so far as the guardsman's cut and thrust differ from the manner in which a savage wields his club." Since the man of science "simply uses with scrupulous exactness, the methods which we all, habitually and at every moment, use carelessly," everyone could be trained, like the guardsman, to develop and perfect the "hewing and poking" of the savage.[29] Huxley rejected the notion, influential before the midcentury, that scientific creativity was an unteachable innate gift of the few. Instead, Huxley presented a seemingly more democratic science, which emphasized how a scientific education could discipline the mind and teach the public to resist such unscientific fancies as table-turning or mesmerism.[30]

Almost twenty years later, in his "On the Study of Biology" (1876), Huxley outlined a position that emphasized his increased commitment to the professionalization of science. Even the title of his earlier essay, "On the Educational Value of the Natural History Sciences," retained a role for the older tradition of natural history, which searched for insight into the order of nature, often from a religious perspective. But in the later essay, Huxley characterized natural history as an outmoded term and unabashedly embraced the term "biology." Interestingly enough, Huxley never liked Whewell's coinage "scientist," as he considered it to be an unscholarly Americanism.[31] In 1894 he remarked, "to any one who respects the English language, I think 'Scientist' must be about as pleasing a word as 'Electrocution.'"[32] But he had no problems with "biology," arguing that it was not "simply a new-fangled denomination" for what used to be known under the title of natural history. On the contrary, Huxley declared, "the word is the expression of the growth of science during the last 200 years, and came into existence half a century ago." Realizing that natural history included "very heterogeneous constituents," discerning men decided to group botany and zoology together since it was possible to obtain an "extensive knowledge of the structure and functions of plants and animals without having to enter upon the study of geology or mineralogy." In his discussion of the origin of the term "biology," Huxley believed that Bichat, Lamarck, and Treviranus, three scientists working at the beginning of the nineteenth century, were among the first to treat all of the sciences dealing with living matter as if they were part of one discipline. He credited Lamarck with the first use of the term *biologie* in 1801 in his *Hydrogéologie*, and asserted that Treviranus's great merit lay in the publication of an entire work titled *Biologie*, the first volume of which appeared in 1802. As a result of the efforts of these men at the beginning of the century, by the 1870s the term "natural history" had come to be seen as "old" and "confusing," and "all clear thinkers and lovers of consistent nomenclature" used "biology" to characterize the study of the totality of living phenomena.[33] In comparison to natural history, Huxley's biology was stripped of all religious content. The study of living phenomena did not involve the analysis of exquisitely designed organs of perfection and it could be conducted without having to deal with any questions concerning the wisdom, power, and benevolence of a divine being. Huxley's work on education reform helped to spread the use of "biology" and led to the decrease of the label "natural history."[34]

In the same essay Huxley laid out the ideal way to study biology. The earlier essay had stressed the democratic nature of science, though common sense required discipline and training. Here Huxley insisted that the study of biology must be analogous to the study of the other physical sciences. It was not enough just to read books and attend lectures. Referring to the way science was studied at Oxbridge, Huxley criticized "the 'paper-philosophers' [who] are under the delusion that physical science can be mastered as literary accomplishments are acquired, but unfortunately it is not so." It was also necessary to perform experiments in the laboratory. The student of science must touch, handle, and see the things symbolized in language in books. Huxley then told his readers how biology was practiced in his laboratory at South Kensington, a newly built facility completed in 1871.[35] But admittance to the lab was restricted to experts and those training to be experts. In his discussion of biology museums a few pages later in the essay, Huxley made it clear that the laboratory was not a public space. The ideal biology museum would be divided in two, one part open to the public, which would give them "easy and unhindered access to such a collection as they can understand and appreciate," and the other part open every day to men of science to give them access "to the materials of science" in a laboratory-type setting.[36] The democratic nature of science became lost in Huxley's stress on expertise developed only in the laboratory.

One more theme emerges in Huxley's essay "On the Study of Biology" that was characteristic of how scientific naturalists read social and political messages into the natural world. In discussing the scope of biology, Huxley claimed that it covered "all the phenomena which are exhibited by living things," including the higher forms. To Huxley this meant that humans and all their ways came under the heading of biology. Huxley gave an evolutionary justification for subsuming psychology, politics, political economy, and civil history within the domain of the biologist. The "rudiments and outlines of our own mental phenomena are traceable among the lower animals" who have "their economy and their polity." If the "polity of bees and the commonwealth of wolves fall within the purview of the biologist proper," Huxley declared, "it becomes hard to say why we should not include therein human affairs." But biologists had, at great sacrifice, given up civil history to a different branch of science, which, following Auguste Comte, Huxley called "sociology," though he warned that it should not seem so surprising if a biologist apparently trespassed "in the region of philosophy or politics" or meddled with "human education; because, after all, that is a part of his kingdom which he has only voluntarily forsaken."[37] Huxley's biological imperialism justified the use of biological theories by professional scientists to solve social problems. It provided Huxley and his allies the cultural authority they craved. Defenders of the Anglican-aristocratic status quo denounced this sweeping notion of biology as reductionist and materialistic.

But Huxley found a new target in the late eighties when he flagrantly trespassed into the social domain in his essay "The Struggle for Existence in Human Society" (1888), which was followed by "Capital—the Mother of Labour" (1890), "On the Natural Inequality of Men" (1890), "Natural Rights and Political Rights" (1890), and "Government: Anarchy or Regimentation" (1890). In these essays Huxley, ever the apologist for middle-class ideals, attacked socialism as an unscientific form of a priori political speculation that sought to upset "the existing arrangements of society on the faith of deductions" from "highly questionable axioms," such as the doctrine of natural rights. To make any progress, political thinkers had to recognize that problems could not be solved a priori and that "the natural order of things," which tended to maintain the war of each against all, had to be taken into account.[38] Huxley joined Spencer and other scientific naturalists in drawing on evolutionary theory to argue that there were natural limits to what could be accomplished through social reform.[39] Kropotkin, as well as other socialist thinkers, responded to Huxley's evolutionary capitalism by calling for a reinterpretation of Darwinian theory, which took into account natural sociability and mutual support within species.[40]

ELITE RESISTANCE TO SCIENTIFIC NATURALISM

The attempts of scientific naturalists to redefine the intellectual and institutional basis of British science were resisted by other members of the intellectual elite. The strongest resistance came from within the Anglican clergy, including many of those gentlemen of science who lingered on into the second half of the century as well as younger Anglicans in science. They continued to look to the Anglican Church as the chief cultural authority within British society and they vehemently rejected the notion that a scientific worldview could provide meaning if divorced from a religious foundation. Materialistic scientific theories missed the spiritual truth that lay behind all natural phenomena. Even as late as the 1890s, Anglican intellectuals such as Arthur Balfour, a future Tory prime minister, were still attacking scientific naturalism. In his *Foundations of Belief* (1895), Balfour carefully dissected the weaknesses of the secular conception of scientific knowledge, denying that scientific naturalism had any intrinsic connection to, or authority over, science.[41] To Balfour, naturalism was but a "poor relation" of science that had forced itself into the "retinue of science" and now claimed to "represent her authority and to speak with her voice."[42] Anglicans were joined by non-Anglican Protestant and Catholic members of the intellectual elite stung by the attacks of scientific naturalists on Christian doctrine.

Effective opposition also came from a group of scientists who from the 1850s to the 1870s constructed the science of energy thereby redrawing the disciplinary map of physics. Like the scientific naturalists, they too had a reform program for the whole range of physical and even life sciences. Bearing the impress of Scottish Presbyterianism, representing Whig and progressive values, and linked to the industrialists of northern Britain, energy physics was founded by a "North British" group composed of Glasgow professor of natural philosophy William Thomson (1824–1907), Scottish natural philosophers James Clerk Maxwell (1831–1879) and Peter Guthrie Tait (1831–1901), and the engineers Henry Fleeming Jenkin (1833–1885) and William Macquorn Rankine (1820–1872). These men found the perceived anti-Christian materialism of the metropolitan scientific naturalists quite distasteful and they were prepared to enter into an alliance with Cambridge Anglicans to undermine the authority of Huxley and his allies. They promoted a natural philosophy (their preferred term for a scientific study of nature) in harmony with, though not subservient to, Christian belief.

Energy physics was an indispensable component of this natural philosophy, since, as constructed by the North British Physicists, it posited that energy flow had a direction, whether expressed as progression or dissipation, which pointed to a universe governed by basic laws ordained by a creative, divine being. Believing that the core doctrine of materialism was reversibility, the North British Physicists denied the scientific validity of the concept of a purely dynamical or mechanical system in which there was no difference between running forward or backward. They resisted the efforts of John Tyndall to deploy the doctrine of the conservation of energy in the service of scientific naturalism within the domain of the physical sciences.[43] Tyndall maintained that the fixed quantity of energy in the universe meant that the mechanism of nature remained closed to all external (read supernatural) interference. In the case of Thomson's debate with Huxley on the age of Earth in the late sixties, the science of energy was harnessed to disrupt the authority of scientific naturalism within the life sciences. The laws of thermodynamics were used by Thomson to raise questions about vast time scale required by Darwinian evolution.[44] Thomson's rejection of Darwinian evolution, with its randomness, was bound up with his commitment to a process that was ordered, law-like, and subject to divine guidance and control. During a heated dispute as to who was the founder of the modern dynamical theory of heat, with Tyndall backing Mayer, and Thomson and Tait supporting Joule, the crucial issue of religion and scientific authority came out into the open. In the *Philosophical Magazine*, a periodical for professional scientists, Tyndall objected that Thomson and Tait had aired their priority dispute in the pages of a Christian journal like *Good Words*. Tyndall charged that Thomson and Tait had acted against the "interests" and "dignity" of science by "taking difficult and disputed points . . . into such a court" as the one composed of readers of *Good Words*. Tyndall rejected the appeal to a court of the people and argued that the case must be tried in a forum made up of "instructed men of science."[45]

The resistance of North British Physicists to the materialist meaning of science defended by metropolitan scientific naturalists can be seen in their conception of natural philosophy. Huxley jettisoned the label "natural history" in favor of "biology" so he could incorporate the laboratory as a crucial component of science and exorcise religious notions from the study of life. But Thomson was content to retain the term "natural philosophy" so he could preserve its traditional religious framework while still introducing experimental research as a key to the progressive development of science. Thomson disapproved of the term "physicist" and as late as 1890 preferred the word "naturalist" to describe his discipline.[46] Thomson was elected as the professor of natural philosophy at the University of Glasgow at the age of twenty-two; he delivered his "Introductory Lecture to the Course on Natural Philosophy" on November 4, 1846. Since this lecture was presented in part to his class at the beginning of nearly every subsequent session, it provides valuable insight into Thomson's program for natural philosophy.[47] As Thomson ranged over the gamut of natural philosophy subjects, such as heat, electricity, magnetism, and optics, he paused from time to time to spell out the religious implications of science, emphasizing the compatibility of the study of nature with a profound belief in the existence of a divine being. In the section on optics, he drew attention to the designed quality of the eye. Though he criticized demonstrations of design in nature by "fanciful or too zealous advocates," he encouraged thoughtful students to trace "the proofs of design in the adaptation of our organs of vision in accordance with the physical laws of light to receive the impressions by which we see external objects." Looking back on the progress that had been made by the human mind in the discovery of truth, Thomson sought to instill the proper attitude of reverence and thankfulness within his students. No progress would be possible unless God had endowed humanity with a mind capable of uncovering the general laws of force that constituted the primary subject of Natural Philosophy. "We feel," Thomson stated, "that the power of investigating the laws established by the Creator for maintaining the harmony and permanence of His works is the noblest privilege which He has granted to our intellectual state." The divine design in nature could only be apprehended by a mind designed by God. This was evidence of divine goodness as compelling as "the various arrangements by which the physical powers of the animal world are produced and maintained." "The earnest student of philosophy must always be impressed" by feelings of gratitude for this gift, Thomson maintained, and thus through his "studies and successive acquirements" he will be led "'through nature up to nature's God.'"[48]

In this lecture Thomson also presented natural philosophy as a progressive, practical study that required key skills in experimental research and the use of precision instruments, rather than as a static, logical structure that relied on illustrative experiments as a teaching device. In comparison to his predecessors, Thomson included far more on experimental topics in his course, including sections on heat, hydrostatics, pneumatics, acoustics, magnetism, and electricity. He also built a research laboratory at Glasgow College where both the professor and his students could make precision measurements in their experiments. Precision measurements reduced to absolute units represented to Thomson the exemplar of the experimenter's art. His approach signified a radical new professionalism distinguished by a clear research orientation. Thomson not only revolutionized the practice of natural philosophy in the local context of Glasgow, he also contributed to a major transformation of British science in terms both of incipient professionalization and of the emergence of laboratory science.[49] The introduction of the laboratory into science in no way involved the elimination of the religious framework of natural philosophy. Nor did Thomson link the professionalization of science with the exclusion of the Christian clergy from the world of science. Huxley and Thomson therefore disagreed violently on the larger implications of professionalizing science for the cultural authority of the Christian church and for religious faith.

Twenty-five years later, another North British Physicist echoed Thomson's thoughts on the compatibility of a professional science based on laboratory research with a profound respect for Christian ideals. James Clerk Maxwell was appointed to the new Cavendish chair at Cambridge in 1871, a telling symbol of the alliance between Cambridge Anglicans and the North British Physicists. Maxwell's chief task was to introduce experimental physics into the mathematical and moral culture of the university. The new course on experimental physics, Maxwell announced, "while it requires us to maintain in action all those powers of attention and analysis which have been so long cultivated in the University, calls on us to exercise our senses in observation, and our hands in manipulation. The familiar apparatus of pen, ink, and paper will no longer be sufficient for us." Distinguishing between experiments of illustration, which were educative, and experiments of research, whose "ultimate object is to measure something which we have already seen," Maxwell affirmed that the latter were "the proper work of a Physical Laboratory." In fact, Maxwell declared, the notion that modern experiments consist principally of measurements was so prevalent that scientists abroad were of the opinion that "in a few years all the great physical constants will have been approximately estimated, and that the only occupation which will then be left to men of science will be to carry on these measurements to another place of decimals."[50]

As if to assure his audience that the combination of experiment and mathematical analysis would not lead Cambridge science students away from Anglicanism, Maxwell asked, "But what will be the effect on the University, if men pursuing that course of reading which has produced so many distinguished Wranglers, turn aside to work experiments?"[51] After discussing the need to adopt statistical methods for dealing with groups of atoms, Maxwell answered the question by pointing to the picture of molecules yielded by the approach to physics that he was proposing. It had been discovered that the mass of any present molecule was absolutely invariable and that it was identical to molecules throughout the universe throughout time. How could the scientist account for this identity in the properties of such a multitude of bodies? Maxwell concluded that he was "forced to believe that these molecules must have been made as they are from the beginning of their existence." The manufactured quality of molecules pointed not only to a maker, but a supernatural maker. "I also conclude," Maxwell proclaimed, "that since none of the processes of nature, during their varied action on different individual molecules, have produced, in the course of ages, the slightest difference between the properties of one molecule and those of another, the history of whose combinations has been different, we cannot ascribe either their existence or the identity of their properties to the operation of any of those causes which we call natural."[52] An experimental approach to natural philosophy led directly to the acknowledgement of the existence of a supernatural cause. While Huxley trained his biology students at South Kensington to see a fully secularized material world through the lens of the microscope, Maxwell and the other North British Physicists were prepared to admit God into their laboratory as the creator who had given purpose to nature and to all scientific activity.

THE POPULARIZERS OF SCIENCE

The vision of science laid out by Huxley and his allies was also resisted by popularizers of science, who managed to build an entire career out of their science journalism, and by members of the vast Victorian reading audience, not just by sections of the intellectual elite, such as the North British Physicists or the Anglican clergy. Without doubt, the latter half of the nineteenth century was marked by the establishment of the ideal of the professional scientist as a replacement for the ideal of the gentleman of science, in part thanks to the efforts of the scientific naturalists. But if it was the age of the professionalization of science, it was also a period when the popularization of science really came into its own and when the audience for scientific books, journals, and lectures grew by leaps and bounds. The revolution in the first half of the century in communications and transportation, both in the vanguard of the economic sectors undergoing industrialization, laid the groundwork for this growth of popular science and its public. As James Secord has argued, "The steam-powered printing machine, machine-made paper, public libraries, cheap woodcuts, stereotyping, religious tracts, secular education, the postal system, telegraphy, and railway distribution played key parts in opening the floodgates to an increased reading public." Driven by technological advances, as well as wider social developments, such as the explosion of the urban population, changes in the book trade, and a rise in literacy rates, the number of book titles increased steadily throughout the first half of the century as the price of books declined.[53]

The establishment of a new market for popular science developed slowly at first. Though not always a commercial success, the appearance of new scientific works in cheap nonfiction series in the 1820s, such as Henry Brougham's Library of Useful Knowledge, sponsored by the Society for the Diffusion of Useful Knowledge, revealed an untapped market and encouraged a series of publishing experiments in the following decade. By the time the *Vestiges of the Natural History of Creation* burst upon the scene in 1844, causing a sensation, the anonymous author (Robert Chambers) could take advantage of the new market for popular science and astutely draw on his vast experience in commercial publishing. With a reputation for being "publishers for the people," the Chambers brothers' publishing house created a new polity of consumers consisting of a family readership among the middle and working classes. Robert Chambers targeted the same audience for his *Vestiges* by domesticating evolutionary theory. He also established a number of important conventions for future popularizers of science, including a successful model for the evolutionary epic with its emphasis on the wonder of the natural world; a new hybrid genre that allowed science journalists to move between a variety of literary forms; and guidelines for establishing a rapport with the public that gave readers a growing sense of power as they worked together with the author to understand the sublime mysteries of nature. Moreover, Chambers challenged the authority of scientists, depicting them as too specialized and narrow, and appealed directly to public judgment to determine the scientific worth of his theories. Despite being denounced by many of the eminent gentlemen of science, not to mention a young Thomas Henry Huxley, in Britain alone, *Vestiges* sold forty thousand copies, the sales boosted by the publication of cheap reprints.[54]

The existence of a robust market for popular science became firmly established in the second half of the nineteenth century, as the structure of communication underwent further changes with the repeal of the remaining taxes on knowledge, the introduction of the rotary press, and a gigantic expansion in newspaper and periodical publishing. New cheap national newspapers slowly replaced expensive provincial ones, and the monthlies superseded the highbrow quarterlies. Secord sums up the results nicely: "The 'People,' imagined by the entrepreneurs of useful knowledge in the 1830s, came into being as the market for the late Victorian mass-circulation press." Though by the end of the century categories like "popular" science and "professional" science became embedded in stable publishing genres, and "popular" science came to be used somewhat dismissively, the power of scientists to control the meaning of science was severely limited by the creation of the new mass media and by the resulting proliferation of alternative points of view.[55] Little did Tyndall and Spencer suspect, as they worked as land surveyors before becoming eminent scientific naturalists, that they were helping to lay the groundwork for the vast railway system that was so essential to the communications revolution that would later confine their ability to dominate science. Ironically, the same social and economic forces, flowing out of the industrial revolution, which favored the development of professional science by bringing the British middle class (with its large Nonconformist component) into prominence during the nineteenth century, created a counterbalance to the influence of scientific naturalism.

The communications revolution opened up a new space in the second half of the century for men and women who were interested in science but who had little or no formal training in the area to forge careers in science journalism. These science journalists created a role for themselves as mediators between the specialized, professional scientist and the members of the reading public interested in the larger religious, moral, and social implications of the most recent discoveries.[56] They had their own set of goals, which did not necessarily coincide with the agenda of those like Huxley, who were determined to professionalize and secularize science. Many popularizers of science, for example, provided their readers with a readily accessible natural theology, updated in light of current scientific developments.[57] Although most popularizers did not openly question the authority of scientists as had Chambers, they insisted that the discovery of new knowledge was not limited to expert professionals. Moreover they conceived of themselves as natural historians, taking on the traditional role of naming, describing, classifying, and uncovering the order of natural objects, while adopting a narrative that retained the explicit use of storytelling as a technique to draw the reader into an appreciation of the wonder of nature.[58] It is no coincidence that many of the Victorians attracted to a career in popularizing science in the latter half of the nineteenth century belonged to those groups that Huxley and his colleagues were trying to edge out of science. Women and the Anglican clergy were particularly well represented among them, even though professional scientists like Huxley did not believe that they had any authority to speak about scientific subjects. If large book sales are any indication, the popularizers were influential in determining the meaning of science in the minds of the reading public.

An Oxford MA and ordained Anglican minister, the Reverend John George Wood (1827–1889) retired from regular clerical work to pursue a writing career as a popularizer of natural history. His most popular contribution, *Common Objects of the Country* (1858), is said to have sold a staggering hundred thousand copies in the very first week, though this is likely an exaggeration. Wood was not the only Christian minister who became involved in popularizing science. He was joined by Ebenezer Brewer (1810–1897), one of Jarrold and Sons' most successful authors, whose *A Guide to the Scientific Knowledge of Things Familiar* (1847) was in its fifth edition by 1850, with 113,000 copies printed by 1874.[59] Charles Alexander Johns (1811–1874), whose *Flowers of the Field* (1853) reached its thirty-fifth edition in 1949, published widely on birds and botany. It was Johns who had encouraged a passion for botany in the influential Broad Churchman Charles Kingsley (1819–1875), his former student at Helston grammar school. Kingsley, of course, went on to publish popular science works of his own, including the charming *Water Babies* (1863). Another popularizer of botany, the Reverend George Henslow (1835–1925), published a series of popular works on flowers in the last three decades of the nineteenth century. The Reverend Henry Neville Hutchinson (1856–1927) capitalized on the interest in dinosaurs and other extinct monsters in his books on popular geology during the 1890s.

If Christian ministers were well represented among the ranks of the popularizers of science, so were women. Mary Somerville (1780–1872) is well known for her popularizations of the physical sciences, but many of her sisters in science are not. Women were turning to popular science in droves, even though Huxley was busily excluding them from professional scientific societies. In his bid to upgrade the professional status of the Ethnological Society in 1868, Huxley, the newly elected president, moved to have women excluded from the "Ordinary Meetings," where the serious scientific discussions took place. Despite the objections of Eliza Lynn Linton, a convert to Darwinism who became famous for her satirical attacks on nineteenth-century feminism, Huxley relegated her and the other women who regularly attended to the larger and popular "Special Meetings."[60] But women could draw on a previously existing tradition of female popularization of science that sanctioned their involvement in science writing. Like their predecessors in this tradition, they took on the role of moral and religious guides and the majority of them claimed, in opposition to Huxley and his allies, that recent scientific developments revealed the workings of a divine creator. They contested the attempt of scientific naturalists to secularize nature. Women wrote about almost every area of scientific knowledge. Rosina Zornlin (1795–1859) produced a series of books on physical geography and geology. Jane Loudon (1807–1858), Anne Pratt (1806–1893), and Elizabeth Twining (1805–1889), as well as Lydia Becker (1827–1890), prior to her involvement with the women's suffrage movement, composed popular works on botany. Arabella Buckley (1840–1929), who had been Charles Lyell's secretary before turning to popular science writing, produced a series of works on evolutionary theory. Sarah Wallis (later Bowdich and then Lee; 1791–1856) and Mary Kirby (1817–1893) published widely in the area of natural history. In addition to her work *British Seaweeds* (1863), Margaret Gatty (1809–1873) published the bestseller *Parables of Nature* (1855), a book of short stories designed to teach children scientific, as well as moral, lessons. Mary Ward (1827–1869) focused in her works on scientific instruments such as the microscope and the telescope. At the end of the century, women's involvement in popular science writing was still going strong. In over a dozen books published from the eighties up until the third decade of the twentieth century, Agnes Giberne (1845–1939) explored both the heavens and the oceans, while Agnes Clerke (1842–1907), tackled a wide range of astronomical topics, ranging from the history of astronomy and biographical studies of famous astronomers to current developments in astrophysics and stellar astronomy. Eliza Brightwen (1830–1906) published a series of works on natural history in the last decade of the century, including her popular *Wild Nature Won by Kindness* (1890).

The career of Richard Proctor (1837–1888) vividly illustrates how many popularizers resisted the ideals of scientific naturalists. Proctor, the prolific popularizer of astronomy who fascinated the Victorian reading public throughout the seventies and eighties with his many works on extraterrestrial life, was neither an Anglican clergyman nor a woman. He actually straddled the worlds of professional and popular science, having been active in the Royal Astronomical Society as honorary secretary in the early seventies. A product of St. John's College, Cambridge, where he studied theology and mathematics, Proctor was forced to live by his pen when he incurred a huge debt as a result of a failed investment. Proctor was involved in almost all aspects of popular science activity. He authored over sixty books, mostly on astronomy; wrote at least five hundred essays that appeared in a wide variety of periodicals such as *Popular Science Monthly*, *Cornhill Magazine*, *Contemporary Review*, *Fortnightly Review*, *Fraser's Magazine*, and *Nineteenth Century*; founded the London scientific periodical *Knowledge* in 1881, which he edited until his death; and lectured extensively in England, the United States, and even Australasia.

Whereas some popularizers expressed their opposition to scientific naturalism by rejecting evolutionary theory on religious grounds or by embracing forms of thought inimical to Huxley and his friends, Proctor's resistance was far subtler. Establishing a close relationship with Alfred Russel Wallace, Arabella Buckley pursued with him a passionate interest in spiritualism behind the backs of her scientific naturalist friends. Margaret Gatty was violently opposed to evolution and wrote a parody of Darwin's theory in her "Inferior Animals" (1860), one of the stories published in the third series of her *Parables*.[61] But Proctor often expressed his respect for Darwin, Huxley, Tyndall, and other scientific naturalists, defending them on several occasions from the charge of atheism.[62] He embraced Spencer's conception of the Unknowable, a shadowy deity behind nature, and accepted Huxley's term "agnosticism" as an accurate label for his own position. He even agreed with the scientific naturalists that the cultural authority of the Anglican clergy needed to be questioned. The clergy, experts in rowing or tennis, were educated only in the "systems of training adopted for ministers of various orders," which limited and narrowed their abilities to "deal with the higher and nobler problems of religion." But the man of science usually "had a special training for discussing questions whose relation to the higher philosophy of religion is somewhat nearer than the relation of whist or cricket, or even Greek and Latin syntax, to theological and doctrinal problems."[63]

However Proctor's attitude toward the professionalization of science would not have endeared him to Huxley. He undercut the scientific naturalists' push for more salaried positions in science in his *Wages and Wants of Science-Workers* (1876), where he was openly critical of schemes to secure state funding for science put forward by the Devonshire Commission (on which Huxley sat). He was particularly harsh on professional astronomers in government observatories, where mechanical and routine work precluded "almost entirely the pursuit of original researches."[64] Moreover, Proctor did not admire "the way in which *'soi-disants'* professional astronomers treat the wonders of the heavens and the grand problems presented by the movements of the celestial orbs. Too often they discuss these as a mere land surveyor might discuss the teachings of the earth's crust." Proctor's disgust with the emphasis on "methods and instruments of observation," rather than the lessons to be learned from careful observations, represented a protest against the obsession with precision measurement that was so important to professionalizing scientists.[65] His use of the phrase "*soi-disants*," or, "so-called," professionals, registers his protest against the notion of the professional scientist, at least when applied to those who worked in government observatories. In fact, Proctor founded his journal *Knowledge* in order to challenge *Nature* for control of the popular science periodical market. Edited by the astronomer Norman Lockyer, one of Proctor's bitterest enemies, *Nature* was founded in 1869 to gain the support of the general public for the agenda of professional science. Lockyer often called on Huxley and his friends to boost the success of his project. The format of *Knowledge* reflected Proctor's aversion for the professionalizing, hierarchical vision of science contained in the pages of *Nature*. His format, which in the early years featured a large correspondence section, drew on an older republican image of scientific community.[66]

In the opening pages of his *Flowers of the Sky* (1879), Proctor quoted John Herschel on how the nature of light provided evidence for the argument for a unity of design and action in the universe.[67] In many ways, Proctor was closer in spirit to the gentlemen of science than to the scientific naturalists. Like Whewell, he pointed to the need for unity in science. Although he realized that "every real worker in science must be a specialist," he also believed that they should have "a correct view of other sciences" that could be found in popular science works. "Every true populariser of science knows," Proctor declared, "that among his readers, if not forming the greater number of his readers, there will be men of science, working in other branches."[68] Like Whewell, he refused to make rigid distinctions between popular and professional science. He argued that writing popular science works was an integral part of the role of the scientist, that it was by no means a secondary activity. Proctor even suggested that popularizing new scientific theories could be a boon to the scientist, as the process of working on how to present them clearly constituted a test of their cogency.[69] Proctor offered a very different definition of the "scientist" at a time when Huxley and his allies were trying to establish their own vision of the professional scientist. Just as Whewell was prepared to allow women into the world of science, Proctor welcomed female contributors to *Knowledge*. Whereas Huxley ranked women's intelligence with that of the "lower" races, Proctor rejected the notion that women were intellectually inferior to men.[70] He poured scorn upon evolutionists who argued that women needed to wear corsets because nature had not yet fitted them for the upright position.[71] As for the attempts of scientific naturalists to draw on scientific arguments to demonstrate the natural limits to social reform, one of Proctor's strictest rules was the exclusion of politics from the columns of *Knowledge*.[72]

But Proctor's affinity with the gentlemen of science is perhaps most apparent in his conception of the cultural and religious meaning of science. To Proctor, science promoted a belief in universal law and the gradual extinction of superstition and fanaticism.[73] This is what Proctor liked about Darwin's theory of evolution. No scientist since Newton had done as much, in Proctor's opinion, "to extend men's recognition of the wideness of the domain of law," and of "the infinitely perfect nature of the laws of the universe." Whereas Newton had demonstrated the lawfulness in space, Darwin had showed that law operated throughout time. Huxley and Proctor could agree that evolutionary theory was important for the way it expanded the reign of natural law. But to Proctor, a devoutly religious soul with an eye for the divine order at the heart of the natural world, the expansion of the laws of nature also enlarged the domain of religion rather than secularizing science. Religion had been "rendered infinitely grander—infinitely more impressive by our new knowledge" and "infinitely more reasonable."[74] Proctor was able to incorporate evolutionary theory into his argument for extraterrestrial life, and then place both into the framework of a theology of nature. In his *Other Worlds Than Ours* (1870), his first bestseller, Proctor extracted lessons from the sun, the planets, the stars, and other heavenly bodies, which pointed to the divine order in nature.[75] A study of the evolutionary process on Earth revealed that nature, according to Proctor, "has a singular power of adapting living creatures to the circumstances which surround them."[76] Darwin's insight held for planets throughout the cosmos, not just Earth. It was a universe teeming with life. Whatever the physical conditions to be found on any particular planet, there were beings that could adapt to the environment. For Proctor, pluralism and natural theology went hand in hand. Since God had created nature to fulfill a certain purpose, and since nature's "great end" was "to afford scope and room for new forms of life, or to supply the wants of those which already exist," the greatest proof of God's existence lay in the discernment of the existence of alien life.[77] Needless to say,

Huxley and his allies would have taken a dim view of Proctor's combination of evolutionary theory, pluralism, and natural theology, especially since it captured the imagination of the Victorian reading public for all the wrong reasons.

VICTORIAN READERS OF SCIENCE

If many professional science journalists involved in popularization activities resisted the scientific naturalists' conception of science, what of their readers? As we have seen, already in the early nineteenth century scientists like Davy tried to establish a relationship with their audience that rendered them passive. The emergence of disciplined, trained cadres of research scientists, such as Huxley's students in his laboratory at South Kensington, widened the gap between the scientist and the wider public. The notion of disciplined expertise was intended to restrict the community active in creating and validating scientific knowledge, and to produce a passive public. Closely connected with these changes in the relationship between the scientist and the public was the creation by professional scientists of the diffusionist notion of "popularization," in which the communication of scientific ideas took place in a linear process from the expert to the public, who passively consumed them. It is no coincidence that the top-down sense of the verb "to popularize" was developed in English at the same time that the word "scientist" came to designate the new experts who sought to position readers as invisible members of a mass audience. This notion of popularization was an important element in the self-fashioning of the professional scientist.[78] Many professional scientists, like Huxley, Tyndall, and Spencer, all tried their hand at writing popular works. They recognized that public support for science was essential to establish the cultural authority of the scientific elite and to free up funding for scientific institutions. Perhaps the most famous attempt at codifying and popularizing scientific knowledge in a systematic fashion to a wide-reading audience, the International Scientific Series appeared in the United States and five European countries in over 120 titles between 1871 and 1910. It was written for the most part by professional scientists and directed in its early years in England by an advisory committee composed of Huxley, Tyndall, and Spencer. Like *Nature*, the series stands as a monument to the efforts of professionals to control the public's understanding of modern science. Though each contributing author was free to present their own definition of science while focusing on their area of specialty, the initial aim was to disseminate an authoritative and collective image of science as stable, secular, and comprehensive (since both the natural and social sciences were included).[79]

The attempt to create and maintain a large, passive, lay audience was not immediately or uniformly successful. More egalitarian popular science journals were only gradually replaced in the second half of the century by journals such as *Nature*. Even as late as the eighties, Proctor was fighting to oppose this trend. But did members of the lay audience for science accept the role provided for them by professional scientists like Huxley? Topham has suggested that we begin to answer this question by examining the patterns of reading of historical actors (who read what and where) and by studying how and why readers read what they read. This will allow us to recover the agency of readers and to see if they subverted authorial intentions and textual strategies.[80] Historians are just beginning to make use of this promising approach. We have a model in Secord's *Victorian Sensation*, an analysis of the diverse readership of the *Vestiges of the Natural History of Creation* that demonstrates how readers actively participated in forging the meaning of Chambers's book, and therefore the meaning of science, for themselves. Secord draws our attention to the varied composition of the vast reading public. We can divide the audience up by such factors as gender, class, age, religion, and region, each of which played a role in how individual readers interpreted the significance of science. Though the emphasis in what follows will be on the agency of members of the public as readers—as consumers of scientific writing—it should be kept in mind that these people affirmed their agency through participation in scientific activities besides reading.[81] Members of the public attended lectures, visited museums and exhibitions, participated in trips to the field to collect specimens, and joined amateur science clubs. As David Allen has reminded us, the Victorian public was swept by a succession of natural history crazes, from fern collecting to seaweed collecting.[82]

Working-class readers constituted an important and large segment of the new reading audience for science created in the latter half of the nineteenth century. How did they read professional scientists like Huxley and science journalists like Proctor? Did they view themselves as invisible members of a passive mass public? What did science mean to them? In her study of working class science in the last three decades of the nineteenth century, McLaughlin-Jenkins treats science as a fundamental component of self-improvement and class emancipation for many working-class Victorians.[83] Science was available to members of the working class in the first half of the century through their family, friends, and neighbors; through community resources; through evening study at mechanics' institutes, community clubs, and pubs; through cooperative societies; and through contact with secularist lecturers and periodicals.[84] By the second half of the century, increased contact with scientific ideas was made possible by the communications revolution, which provided cheap reprints of scientific texts and more affordable periodicals with scientific content, by the establishment of free libraries, by the explosion of middle-class popular science, and by working-class educational initiatives. During the final decades of the century, the Victorian Left "capitalized" on the ideological power of science through the creation of a socialist science. Rejecting the elitist, capitalist science of the scientific naturalists, based on evolutionary notions of a struggle in nature and society, socialist propaganda in periodicals, pamphlets, and books offered a scientific validation for radical social ideals. Darwin had shown that struggle in nature was countered by natural sociability and mutual support within species. Insisting that science was common property, or collectively owned, the proponents of socialist science challenged the cultural authority of Huxley and his allies.

No doubt, members of the working class reacted in diverse ways to the science they encountered, whether it originated from scientific naturalists like Huxley, science journalists like Proctor, or even purveyors of socialist science. But in many cases, they exercised their agency as readers and declined to accept the role of passive consumer of knowledge. For example, consider Tom Mann (1856–1941), engineer, trade unionist, and socialist.[85] The son of a clerk at the Victoria Colliery, Mann started work in a mine at the age of nine, dragging boxes of coal. When the mine had to be closed due to an accident, the family moved in 1870 to Birmingham, where Mann became apprenticed to engineering in a tool-making firm. After completing seven years' apprenticeship, he moved to London in 1877 to take on various jobs in engineering. While employed at Cubitt's engineering works he was given the task of cutting a large meteorite into three pieces for exhibition in different museums. The meteorite's strange composition intrigued him, and from then on astronomy became a serious recreational activity. In the early eighties, Mann was working near London on engines for torpedo boats. "At this period," Mann recalled in his autobiography, "Mr Richard A. Proctor, the astronomer, was lecturing at Kensington on, 'The Birth and Death of Worlds,' I was a regular reader of Proctor's magazine *Knowledge*." Mann convinced two of his workmates to accompany him to the lecture. Proctor began the lecture by exhibiting a series of pictures of nebulae "and the resultant worlds, their life and decay." Though Mann was engrossed by the lecture, his two companions fell asleep. Mann's chief exposure to science seems to have come through a popularizer of science rather than a scientific naturalist.[86]

Mann's subsequent intellectual development raises further questions about the ability of scientific naturalists to control the meaning of science and underlines the agency of the working class reader. Mann became politically active about the same time that he became interested in astronomy. He declared that in London the only two subjects that attracted him apart from workshop affairs were "social problems and astronomy." Proctor's embargo on politics in *Knowledge* did not prevent him from yoking together an interest in astronomy and social problems. In 1884 he joined the Social Democratic Federation (which based its program in Marxist Scientific Socialism), became a strong advocate of shorter working hours, and in 1889 became deeply involved in the London dock strike. Greatly in demand as a speaker due to his articulate style and infectious enthusiasm, he gave up his career as engineer and worked in the labor movement in England and Australia, later helping to found the British Communist Party in 1920. Mann's reading of evolutionary theory was profoundly affected by his experiences as a member of the working class and by his study of left-wing thinkers such as Henry George. If he encountered evolution in the writings of Huxley, his rejection of Malthusianism colored his interpretation of its social significance. Evolutionary theory and socialism were by no means in opposition. He brought all of his working-class categories to his reading of science and his understanding of its practical consequences.[87] He did not recognize the cultural authority of scientific naturalists like Huxley and he contested their interpretations of the meaning of science.

There are parallels in the case of Percy Redfern (1875–1958), clerk and cooperative journalist. Born to an unmarried woman of a proudly Nonconformist and puritan family, Redfern attended a variety of schools as a child.[88] At the age of thirteen he was instructed to abandon his love of sensationalist fiction for "scientific, informative books." "I found beckoning me the popular and romantic introductions to astronomy of R. A. Proctor," Redfern recalled. "For the first time in my life, it seemed, I looked at the stars." Proctor's books provided an escape from his squalid circumstances in Leicester, dominated by its "garish and mean" shops, and offered him a grand vision of the universe. Redfern was enthralled by Proctor's pluralism. "Mars had its canals," he recollected, "and beyond that hopeful sphere there was nothing to discourage imagination from roaming an infinitely-peopled universe." Although Proctor intended his lectures and books to guide the reader toward an appreciation of divine law in the universe, Redfern's interest in astronomy "at once took me away from the Bible." There was "no earth-centred cosmos, no skies that could make a heaven, and stars that were no lesser lights," and this "discredited" the Bible and cast doubt on the existence of a God. Redfern was now receptive to agnosticism and secularism.[89]

Finding school unsatisfying, Redfern was apprenticed to a draper in Nottingham at the age of fourteen but left to seek employment in London and Coventry. In Coventry, where he was employed as a shop worker, he was plagued by the tension that he saw between his scientific secularism and his faith in the religious ideal "love thy neighbour." Secularists firmly believed in science "and science questioned or contradicted the faith. Man was of Nature, and Nature demanded the evolutionary struggle." But the more Redfern was drawn into the world of radical politics, the more he began to question evolutionary naturalism. He joined the Social Democratic Federation, where the ideal of "fraternal happiness" was seen as the antidote to "those ever-depressing doubts created by scientific agnosticism." In 1895 he moved to London and joined the Hammersmith Socialist Society, and then in 1896 he became a member of the Independent Labour Party when he moved to Huddersfield. Shortly afterwards, he read Huxley's "Evolution and Ethics" and was appalled even more by the scientific naturalist's notion that there were natural limits to social reform. "So it was the message of science," Redfern wrote, "that we could change the world only here and there, easing this and that, but altering nothing radically. The last word was death; and the intermediate hope, some small mitigation of the old, animal struggle." Redfern was not prepared to accept Huxley's reading of evolutionary theory as "the last word." Exercising his agency as reader in order to escape the "paralysis of agnosticism," he questioned the authority of scientists like Huxley, limited the role of science in legitimating leftist ideology, and maintained that scientific data had nothing of value to say about the true "meaning" of human life. Though he continued to work in the Labour movement throughout the twentieth century, becoming editor of *Wheatsheaf*, the Co-Operative Wholesale Society's monthly magazine, he was critical of the abandonment of religion by British socialism.[90] Redfern, then, began his active engagement with science by rejecting the religious dimensions in Proctor's astronomy, became attracted to the secularized vision of nature offered to him by scientific naturalism, and ended up returning to a vague religiosity when evolutionary naturalism conflicted with his socialist ideals. Working class readers of science like Mann and Redfern were not passive consumers of science, whether it came from science journalists such as Proctor or scientific naturalists like Huxley.

PATTERNS OF SCIENCE

The nineteenth century is a particularly important period in the history of science, as during this era many distinguishing characteristics of contemporary science began to take shape. A number of key theories were incorporated into science, such as evolutionary theory and thermodynamics, which transformed both the life sciences and the physical sciences. It was a time when scientists changed the way they did science and where they did it, as the use of precision instruments in laboratory experiments became the defining scientific activity, and the work of collecting, naming, describing, and classifying in the field became secondary. The titles used to describe scientific disciplines reflected this shift, as terms such as "biologist," "physicist," and "scientist" became more widely used, edging out the older labels "natural history" and "natural philosophy." As science became more professionalized, some believed that the privilege of making knowledge should be restricted to those with the proper training and expertise. It was also a time when scientists gained unprecedented cultural authority at the expense of the Christian clergy, as science came to be seen as providing a model for obtaining truth. Those deemed the legitimate interpreters of science could determine its implications for the religious, social, and political issues of the day. Driven by the twin engines of empire and industrialization, science became a cultural and social force to be reckoned with. Exotic flora and fauna from around the world and technological breakthroughs based on scientific know-how fascinated the public, who flocked to see them on display at the London Zoo, Kew Gardens, the Great Exhibition of 1851, the Royal Institution, and the British Museum (Natural History), which first opened in 1881. Popular science also thrived in print form, as the revolution in communications provided the conditions for the development of a vast cadre of professional science journalists whose writings filled the pages of new popular science journals and books. Science was the "in" thing for much of the second half of the nineteenth century. It was a "hot property," and everyone wanted a piece of it. Exactly who "owned" science was vigorously disputed and there were those who were firmly opposed to the entire metamorphosis of science.

The story of the contested meanings of mid- and late-Victorian science began with the London-based scientific naturalists and their attempt to reform and then control the scientific scene. But as we move from the scientific naturalists to other groups, seemingly from the center of the scientific community to the periphery, we discover intense opposition to Huxley and his allies. By fanning out to another group of reform minded, professional scientists, the North British Physicists, the resistance to scientific naturalism within the scientific elite became apparent. Then, by moving on to groups normally seen as outside professional science, the science journalists who popularized science, and the reading public who consumed science, it becomes clear that there was strong resistance to the agenda of scientific naturalism from those who were supposed to be passive mediators or consumers of science. The entire analogy of center and periphery becomes questionable as an adequate model for understanding Victorian science. Of course the picture could have been complicated even more had we included provincial science and even science in the colonies. But the story is complicated enough to remind us of the importance of studying the local settings of science that, as Jim Secord argues, "has made it possible to escape the old image of science as dominated by a handful of great theorists and simultaneously to understand theory making as a form of practice."[91] Looking back at the nineteenth century from the vantage point of the twenty-first, we may be tempted to conclude that the scientific naturalists won the contest to determine the meaning of science, since our contemporary conceptions seem to share much in common with Huxley's vision of a secular and professional science. But victory was never assured during Huxley's lifetime, and we must be careful not to treat the professionalization of science as if it were an inevitable process—akin to a law of nature—just because we think we know the end of the story.

Coming to any overarching conclusions about similar contests during the nineteenth century over the meaning of science in other national contexts, involving professional scientists, science journalists, and the public, will have to await further research by historians of science.[92] Though a process of professionalization played a significant role at some point in the nineteenth-century in Europe and the United States, the intricate interplay between the social and cultural context and the development of scientific institutions and ideas varied from country to country. In the case of nineteenth-century German popular science, for example, where Andreas Daum has done some work, there are fascinating parallels with Britain. As in Britain, the genre of popular science was established in Germany as a specific field of cultural activity in the middle of the century. Figures such as Emil Adolf Rossmässler came from outside the scientific establishment, established journals devoted to popularizing science, revered the natural history tradition, eschewed the emphasis on research, and were more likely to adhere to Humboldt's cosmic concept of nature rather than to scientific materialism. The differences between the German popularizers and their British counterparts are just as interesting and, in large part, are related to the different contexts in which they worked.

The failed revolution in 1848 became the impetus for the development of popular science in Germany, as men such as Rossmässler believed that a democratic Germany was more likely to emerge through the impact of popularizing activities than by political means.[93]

Only in-depth comparative studies of professional and popular science in other national contexts in this period will reveal larger patterns of significance. They will also help us to avoid the temptation of treating the British story as paradigmatic for our understanding of science and its contested meanings and for our analysis of the participants involved in the contest.[94] To determine what is peculiarly British about the story told here, future studies could explore a series of interrelated questions. Were there equivalents to the gentlemen of science, the scientific naturalists, the North British Physicists, the popularizers, and the reading public in other countries? If so, then did these groups relate to each other in a similar way, raise identical issues, or command the same amount of power? If different groups existed, what did the geography of the scientific landscape look like and how did it affect the contest over the meaning of science? Is there a satisfactory explanation for the absence of important groups of players, if, for example, women or Christian clergymen are not well represented among the popularizers? Were there tensions within the group of professional scientists akin to the rivalry between the scientific naturalists and North British Physicists? Did professional scientists attempt to control popular science and, if so, how resistant were readers to the authority of self-proclaimed scientific experts?

This analysis of the contested meanings of science in Victorian Britain has only complicated the task of writing a history of "science." There is no single, univocal meaning for science in a particular space, which rules out the possibility of locating one over time. Historians have come to be suspicious of "essentialism"—the notion that science has an essence, that it can be considered to be one sort of thing, unambiguously identifiable in every historical era. It is a real inconvenience for the historian bent on writing the definitive master narrative. But science is inevitably altered as it develops within a constantly shifting set of cultural and social contexts which subtly shape its meaning. At any one time in history, the scientific world is composed of an intricate quilt pattern of local communities with varying definitions of science, usually in line with their social aspirations. The challenge is to explore the varied meanings of science in each of the patterns, to specify them in their local settings, and to identify those sites of contestation—scientific societies, journals, lecture halls, museums, and exhibitions, to name just a few—where the seams between patterns have become visible because they are loose, worn, or newly sewn. Made by human hands from materials supplied by nature, the fabric of science is delicate and the quilt must be handled carefully. But if we can pick apart the threads without unraveling the whole, we will understand how the patterns have been formed, how they fit, and, ultimately, what holds them all together in the fragile tapestry of history.

NOTES

The author would like to acknowledge those scholars whose helpful suggestions strengthened this paper immeasurably. Paul Farber gave me the reference to Huxley's essay on biology and natural history. In addition to answering several questions about the North British Physicists, Crosbie Smith drew my attention to lectures by Thomson and Maxwell on natural philosophy. Erin Jenkins guided me toward Redfern and Mann as good case studies for working class intellectuals who encountered science. Sally Kohlstedt supplied me with some useful references to nineteenth-century American science. With Andy Daum's permission, Lynn Nyhart kindly sent me his then-unpublished paper on German popular science. I am also indebted to Adrian Desmond, Don Opitz, Lynn Nyhart, and the editors for sharing their stimulating comments and reactions to earlier drafts of this chapter. Finally, a version of this chapter appeared as "Science, Scientists and the Public: The Contested Meanings of Science in Victorian Britain," in Bernard Lightman, *Evolutionary Naturalism in Victorian Britain: The 'Darwinians' and their Critics* (Farnham, UK: Ashgate, 2009), 1–40. The editors are grateful to Ashgate for permission to republish it here.

1. [William Whewell], "Mrs. Somerville on the Connexion of the Sciences," *Quarterly Review* 51 (1834): 59.

2. Ibid., 54–56.

3. Sydney Ross, "'Scientist': The Story of a Word," *Annals of Science* 18 (1962): 65–66.

4. Frank M. Turner, *Between Science and Religion: The Reaction to Scientific Naturalism in Late Victorian England* (New Haven, CT: Yale University Press, 1974), 12–13.

5. John Tyndall, *Fragments of Science*, 8th ed. (London: Longmans, Green, & Co., 1892), 2:197.

6. [John Tulloch], "Modern Scientific Materialism," *Blackwood's Edinburgh Magazine* 188 (Nov. 1874): 520.

7. James A. Secord, *Victorian Sensation: The Extraordinary Publication, Reception, and Secret Authorship of* Vestiges of the Natural History of Creation (Chicago: University of Chicago Press, 2000), 405.

8. Ibid., 406.

9. William Paley, *The Complete Works of William Paley, D.D.* (London: J. F. Dove, 1825), 1:429.

10. Humphry Davy, *The Collected Works of Sir Humphry Davy* (ed. John Davy, [London: Smith, Elder, 1839–1840], 2:326), as cited in Jan Golinski, *Science as Public Culture:*

Chemistry and Enlightenment in Britain, 1760–1820 (Cambridge: Cambridge University Press, 1992), 197.

11. Golinski, *Science as Public Culture*, 285.

12. Alison Winter, "The Construction of Orthodoxies and Heterodoxies in the Early Victorian Life Sciences," in *Victorian Science in Context*, ed. Bernard Lightman (Chicago: University of Chicago Press, 1997), 24–50.

13. Secord, *Victorian Sensation*, 223.

14. Richard Yeo, *Defining Science: William Whewell, Natural Knowledge and the Public Debate in Early Victorian Britain* (Cambridge: Cambridge University Press, 1993), 8.

15. Adrian Desmond, *The Politics of Evolution: Morphology, Medicine, and Reform in Radical London* (Chicago: University of Chicago Press, 1989).

16. Richard John Noakes, "'Cranks and Visionaries': Science, Spiritualism and Transgression in Victorian Britain" (PhD diss., Cambridge, 1998), 23.

17. Iwan Rhys Morus, *Frankenstein's Children: Electricity, Exhibition, and Experiment in Early-Nineteenth Century London* (Princeton, NJ: Princeton University Press, 1998).

18. Noakes, "'Cranks and Visionaries,'" 25.

19. Beatrice Webb, *My Apprenticeship*, 2nd ed. (London: Longmans, Green & Co., 1950), 112.

20. W. K. Clifford, "On the Aims and Instruments of Scientific Thought," in *Lectures and Essays*, 2nd ed., ed. Leslie Stephen and Frederick Pollock (London: Macmillan & Co., 1886), 86.

21. Thomas H. Huxley, *Method and Results* (New York: Appleton & Company, 1897), 66.

22. Turner, *Between Science and Religion*, 9, 24.

23. Ruth Barton, "The X Club: Science, Religion, and Social Change in Victorian England" (PhD diss., University of Pennsylvania, 1976); Ruth Barton, "'Huxley, Lubbock, and Half a Dozen Others': Professionals and Gentlemen in the Formation of the X Club, 1851–1864," *Isis* 89 (1998): 410–44; Roy M MacLeod, "The X-Club: A Social Network of Science in Late-Victorian England," *Notes and Records of the Royal Society of London* 24 (1970): 305–22; J. Vernon Jensen, "Interrelationships within the Victorian 'X Club,'" *Dalhousie Review* 51 (Winter 1971–72): 539–52.

24. Frank M. Turner, "The Victorian Conflict between Science and Religion: A Professional Dimension," in *Contesting Cultural Authority: Essays in Victorian Intellectual Life* (Cambridge: Cambridge University Press, 1993), 171–200.

25. Adrian Desmond, "Redefining the X Axis: 'Professionals,' 'Amateurs' and the Making of Mid-Victorian Biology A Progress Report," *Journal of the History of Biology* 34 (2001): 3–50.

26. Secord, *Victorian Sensation*, 403; Alison Winter, *Mesmerized: Powers of Mind in Victorian Britain* (Chicago: The University of Chicago Press, 1998), 300.

27. George H. Daniels, "The Process of Professionalization in American Science: The Emergent Period, 1820–1860," in *Science in America Since 1820*, ed. Nathan Reingold (New York: Science History Publications, 1976), 63, 66, 70; see also Robert V. Bruce, *The Launching of Modern American Science 1846–1876* (New York: Alfred A. Knopf, 1987).

28. Marc Rothenberg, "Gould, Benjamin Apthorp," in *American National Biography*,

ed. John A Garraty and Mark C. Carnes (New York, Oxford: Oxford University Press, 1999), 337–38.

29. Thomas Henry Huxley, *Lay Sermons, Addresses, and Reviews* (New York: D. Appleton and Company, 1895), 77–78.

30. Winter, *Mesmerized*, 296–303.

31. Ross, "'Scientist,'" 78, 80.

32. "The Word 'Scientist,'" *Science-Gossip* 1 (Jan. 1895): 242.

33. Thomas H. Huxley, *Science and Education* (New York: H. L. Fowle, [1893]), 263, 267–68.

34. Paul Lawrence Farber, *Finding Order in Nature: The Naturalist Tradition from Linnaeus to E. O. Wilson* (Baltimore, MD: The Johns Hopkins University Press, 2000), 85–6.

35. Huxley, *Science and Education*, 281–84. For a discussion of how Huxley trained his students to use the microscope in his lab see Graeme Gooday, "'Nature' in the Laboratory: Domestication and Discipline with the Microscope in Victorian Life Science," *British Journal for the History of Science* 24 (1991): 333–40.

36. Huxley, *Science and Education*, 288.

37. Ibid., 270–1.

38. Huxley, *Method and Results*, 338, 427.

39. Leon Stephen Jacyna, "Scientific Naturalism in Victorian Britain" (PhD diss., University of Edinburgh, 1980), 276–311.

40. Erin McLaughlin-Jenkins, "Common Knowledge: The Victorian Working Class and the Low Road to Science 1870–1900" (PhD diss., York University, 2001).

41. Bernard Lightman, "'Fighting Even with Death': Balfour, Scientific Naturalism, and Thomas Henry Huxley's Final Battle," in *Thomas Henry Huxley's Place in Science and Letters: Centenary Essays*, ed. Alan Barr (Athens: The University of Georgia Press, 1997), 323–50.

42. Arthur James Balfour, *The Foundations of Belief: Being Notes Introductory to the Study of Theology* (London: Longmans, Green, 1895), 135.

43. Crosbie Smith, *The Science of Energy: A Cultural History of Energy Physics in Victorian Britain* (Chicago: University of Chicago Press, 1998), 7, 171, 307, 311–12.

44. Joe D. Burchfield, *Lord Kelvin and the Age of the Earth* (London: The Macmillan Press Ltd., 1975), 81.

45. John Tyndall, "Remarks on an Article Entitled 'Energy' in 'Good Words,'" *Philosophical Magazine* 25 (1863), 221.

46. Ross, "'Scientist,'" 73. According to the *Oxford English Dictionary*, the term "physicist" was occasionally used in the eighteenth century to refer to an individual knowledgeable about medical science. In 1840, Whewell suggested in the preface to his *Philosophy of the Inductive Sciences* that a distinction be made between "physician," to be used as the equivalent of the French *physicien*, and "physicist," who "proceeds upon the ideas of force, matter, and the properties of matter." See William Whewell, *The Philosophy of the Inductive Sciences* (London: John W. Parker, 1840), 1:lxxi.

47. Crosbie Smith and M. Norton Wise, *Energy and Empire: A Biographical Study of Lord Kelvin* (Cambridge: Cambridge University Press, 1989), 120.

48. William Thomson, "Introductory Lecture to the Course on Natural Philosophy," in *The Life of William Thomson*, by Silvanus P. Thompson (London: Macmillan and Co., 1910), 1: 244–46, 250–51.

49. Smith and Wise, *Energy and Empire*, 117–28.

50. James Clerk Maxwell, "Introductory Lecture on Experimental Physics," in *The Scientific Papers of James Clerk Maxwell*, ed. W. D. Niven (1890; New York: Dover Publications, Inc., 1965), 2: 241, 243–44.

51. "Wrangler" was the name given to each Cambridge undergraduate placed in the first class in the mathematical tripos.

52. Maxwell, "Introductory Lecture on Experimental Physics," 247, 254.

53. Secord, *Victorian Sensation*, 30.

54. Ibid., 48, 69, 461, 41, 100–9, 3; Richard Yeo, "Science and Intellectual Authority in Mid-Nineteenth-Century Britain: Robert Chambers and 'Vestiges of the Natural History of Creation,'" *Victorian Studies* 28 (Autumn 1984): 5–31.

55. Secord, *Victorian Sensation*, 523–25.

56. Steven Shapin, "Science and the Public," in *Companion to the History of Modern Science*, ed. R. C. Olby, G. N. Cantor, J. R. R. Christie, and M. J. S. Hodge (London: Routledge, 1990), 990–1007.

57. Bernard Lightman, "'Voices of Nature': Popularizers of Victorian Science," in Lightman, *Victorian Science in Context*, 187–211; Bernard Lightman, *Victorian Popularizers of Science: Designing Nature for New Audiences* (Chicago: University of Chicago Press, 2007).

58. Bernard Lightman, "The Story of Nature: Victorian Popularizers and Scientific Narrative," *Victorian Review* 25, no. 2 (1999): 1–29.

59. Lightman, *Victorian Popularizers of Science*, 66.

60. Evelleen Richards, "Huxley and Women's Place in Science: the 'Women Question' and the Control of Victorian Anthropology," in *History, Humanity and Evolution*, ed. James R. Moore (Cambridge: Cambridge University Press), 253–84.

61. Suzanne Le-May Sheffield, *Revealing New Worlds: Three Victorian Women Naturalists* (London: Routledge, 2001), 58.

62. [Richard A. Proctor], "Science and Art Gossip," *Knowledge* 2 (Oct. 27, 1882): 349; Richard A. Proctor, "The Unknowable. A Digression Concerning the Name and Purpose of These Papers," *Knowledge* 9 (June 1, 1886): 233–34; Richard A. Proctor, "The Unknowable; or, The Religion of Science," *Knowledge* 9 (Nov. 1, 1885): 1–3.

63. [Richard A. Proctor], "Science and Religion," *Knowledge* 10 (June 1, 1887): 172–73.

64. Richard A. Proctor, *Wages and Wants of Science-Workers* (London: Frank Cass and Company Limited, 1970), 38.

65. [Richard A. Proctor], "Popular Astronomy," *Knowledge* 1 (Feb. 17, 1882): 336.

66. Bernard Lightman, "'Knowledge' Confronts 'Nature': Richard Proctor and Popular Science Periodicals," in *Culture and Science in the Nineteenth-Century Media*, ed. Louise Henson et al. (Aldershot, UK: Ashgate Publishing Limited, 2004), 199–210.

67. Richard A. Proctor, *Flowers of the Sky* (London: Chatto and Windus, 1883), 2.

68. Proctor, *Wages and Wants of Science-Workers*, 34–5.

69. Ibid., 31.

70. Evelleen Richards, "Redrawing the Boundaries: Darwinian Science and Victorian Women Intellectuals," in Lightman, *Victorian Science in Context*, 126; Richard A. Proctor, *The Universe of Suns and Other Science Gleanings* (London: Chatto and Windus, 1884), 338.

71. Proctor, *The Universe of Suns*, 346–8.

72. [Richard A. Proctor], "Letters Received and Short Answers," *Knowledge* 6 (Nov. 28, 1884): 452.

73. Proctor, *Wages and Wants of Science-Workers*, 18, 22.

74. Richard A. Proctor, "Newton and Darwin," *Knowledge* 1 (April 28, 1882): 545–46.

75. Bernard Lightman, "Astronomy for the People: R. A. Proctor and the Popularization of the Victorian Universe," in *Facets of Faith and Science*, ed. Jitse van der Meer (Lanham, New York: The Pascal Centre for Advanced Studies in Faith and Science, Redeemer College, Ancaster, Ontario, and University Press of America, Inc., 1996), 3:31–45.

76. Richard A. Proctor, *Other Worlds Than Ours: The Plurality of Worlds Studied Under the Light of Recent Scientific Researchers* (New York: A. L. Fowle, 1870), 25.

77. Ibid., 18.

78. Jonathan Topham, "Scientific Publishing and the Reading of Science in Nineteenth-Century Britain: A Historiographical Survey and Guide to Sources," *Studies in History and Philosophy of Science* 31 (2000): 560–61.

79. Roy M. MacLeod, "Evolutionism, Internationalism and Commercial Enterprise in Science: The International Scientific Series 1871–1910," in *Development of Science Publishing in Europe*, ed. A. J. Meadows (Amsterdam: Elsevier Science Publishers), 63–93; Leslie Howsam, "An Experiment with Science for the Nineteenth-Century Book-Trade: The International Scientific Series," *British Journal for the History of Science* 33 (2000): 187–207. The North British Physicists were also involved in popularizing efforts.

80. Topham, "Scientific Publishing and the Reading of Science in Nineteenth-Century Britain," 568.

81. The emphasis on texts in this paper creates something of a problem in getting across the point that members of the public contested the meanings of science offered to them by elite scientists, and even by popularizers of science. It inevitably focuses attention on battles over who gets to write the texts with the authority of science behind them and it places the public in the role of the consumer of scientific writing. Since it is sometimes difficult to see any real agency in the activity of reading, it is important to draw attention toward the ways in which the public may have shaped the meaning of science through their participation in scientific activities—their scientific practice—above and beyond their role as critical readers. Lynn Nyhart brought this point to my attention.

82. David Allen, *The Naturalist in Britain: A Social History*, 2nd ed. (Princeton, NJ: Princeton University Press, 1994); David Allen, *The Victorian Fern Craze: A History of Pteridomania* (London: Hutchinson, 1969). In a striking instance of the shaping of professional science by members of the public, Alberti has discussed how amateurs and professionals redefined their identities in response to each other in late Victorian Yorkshire. See Samuel J. M. M. Alberti, "Amateurs and Professionals in One County: Biology and Natural History in Late Victorian Yorkshire," *Journal of the History of Biology* 34 (2001): 115–47.

83. McLaughlin-Jenkins, "Common Knowledge."

84. Anne Secord, "Science in the Pub: Artisan Botanists in Early Nineteenth-Century Lancashire," *History of Science* 32 (1994): 269–315.

85. J. S. Middleton, "Mann, Thomas (1856–1941)," in *Dictionary of National Biography 1941–1950*, ed. L.G. Wickham Legg and E. T. Williams (London: Oxford University Press, 1959), 568–570.

86. Tom Mann, *Tom Mann's Memoirs* (London: Macgibbon & Kee Ltd., 1967), 13, 19.

87. Ibid., 13, 16.

88. Joyce Bellamy, "Redfern, Percy (1875–1958)," in *Dictionary of Labour Biography*, ed. Joyce M. Bellamy and John Saville (London: Macmillan Press Ltd., 1972), 1:280–81.

89. Percy Redfern, *Journey to Understanding* (London: George Allen & Unwin Ltd., 1946), 12–13

90. Ibid., 29, 32, 41, 40, 68, 70–71.

91. Secord, *Victorian Sensation*, 520.

92. For an important treatment of the relevant issues see Jonathan R. Topham et al., "Focus: Historicizing 'Popular Science,'" *Isis* 100 (2009): 310–68.

93. Andreas Daum, "Science, Politics, and Religion: Humboldtian Thinking and the Transformations of Civil Society in Germany, 1830–1870," in *Science and Civil Society*, ed. Tom Broman and Lynn Nyhart (Chicago: University of Chicago Press, 2002), 107–40.

94. I am indebted to Lynn Nyhart for helping me to develop my thoughts on this point.

CHAPTER 14

Science and Place

David N. Livingstone

OF PERIOD AND PLACE

In one of those grand generalizations for which he remains justly notorious, Michel Foucault once proclaimed that the "great obsession of the nineteenth century . . . was history: with its themes of development and suspension, of crisis and cycle, themes of the ever-accumulating past." As he read the signs of his own day, however, Foucault divined that the hegemony of time over space—the triumph of period over place—was beginning to invert. "The present epoch will perhaps be above all the epoch of space," he surmised. "We are in the epoch of simultaneity: we are in the epoch of juxtaposition, the epoch of the near and far, of the side-by-side, of the dispersed."[1]

No doubt Foucault's prognostication can be read, at least in part, as a commentary on what might be termed—no doubt rather sloppily—the condition of postmodernity. In an age when the particular is taking its revenge on the universal, when generality recedes before singularity, and when local stories replace global theory, it is understandable that the eye of scholarly attention might divert from periodicity to spatiality. Whether this diagnosis approaches accuracy, or indeed whether the complexities of Foucault's hermeneutic undertakings are either persuasive or coherent, are not germane to my basic contention here, namely, that hitherto history has taken precedence over geography in our understanding of the scientific enterprise. Accounts of that imagined singularity "science" have conventionally been cast in terms of historical epochs and temporal change, frequently—though not invariably—touched with a progressivist brush. Such readings, moreover, have routinely been placed at the service of philosophical argument or social policy in order to provide grounds for investment in such cultural capital as intellectual advancement, technical control, and instrumental progress. Both cognitively and socially, the study of science as a human enterprise has fitted comfortably into a linear eschatological schema.

In this environment, the idea of a geography of science runs against the grain. That science has a history and a philosophy makes sense. But the suggestion that scientific inquiry has been influenced in any significant way by location seems counterintuitive. Science, we have long been told, is an enterprise untouched by the particularities of place; it is a transcendental undertaking, not a provincial practice. Of all the human projects devoted to getting at the truth of how things are, that venture we call science has surely been among the most industrious in its efforts to transcend the parochial. To bring science within the domain of geographical scrutiny disturbs settled assumptions about the kind of enterprise science is supposed to be and calls into question received wisdom about how scientific knowledge is acquired and stabilized.

And yet if we want to take seriously the geographical adjectives in such standard designations as "Chinese science" under the Sung emperors, or "Arabic science" under the patronage of Abassid Caliph al-Mansur, or "American science" in the age of Jackson, or "French science" in the late Enlightenment, we will have to attend to the significance of locational coordinates. Equally, urban particularities will have to be registered if we are plausibly to refer to "Edinburgh science" in Enlightenment Scotland, "London science" in the early Victorian period, or "Charleston science" in antebellum America. At one scale of operations, then, "science" is always a Renaissance French, a Jeffersonian American, an Enlightenment Scottish thing—or some other modifying variant. At another, it is always laboratory science, or field science, or museum science.

In many ways the attempt to cultivate a greater awareness of spatial sensibilities resonates with some observations on de-essentializing the term "science" that Dave Lindberg makes in the early pages of *The Beginnings of Western Science*. Language, he reminds us, "is not a set of rules grounded in the nature of the universe, but a set of conventions adopted by a group of people; and every meaning of the term 'science' . . . is a convention accepted by a sizeable community . . . Or to put the point in a slightly different way, lexicography must be pursued as a descriptive, rather than a prescriptive, art. We must acknowledge, therefore, that the term 'science' has diverse meanings, each of them legitimate."[2] My suggestion is that what passes as "science" is different not only from time to time, but also from place to place. The kinds of activities that go on in an astronomical observatory are different from those carried out in a museum or a botanical garden. Lindberg makes the comment that if "the historian of science were to investigate past practices and beliefs only insofar as those practices and beliefs resemble modern science, the result would be a distorted picture."[3] The same sentiments can appropriately be given a spatial rendering. That is, if the student of science were to investigate practices and beliefs in different spaces only insofar as those practices and beliefs resemble what passes for science in one place—say, the laboratory—the result would be a distorted picture.

From a geographical standpoint, it is pleasing to note the degree to which historians and sociologists of science have begun to "put science in its place." By this I mean that the place-specific character of scientific inquiry has recently become a focus of attention. Allow me to identify what I consider three diagnostic moments in this spatial turn, though others could certainly be isolated. First, the 1991 thematic issue of *Science in Context*—devoted to "The Place of Knowledge: The Spatial Setting and its Relations to the Production of Knowledge"—firmly placed space on the agenda. Conceived as an antidote to idealist conceptions of science as disengaged and transcendental, the editorial introduction emphasized the situatedness of science due to how particular venues conditioned ways of knowing, installed means of securing credibility, established standards of justification, and shaped meaning.[4] To Ophir and Shapin, the traditional resort to local factors only as an explanation for deviations from universal objectivity was hopelessly inadequate.[5]

Three years later, in March 1994, the British Society for the History of Science sponsored a conference with the theme "Making Space: Territorial Themes in the History of Science," selected papers from which were published in 1998. Here again the signal was clear: as the editors insisted in their introductory remarks, if a science "is to be produced and sustained" it has to be "credible and embodied" and "spatially located." This meant that what they called "spatialized approaches" to understanding scientific practice could provide insights and perspectives otherwise not available.[6] Of key significance here was the way in which the interlacing cords of knowledge, power and place, delivered cognitive, instrumental and moral authority to privileged scientific sites. Throughout, the location of scientific activity was shown to have been intimately connected with such matters as who engaged in scientific pursuits, what was investigated, how inquiries were pursued, and what passed as good practice.

Finally, the conversation between geographers and historians of science was further advanced at the Annual Conference of the Royal Geographical Society in 1996. Here, Steven Shapin fastened on matters having to do with the circulation of scientific knowing. Even if it were "established beyond doubt that science is indelibly marked by the local and spatial circumstances of its making," he insisted, there were still important issues to address about how "transactions occur between places." All of this was a way of insisting that "the problem is not that geographical sensibility has been taken too far but that it has not been taken far enough."[7] This challenge effectively broadens the scope of the "place of science," for it brings into view not only how scientific knowledge is made in particular places but also how it migrates from site to site, from locality to locality, from region to region. To put it another way, it opens up to geographical purview the *circulation*, as well as the *production*, side of scientific enterprises.

These three moments do not exhaust the scope of geographical engagements with scientific inquiry.[8] A host of studies of particular scientific spaces have recently become available as the particularities of place have begun to be probed with much greater sophistication by historians of science. At the same time, scrutiny of the significance of architectural arrangements for scientific pursuits has opened up intriguing questions about links between what we might call the buildings of science and the building of scientific knowledge.[9] But these three interventions are nonetheless emblematic of a growing sensitivity to the role of space in the production and reproduction of scientific cultures. What follows is an attempt to map out something of this terrain. Before turning expressly to what I am calling "Places of Practice" and "Places of Reception," however, it will be profitable to reflect a little on the nature of space itself. And to do so, I want to make use of the idea of "the production of space" drawing selectively on the judgments of the French philosopher and social theorist Henri Lefebvre.

THE PRODUCTION OF SPACE

> What is an ideology without a space to which it refers, a space which it describes, whose vocabulary and links it makes use of, and whose code it embodies? What would remain of a religious ideology . . . if it were not based on places and their names: church, confessional, altar, sanctuary, tabernacle? What would remain of the Church if there were no churches? The Christian ideology . . . has created the spaces which guarantee that it endures.[10]

With these comments, Lefebvre stakes out what he sees as the constitutive role of spaces in the reproduction of social life. I certainly do not intend to expound anything of the filigree intricacy of Lefebvre's project, not least because it has attracted many different readings, and critics have not been slow to identify stresses and strains in its architecture. Moreover, while Lefebvre's take on the whole matter is resolutely materialist, I do not consider that philosophical materialism is a necessary concomitant of the perspective he develops. What appeals is the general idea rather than the specific formulation, namely, that spaces must not be thought of as "givens," as mere "containers" inside which human activities take place. No. Spaces are to be thought of as social productions, that is, as constituted by social life in such a way that "mental space" and "material space" are elided. As he himself styles it, his design is to elaborate an account of social life "which would analyze not things in space but space itself, with a view to uncovering the social relationships embedded in it."[11] Because it signifies "dos and don'ts," space "is at once result and cause, product and producer."[12] Here we sense something of the dynamic reciprocity of the spatial and the social, their mutual making, and their inextricable intertwining. Such a way of thinking stands defiantly against the dichotomizing of social space into spatial form on the one hand, and social action on the other. Instead Lefebvre intends that this abstracted polarity is reconceptualized dialectically as a single, mutually constitutive process.

For Lefebvre, the production of space is intimately bound up with what he calls "spatial practices," namely, the activities by which actors constitute the arenas in which they participate. "Everyone knows what is meant when we speak of a 'room' in an apartment, the 'corner' of the street, a 'marketplace,' a shopping or cultural 'centre,' a public 'place,' and so on" he begins. "These terms of everyday discourse serve to distinguish, but not to isolate, particular spaces, and in general to describe a social space. They correspond to a specific use of that space, and hence to a spatial practice that they express and constitute."[13] Each of these arenas embodies a suite of performances, or "spatial practices," that are both cause and consequence of the social formations that are appropriate to those spaces. In this way practice is located at the very center of Lefebvre's social theory of spaces. Thus, as he observes, "Like all social practices, spatial practice is lived directly before it is conceptualized."[14] This immediately directs attention away from what social actors *say* about their activities, from how they *interpret* them, and from what they *think* about them, to how they *act*. As he puts it: "space was *produced* before being *read*; nor was it produced in order to be read but to be *lived* by people with bodies and lives in their particular context."[15]

The implications of these considerations for scientific enterprises are considerable. For example, if we substitute "laboratory," say, for "room," or "museum" for "street corner," or "field station" for "marketplace" in the gobbet above and ask what kind of spatial practices and social formations are constituted in these venues, we can begin to appreciate that different sorts of inquiry—different forms of scientific life—find expression in each. Moreover, the relevance of the idea of spatial practice to scientific engagement is not difficult to discern. For science is about performances as well as principles, about practices as well as theories, about institutions as well as ideas. Science thus understood is a "spatial practice." Recall Clifford Geertz's quip: "If you want to understand what a science is, you should look in the first instance not at its theories or its findings, and certainly not at what its apologists say about it; you should look at what the practitioners of it do."[16]

To Lefebvre, then, places are not simple slabs of earth stuff. Rather they are a multilayered mosaic of social spaces which are both cause and consequence of human agency. Spaces, produced and reproduced by social actors, are an elision of the material and the mental. And for this very reason, certain spaces are privileged sites because from them emanate discourses that exercise immense power in society. I do not think that one has to endorse the entire outlook of figures like Michel Foucault and Edward Said to see their relevance here.[17] From the church, the courtroom, the hospital, the asylum, and so on flow utterances that authorize the labeling of people in all sorts of powerful ways. Saint and sinner, innocent and guilty, healthy and sick, sane and insane, rational and irrational—just a few of the bipolar categories that have shaped our culture—have all been named and brought into cultural circulation from particular sites. In such spaces, people are brought under the authority of medical, religious, and legal knowledge for different purposes. To understand the *history* of medicine, or religion, or law, or science, then, necessarily requires us to grasp the *geography* of medical, religious, legal, or scientific discourses. It is critically important, then, to attend to those sites that have generated knowledge claims and then wielded them in different ways. No less is this the case on the reception side of the equation. *Where* ideas and theories are encountered conditions *how* they are received. Like objects and commodities, as Said has noted, "ideas and theories travel—from person to person, from situation to situation, from one period to another."[18] But this migration process is never mere replication; traveling involves translation, and translation is transforming. Precisely because arguments are the product of time and place, for that very reason they are always appropriated in time and place. At every scale, knowledge, space, and power are tightly interwoven.

PLACES OF PRACTICE

Initially, then, we direct our attention to sites where scientific inquiry is activated and scientific claims are generated, that is, to the production sector of the knowledge circuit. Science, of course, has been practiced in a remarkably wide range of places. Venues like the laboratory, the observatory, the museum, the field, the botanical garden, and the hospital, spring readily to mind. But such standard locations do not exhaust the range of possibilities. The ship, for example, has not only been a space in which scientific gadgetry was transported; it itself has been—as in the case of James Cook's vessel, the *Endeavour*—a scientific instrument delivering to surveyors, through the plotting of its own geodetic position, a sort of cartographic shadow of the coastline.[19] During the middle decades of the nineteenth century, the English village inn became a key site in the pursuit of artisan botany. In the public house, as Ann Secord has tellingly shown, amateur horticulturalists would congregate on Sunday mornings to share plant information, swap specimens, and pore over the botanical textbooks that the club kept in the local hostlery. Indeed gentlemen scientists, like J.D. Hooker of Kew Gardens, would resort to these gatherings for samples and skills alike. Here the chasm between philosopher and craftsman, head and hand, began to be bridged as working-class practitioners, long denied access to the elite sites of scientific investigation, pursued botanical inquiries in a congenial atmosphere.[20] And then there is the library. As Adrian Johns has reminded us, "far from emerging from a rejection of words, in fact science originated partly from a need to master as many of them as possible."[21] The research library, only spatially differentiated from museum and laboratory space during the European Enlightenment, installed various bureaucratic practices that effectively regulated access to crucial scientific texts. The remarkable library at the University of Göttingen for example, itself "oriented on a pragmatic, rationalising view of knowledge" from its opening in 1737, stamped an Enlightenment philosophy of utility on its collections by becoming as renowned for what William Clark calls "its reader-friendly atmosphere," as for its effective cataloguing system. Such arrangements, expressed in the spatial layout of stock, went hand in hand with the systematization of knowledge.[22] To these we might also add such locations as cathedrals, coffee houses, tents, breeding clubs, royal courts, stock farms, exhibition stages and, no doubt, many more.[23]

There are indeed important things to be said about each of these spaces as sites of scientific practice and the production of scientific knowledge, just as there are about individual cases of each. The stories of, say, the

Jena library under Goethe's rule, the Philoperisteron pigeon breeding club with which Darwin was familiar, or the court of the Medici in Galileo's time, and their significance for scientific endeavor, are now beginning to be heard. "Thick description" of the ethnographic variety, however, is not my quarry here. Instead I want to dwell on three *classes* of scientific space in order to tease out something of the different kinds of inquiry they embody and the ways in which location matters in the claims they deliver.

SPACES OF EXPERIMENTATION

The securing of space within which to carry out experimental investigations of nature was the outcome of historical negotiations revolving around the appropriate location in which to acquire dependable knowledge. In the West, it was long believed that retreating from the turmoil of everyday life was imperative for the acquisition of authentic knowledge.[24] Sages and seers sought spaces of solitude. To obtain knowledge that was true *everywhere*, they had to go *somewhere*, to find wisdom that bore the marks of *nowhere*. But such places as the wilderness or the hermitage were not equipped for the pursuit of experimental knowledge. A new sort of space for withdrawal was needed for that activity, and this was often sought within the confines of the home. The instrument maker and alchemist John Dee, for example, found it a difficult task to carve out a piece of domestic space for his investigations in a house cram-full of servants, and at a time when the home was considered female space.[25] Robert Boyle, gentleman-philosopher had his laboratory in the basement of the home of his sister Katherine.[26] Circumstances like these posed important questions about who had right of access to such quarters, and not least because experimental activities, even if carried out in a *private* space, had to be, in some significant way, a *public* affair. In order to achieve the status of warranted knowledge, experimental claims had to undergo public attestation. But this did not mean public in today's sense. As Steven Shapin has been at pains to demonstrate, the "experimental public" was composed of gentlemen whose supporting presence was essential to the confirmation of experimental findings.[27]

I do not want to adjudicate here on the cogency of Shapin's arguments about the constitutive ties between the making of science, gentlemanly codes of conduct, and testimonial attestation, in seventeenth century England. Rather it is the significance of *space* in his account of the generation of knowledge that presently appeals. Consider. An experiment might "work" perfectly well in the private recesses of the scientist's workplace.

But that was not enough to establish it as a piece of genuine knowledge. To secure *that* level of epistemic standing it had to receive the approval of the relevant experimental public. A chasm thus opens up between the private "trying" of an experiment, and its public "showing." Only when it had successfully traveled from private to public space could a scientific claim enjoy the status of warranted knowledge. Through public *demonstration*, private *experimentation* achieved open *confirmation*. The shift from "trying" to "showing," from delving to demonstrating, we might say, is a spatialisation of the move from the context of scientific discovery to the context of justification. It is important to note here that it was only what Shapin calls "geographically privileged persons" who could deliver warranted credibility. But this did not involve just *anybody* who crossed the material threshold. A variety of visitors could, under certain conditions, enter the lab. It was only men of a certain standing that could provide trustworthy testimony. Occupying the laboratory's physical space was one thing; occupying its discursive space quite another. Casual callers inhabited a different epistemic space from those socially and cognitively sanctioned to adjudicate on experimental reliability. Here were boundaries that, though unmarked in physical space, were prominently displayed in the laboratory's mental cartography.

As a place of obtaining scientific knowledge of nature, the laboratory is, in Lefebvre's terms, a produced space, a distinct material and mental space, where knowledge is made in very particular sorts of ways. And the kind of place the laboratory is, turns out to be significant in the cognitive claims that emerge from it. To be sure, experimental knowledge can be secured in other spaces too. The body, for example, has itself often been a site of experiment. Take Alexander von Humboldt. He frequently used his own body as an instrument for acquiring experimental knowledge of the environments through which he traveled. It was, as he put it, "a kind of gauge" for registering different atmospheres and conditions. He even applied electrodes to himself to determine the effects of an electric current on a secretion of blood and serum derived from deliberately raised blisters on his back. And, indeed, he and Aimé Bonpland used their own bodies as virtual Leyden jars to test the discharge from electric eels—with fairly nasty results.[28] Such procedures point to the body as an experimental space. But such performances raised intriguing questions about which bodies—and which minds—could be trusted to deliver empirical truths. Typically, during the eighteenth century it was only the genteel whose word could be taken seriously. Only they had sufficient composure to distinguish what was fact from what was fabulous. Thus the French student of electricity, Jean-Antoine Nollet, debarred from the republic of learning "either Children, Servants, or People of the lower Class."[29] Their testimony, even to the experiences of their own bodies, could not be trusted.[30] Once again, in determining the reliability of experimental testimony, the space of its practice was a significant element in the acquisition of warrant. Moreover, using the body as an experimental site raised moral questions. Those grotesque experiments in what was called Racial Hygiene during the 1940s in concentration camps like Auschwitz and Buchenwald—which delivered such experimental knowledge as how long a person can survive in icy waters or in very low pressures—reminds us that in these sinister spaces, the cost of scientific enlightenment was moral darkness.[31] Physical spaces are moral spaces too.

SPACES OF EXPEDITION

Recalling attention to Humboldt as the embodied site of experimentation reminds us that it is not only in the laboratory that scientific inquiry, even of the experimental variety, is conducted. Indeed, as Dorinda Outram has shown, it was around the persona of Humboldt that disputes revolved about the proper place of "real" scientific investigation. The debate found its animus in Georges Cuvier's contrast between the methods of a scientific traveler like Humboldt and a sedentary naturalist like himself.[32] Because the experience of explorers was inescapably fragmentary and fleeting, Cuvier felt it could not be compared with the stable and thus reliable vision of the bench naturalist who saw nature steadily and saw it whole. Only when the samples harvested by the traveler were spread out before the laboratory man could impressionism give way to information. The expeditionary naturalist passed too quickly over terrain to do good science. For Cuvier, the most successful scientific voyages never weighed anchor for the simple reason that they never left the workshop. So, while the sedentary naturalist insisted that it was *absence* from wild nature that secured epistemic privilege, this was precisely what field naturalists denied. To them *presence* in the wide-open spaces of the field was required to deliver warranted credibility. Here again the space of inquiry was a crucial ingredient in disputes over the appropriate means of making natural knowledge.

Knowledge acquired in the field, however, raised other questions about credibility. How could metropolitan scientists be assured that the information brought home from afar was reliable? And how could scientific travelers justify their reports? A moment late in the motion picture *Mountains of the Moon* provides some intriguing hints about what I want to call "embodied warrant."[33] The scene is staged at the September

1864 meeting of the British Association for the Advancement of Science in Bath. The occasion was to have been a public debate between Richard Burton and John Hanning Speke, both African explorers, about whether or not the latter had succeeded in finding Lake Victoria and the source of the Nile. England's favorite missionary-explorer, David Livingstone, had been invited to adjudicate. The debate never materialized, of course, due to the untimely death of Speke in a hunting accident. The circumstances occasioning this controversy, however, are not my concern here. Instead I want to recount the meeting—no doubt cinematically overdramatized—between Burton and Livingstone prior to the scheduled public discussion. "Risky profession we're in," observes Livingstone, beginning to open his shirt when he and Burton are left alone. "You know of course that I was mauled by a lion. He only chewed my shoulder," he went on, pointing to the spot. Burton, not to be outdone, pulls back his shirt and points: "Bullet hole. Single bore." Livingstone now takes to unbuttoning his breeches to display a scorpion bite. "Cellulitis," Burton replies, pulling up one trouser leg. And so it continues, a ritual exchange of wounds. What is going on here is an exercise in the establishment of credibility. Each explorer could be trusted because they bore in their bodies the authenticating marks of their expeditions.

Field work, it is clear, raised particularly acute questions about who could be trusted and about what might be called "the moral economy of wounds" as the embodied insignia of testimony. As Michael Heffernan has compellingly shown, precisely this species of warrant was at the heart of the controversy about who should be credited with the distinction of being the first European to enter the African city of Timbuktu in the 1820s.[34] In adjudicating the competing claims of a young Frenchman and a Scottish soldier, John Barrow, permanent secretary to the Admiralty, considered that the lengthy list of wounds that the Scotsman has sustained—multiple saber slashes to the head, left temple, and right arm; a variety of fractures; and a musket ball in the hip—bore witness to the genuineness of the geographical knowledge he had so painfully acquired. The injuries he had sustained were nothing less than the signs of truth imprinted in the flesh.

Field science, it seems, raises rather different epistemological questions from, say, the laboratory about what is known, how it is known, and how witnesses secure warrant. Besides, because the field is an inherently unstable scientific site, practical reason is at a premium. Here, the good scientist is the skilled hand, the resourceful practitioner. Certainly these aptitudes are crucial in the laboratory too. But in the field, replication is not so easily effected, the environment is less readily controlled, and impromptu ingenuity is in correspondingly greater demand. Yet however innovative in situ practices may be, the techniques deployed in the field are typically acquired at home. Encounters with the extraordinary are routinely construed in customary ways, for field scientists "travel with their domestic habits of mind and behaviour."[35] To this extent the homeland is always present with the scientific traveler.

Like the laboratory, the field is a produced space. On the surface this seems a strange way to put it. Surely the field is just *there* in a taken-for-granted kind of way. Not so. It is constituted *as the field* by the activities of scientific investigators. In fact in some academic disciplines, notably anthropology, the place of the field has been crucial to the normalization of the discipline's inquiries, empowering some, impeding others. Malinowski's role was crucial here; by installing fieldwork as central to the institutionalization of the discipline he effected a move away from the worldview of Victorian gentlemen-scholars who considered going to the field to be beneath their dignity.[36] The field methods Malinowski had deployed in the Trobriand Islands rapidly became the legitimating insignia of the profession—"the central ritual of the tribe."[37] The *place* of anthropological practice was thus crucial to its own sense of identity. Moreover, anthropological investigators have the power to group individuals into some abstract collective, name them, and bring them into scholarly circulation. By categorizing human subjects as slum dwellers, primitive peoples, middle-class fundamentalists, or some such, social scientists delineate boundaries and define who is in and who is out of the empirical circle.

The field, as a place of scientific investigation, turns out to be anything but the obvious site it might initially seem. Not only is it constructed by the activities of the academy, but it has provided—at least for some scientific traditions—an *operational* answer to questions about appropriate ways of knowing. Absence from home and presence in the field as the necessary precondition of genuine knowledge was the outcome of an historical settlement that gave the field sciences their characteristic place in the scientific division of labor. Here, epistemological warrant was built upon the foundations of spatial practices.

SPACES OF EXHIBITION

In the production of scientific knowledge, spaces of exhibition take their place alongside sites of experimentation and expedition. Whether in botanical gardens, museums, or zoos, scientific knowledge is intimately bound up with display. In such places, the aim is less to *manipulate* the natural world by experiment or to *participate* in it through travel, but to *arrange* it through classification. What unites these venues is an emphasis on collecting items and ordering them on the basis of some taxonomic system. Take, for example, the museum, the origins of which go back to those sixteenth-century "cabinets of curiosities" into which gentlemen heaped oddities of all sorts.[38] In this space of accumulation a new form of scientific knowing began to emerge. The stunning variety of natural and cultural objects that the museum amassed did a good deal to feed the hunger for facts. Thus, in his *Novum Organum*, Francis Bacon called for the acquisition of what he called particulars and prerogative instances precisely because such practices challenged deductive reasoning and a priori syllogism.[39] In this way the culture of collecting became established as a valid way of knowing.

But museums were not simply about acquisition. As commodities flowed in from near and far they were reshuffled, placed, and displayed according to the thinking of the curator. Thus even while museums exhibited real-world objects, they refashioned reality through classification, location, and genealogy. Museums have thus always been, in a crucial sense, hermeneutic practices in which the spatial ordering of phenomena fundamentally realigns the world of nature. This means that the internal spatial configuration of the museum has always been crucial to the cognitive claims it delivers. And as such, museum space has often been a site of controversy. The different perspectives on evolution taken by William King Gregory and Henry Fairfield Osborn, for example, found expression in the spatial layout of their respective halls at the American Museum of Natural History during the 1930s.[40] In Gregory's Hall of the Natural History of Man, evolutionary continuity between the different human races was the driving force, whereas Osborn's Hall of the Age of Man sought to undermine the theory of ape ancestry and to portray the different human races as discrete "species." Their respective displays gave visual expression to the different social, political, and religious convictions of the two scientists. In ways like this, the museum voiced the values of its curators.

In other places too, where scientific claims were bound up with exhibition, space crucially mattered, and disputes often arose. Take the case of the special exhibit at the Bronx Zoological Gardens in September 1906. Housed in the ape enclosure was a specimen designated on the posted sign as: "The African Pygmy: Ota Benga. Age 23 years. Height 4 feet 11 inches. Weight 103 pounds. Brought from the Kasai River, Congo Free State, South Central Africa by Dr. Samuel P. Verner. Exhibited each afternoon during September."[41] Prior to being presented to William Hornaday, director of the zoo, Ota had already been put on display by WJ McGee in the anthropology wing of the St. Louis World's Fair in 1904, along with other pygmies, as "emblematic savages." Here, as later in New York, the aim was to illustrate the stages of human evolution. Crowds came and controversy flared. Ota was poked and taunted to such a degree that he made a bow and fired arrows at particularly obnoxious visitors; the whole episode offended a range of black clergymen; and he was eventually removed from the zoo for good.

Just where to *place* certain races in the natural order had long been part of a Western exhibitionary impulse. Carl Hagenbeck, for example, who spearheaded the idea of the zoo "panorama," where animals were exhibited in open spaces rather than in cages, introduced what he called "anthropological-zoological" displays into his Hamburg Tierpark in 1874.[42] That year he had Lapps acting out daily life with reindeers before enthusiastic audiences. And in the years that followed he orchestrated some seventy ethnographic performances—Oglala Sioux performing ritual dances in the shadow of a constructed mountain proving to be among the most popular.

Zoos, of course, were also about the transfer of animals from one place to another. And it is not surprising that many had their origins in acclimatization societies aiming to domesticate foreign animals to new climatic regimes. In France, the scientific community had a long-standing interest in acclimatization, not least because it bore directly on matters of adaptation, inheritance, and evolutionary change. In fact the zoological garden was in large measure the public laboratory of the Société Zoologique d'Acclimatation which had come into being in 1854.[43] If successful long-term adaptation of species to new environmental niches could be effected, that would do much to confirm the doctrine of the inheritance of acquired characteristics and thus the biological transformism rooted in the ideas of Buffon and Lamarck. So in nineteenth-century Paris, the breeding and dealing of exotic animals was seen as contributing to agriculture and industry, scientific advancement, and commercial success alike.

Much the same was true of botanical gardens. Prior to the nineteenth century, they had performed several different roles.[44] By striving to gather together the botanical riches of an expanding globe they hankered after the Garden of Eden. Gardeners like John Tradescant in the mid-seventeenth century were latter-day Noahs engaging in a task of spiritual and scientific retrieval and reversing the global effects of negligence and depravity. At the same time, as physic gardens, they acquired plants with medicinal powers and were thus often connected with medical faculties. Now, in an age of high empire, places like Kew Gardens became the repository of what might be called botanical imperialism.[45] Burgeoning under the vegetative booty brought back by men like Joseph Banks, they became centers of a worldwide network of plant acquisition and exchange no less for commercial gain as for scientific inquiry.

In these different sites, scientific inquiry rotated around matters to do with the acquisition of specimens—whether plant, animal, or human—and their exhibition in specified spaces. Science in these locations was about the ordered display of the natural world shaped by the dictates of scientific theory and taxonomy. The kinds of investigation that went on there were different in crucial respects from those in either the laboratory or the field. But in all three, the *place* of investigation mattered to the activities engaged in. This does not mean that experimental, expeditionary, and exhibitionary spaces were mutually exclusive arenas. To the contrary, botanical gardens could have their own laboratories, and field workers often took mobile labs with them on their expeditions. To that degree a single site might be polysemic—a composite of different spatialities. But taken overall, laboratory science, museum science, and field science name rather different species of scientific endeavor. In important ways, this is because the production of these spaces has necessarily gone hand in hand with the production of scientific knowledge. And the same is true of the consumption sector of scientific enterprises—a subject to which we now turn.

REGION, READING, AND RECEPTION

Just as scientific knowledge is produced in a variety of places, so too the results of scientific inquiry are received in different venues. Here I want to advertise something of the significance of place in the consumption of science by dwelling on the geographies of reading and the management of textual space.

Ideas and instruments, texts and theories, individuals and inventions all diffuse across the surface of Earth. But they do not diffuse evenly across the globe. Thanks to the research of Owen Gingerich, we are beginning to learn something of the diffusion of Copernicanism across seventeenth-century Europe.[46] By tracing both the 1543 Nuremberg and 1566 Basel editions of *De Revolutionibus* and locating copies that were altered on account of the papal decree of 1616, Gingerich has been able to identify where unexpurgated and censored versions of the treatise turned up. The results disclose an uneven geography. While most Italian copies were censored, the Decree of the Holy Congregation had relatively little impact elsewhere. Even in France, where most copies were located in Jesuit libraries, there is little evidence of censorship.

Darwin's theory of evolution also experienced a different fate in different national settings, religious spaces, and institutional arenas.[47] Because Darwin was heard to say different things in different places, different rhetorical strategies were deployed in these theaters of operation to meet the challenges he was taken to be provoking. Prominent Protestant Church leaders in Edinburgh, Belfast, and Princeton, for example, developed markedly different tactics in their reading of the Darwinian challenge. In each case local circumstances were crucial to the stances they adopted and to the vocabulary—whether irenic or bellicose—they chose to express their sentiments. The reading of Darwin, indeed, often became an occasion for engaging in culture wars seemingly far distant from scientific matters.[48]

Elsewhere other factors shaped Darwinian hermeneutics. In the American South, for instance, the monogenetic implications of Darwin's account of human origins did not sit at all comfortably with the widespread belief that the human race was composed of entirely different species. Here racial politics conditioned the response to Darwin precisely because scientific language had long been the bearer of local ideological freight.[49] In New Zealand, by contrast, racial politics tended to favor the espousal of Darwinism because it was seen as justifying an ethnic struggle for life and legitimizing the settlers' routing of the Maori.[50] In Russia, the story was again different, and here the vast expanse of a harsh, sparsely populated environment was crucial. The metaphor of a struggle for existence was resisted by leading members of the Russian scientific intelligentsia. Beketov, Bogdanov, Dokuchaev, Kessler, Kropotkin, and Poliakov were all deeply skeptical of the Malthusian elements in the Darwinian scheme. A meager population and extreme climatic severity did not fit at all well with Darwin's picture of teeming life-forms or Wallace's lush tropical vegetation. For Russian evolutionists, *On the Origin of Species* seemed a theory inspired in, and by, the tropics.[51]

Scientific ideas rarely circulate as immaterial entities. Rather, as they move from place to place, they are routinely embodied in written texts. At a fundamental level, it is print rather than thought or theory that is let loose upon the world. Scientific forms of reasoning crystallize in texts of all kinds, including—according to Nick Jardine—"routinely authored works—instrument handbooks, instruction manuals, observatory and laboratory protocols" that are crucial to the regulation of empirical practices.[52] All of this brings the book to the forefront of our considerations and to the importance of textual space, namely those arenas of engagement where moments of hermeneutic encounter are effected.[53]

Textual meaning is mobile and shifts from place to place at a range of different scales. Distinctive cultures of reading can be detected within regions and between them, within cities and between them, within neighborhoods and between them. Something of the dynamic of these readerly spaces can be appreciated by turning to James Secord's remarkable account of how, in different venues, the controversial *Vestiges of the Natural History of Creation*, which first appeared in 1844, was read. An anonymous text, later acknowledged to be the work of the Scottish publisher Robert Chambers, it caused a sensation at the time. Embraced by some, vilified by others, it at once bemused, infuriated, consoled, and revolted readers in its bold portrayal of the drama of evolution. One thought it a "priceless treasure," another dismissed it as materialist "pigology." Some found it manly; others were sure they could detect a womanly hand behind its anonymity. Some thought it daring; still others found it melancholic. Moreover, writers outdid one another in the metaphors they devised to stage-manage the text for readers. The book's striking red binding prompted one to "attribute to it all the graces of an accomplished harlot." Its continuing anonymity drew from another the exclamation: "Unhappy foundling! Tied to every man's knocker, and taken in by nobody; thou shouldst go to Ireland!" Because its anonymity ruptured the long-established suture lines between author, narrative, and reader, it stimulated a lengthy list of authorial suspects. Significantly, a differential geography of suspicion surfaced. As Secord puts it, "Names that seemed likely in Liverpool or Edinburgh were barely canvassed in Cambridge or Oxford; those that were common in London's fashionable West End were barely known in the Saint Giles rookeries only a few blocks away in the squalid dens of Holywell Street, notorious for atheism and pornography."[54] Yet speculation was intense. Why? Because aligning an author was required to fix a reading.

In different London salons, the book entered fashionable conversation in different ways and found itself very differently treated. Among aristocratic readers, it was regarded as poisonous. In the homes of progressive Whigs, it was boldly visionary. In Unitarian drawing rooms, its emphasis on change from below was seen as a telling blow against a smug ecclesiastical establishment. Outside London, the book also fared differently. In Liverpool, for example, the way it was read reflected the social topography of the city. Here it sold briskly among those pressing for urban reform because it could be taken as scientific justification for social improvement. In Europe too its fortunes differed from place to place in various translations, as Nicolaas Rupke has shown.[55] The German version translated by Adolf Friedrich Seubert, for instance, strangely incorporated material—from William Whewell's *Indications of the Creator* (1845)—originally intended to refute the *Vestiges*. But by interlacing the two texts, Seubert succeeded in making the book into a treatise supporting the conviction that evolutionary development took place according to divinely ordained laws. All in all, very different messages were read in, and read into, *Vestiges* depending on local circumstances. Textual meanings are mobile. Books exhibit their own "geographies of reading."

MANAGING TEXTUAL SPACE

Something of the significance of location in the staging of texts can be discerned from a brief case study. In 1922, nearly two hundred years after its first appearance, Isaac Newton's work on the biblical books of Daniel and Revelation was made available again to the reading public.[56] The text—reissued under the editorship of Sir William Whitla, emeritus professor of *Materia Medica* at the Queen's University, Belfast, former president of the British Medical Association, and honorary physician to the King in Ireland—affords an opportunity for reflecting on the mobilization of intellectual resources for cultural projects, and in particular, on the enlistment of Newton as an ally in local campaigns of various stripes.[57] Scrutinizing Whitla's Newton text in the context of early 1920s Belfast serves to highlight the salience of textual space by furnishing an opportunity to disentangle one individual's editorial tactics in the management of textual encounter.

Whitla's concern to retrieve Newton's prophetic treatise had initially been stimulated by an invitation to deliver a lecture on the book of Daniel to his own Methodist church. In a short space of time his preliminary thoughts had lengthened out into a lecture series which in turn became a ten-chapter introduction to his reprint of Newton's treatise. Right from the start, Whitla made it perfectly clear how he intended Newton to be read. In a day when critics were undermining the authority of the Bible, inspiration could be derived from Newton who "in strong and childlike faith lent his mighty intellect to the study of this fascinating record."[58] The aim was to muster biblical prophecy Newton-style in the conduct of current culture wars. In the religious world of the early twentieth century, old certainties seemed to be evaporating in the face of the erosion of traditional virtues and the new literary criticism. In Northern Ireland itself, the 1920s heresy trial of J. Ernest Davey rotated around the issue of whether higher criticism and orthodox faith were compatible.[59] In these circumstances, Whitla's turning to Newton was precisely to enlist his reputation in the cause of religious and social stability.

Central to Whitla's diagnosis of Europe's current crisis was the "moral leprosy" that he believed had infected the Church itself. Unperturbed by evolutionary biology and other scientific theories, he vigorously denounced the "deadly" and "poisonous" Graf-Wellhausen theory of the Pentateuch. It was no accident that such doctrines had emanated from Germany. That old enemy stood highest in the international circle of sacrilege. The result was that Germany exhibited widespread "moral myopia" and a craving after "world-empire."[60] And it was to this cause that he later traced "the chaos and desolation of Europe resulting from the late world-war."[61] The fractured geography of postwar Europe was a direct result of the spiritual bankruptcy of Germanic space. It was in pursuit of his campaign to rout these "diabolical" forces of rationalism that Whitla called upon Newton. *This* was the context in which Newton needed to be read. When readers witnessed for themselves how carefully Newton interpreted the symbolic language of the Bible, Whitla wanted them to recognize an ally in a continuing campaign against the evils of Teutonic intellectual arrogance. Newton was the best possible antidote to those Higher Critics who rewrote chronology, deconstructed authors, and mythologized history.

While Whitla staged Newton as a partner in the culture conflicts of postwar Europe, he was not blind to the potential uses of Newton's text in the even more local political context of 1920s Belfast. His introductory reflections were thus intended to manage the local space of textual rendezvous. Whitla himself, it turns out, had an active political career. He was a strong Unionist; signed Ulster's Solemn League and Covenant in 1912, which opposed Home Rule and affirmed loyalty to the Crown; and was elected as a member of the Westminster Parliament in 1918. Indeed, in March 1912, speaking from the chair at a mass rally of Methodists in Belfast's Ulster Hall, he called his co-religionists to unite "as one man in the deliberate conviction that Home Rule means disaster and ruin to our native land, and irreparable injury to our Church and to the civil and religious liberty which we and our fathers have enjoyed under the impartial freedom of the British flag."[62]

During the winter of 1921–1922, when he delivered his prophecy lectures, the new Northern Ireland state was struggling to life. And his first address took place in the midst of a tumultuous week.[63] "Week of Terror in Belfast," screamed a newspaper headline. "Dastardly Bomb Outrages. Sinn Fein Snipers Active. 20 Persons Killed. 76 Wounded."[64] The week before, this same paper had run a piece titled "Dangers Confronting Us Viewed in the Light of Church History," in which the author fastened on the role of the Catholic Church. Because of the papacy's claims to "power in temporal things, over all kingdoms and empires," he could only conclude that "In the Papacy we are face to face with an autocracy of the most satanic character. Judging it by what is written, we can easily understand what it would be for this country to come under its domination."[65] Indeed, Whitla had himself already given voice to precisely this species of politico-religious fear. The Catholic Church, he insisted, "never falters, never wavers in her endeavour to establish supreme ascendancy in the domains of both politics and religion in whatever part of the world on which she obtains a foothold." Such were the only "lessons to be read in the conditions which at present exist in Spain, Portugal, the South American Republics . . . our own Province of Quebec."[66] Biblical exegesis, religious history and contemporary geopolitics thus convened to confirm the legitimacy of Ulster Unionist alarm.

It was with sentiments such as these in the atmosphere that Whitla introduced Newton to his Belfast hearers. The restaging of Newton's anti-Catholicism, which found unambiguous expression in his prophetic work, is thus hardly insignificant. As Newton read the political landscape foretold in the biblical documents, he was certain he could discern the ways in which papal dominion had spread beyond the ecclesiastical sphere and into the political realm. For Whitla, salvaging Newton's text was a means of valorizing the spiritual space Ulster Protestants occupied, even if their political geography was altogether precarious. Moreover, by readvertising Newton's exegesis, speculating about the date on which "the Papacy became first established as a temporal power," and hinting at the eventual "extinction of Papal Rome," his inclinations were far from ambiguous.[67] To oversee the republication of Newton's anti-Catholic reading of Daniel during the earliest days of the Northern Ireland state was hardly fortuitous at a time when the mantra "Home Rule means Rome Rule"—an expression to be found on Whitla's own lips—was reverberating throughout the society.[68]

At every stage in the cycle of scientific culture, place matters. *Where* scientific work is conducted and *where* its wares are encountered make a difference to both the production and consumption sides of the enterprise. This means that if we are to understand the place of science in our culture we will need to attend more carefully to the places of scientific culture and to how these spaces have historically come into being.

NOTES

1. Michel Foucault, "Of Other Spaces," *Diacritics* 16 (1986): 22–23.
2. David C. Lindberg, *The Beginnings of Western Science: The European Scientific Tradi-*

tion in Philosophical, Religious, and Institutional Context, 600 B.C. to A.D. 1450 (Chicago: University of Chicago Press, 1992), 2.

3. Ibid., 3.

4. Adi Ophir and Steven Shapin, "The Place of Knowledge: A Methodological Survey," *Science in Context* 4 (1991): 3–21.

5. Here I leave aside the question of whether seeing science as a local practice is inevitably implicated in wholesale relativism. For myself, I am not convinced that there are necessary links between the relativity of warranted credibility and relativism over substantive concepts of truth.

6. Crosbie Smith and Jon Agar, "Introduction: Making Space for Science," in *Making Space for Science: Territorial Themes in the Shaping of Knowledge*, ed. Crosbie Smith and Jon Agar (London: Macmillan Press, 1998), 2, 3.

7. This was published as Steven Shapin, "Placing the View from Nowhere: Historical and Sociological Problems in the Location of Science," *Transactions of the Institute of British Geographers* 23 (1998): 6, 7.

8. For geographical engagements with the history and sociology of science, see David N. Livingstone, *Putting Science in its Place: Geographies of Scientific Knowledge* (Chicago: University of Chicago Press, 2003); David Demeritt, "Social Theory and the Reconstruction of Science and Geography," *Transactions of the Institute of British Geographers* 21 (1996): 484–503; Trevor J. Barnes, *Logics of Dislocation: Models, Metaphors, and Meanings of Economic Space* (New York: Guildford Press, 1996); Charles W. J. Withers, "Towards a History of Geography in the Public Sphere," *History of Science* 34 (1999): 45–78.

9. So, for instance, Peter Galison and Emily Thompson, eds., *The Architecture of Science* (Cambridge, MA: MIT Press, 1999).

10. Henri Lefebvre, *The Production of Space*, trans. by Donald Nicholson-Smith (Oxford: Blackwell, 1991), 44.

11. Ibid., 89.

12. Ibid., 142.

13. Ibid., 16.

14. Ibid., 34.

15. Ibid., 143 (original emphasis).

16. Clifford Geertz, "Thick Description: An Interpretive Theory of Culture," in *The Interpretation of Cultures: Selected Essays* (New York: Basic Books, 1973), 5.

17. I am thinking here especially of Michel Foucault, *The Birth of the Clinic: An Archaeology of Medical Perception*, trans. by A. M. Sheridan Smith (New York: Pantheon, 1973); Foucault, *Discipline and Punish: The Birth of the Prison* (New York: Vintage Books, 1977); and Edward W. Said, *Orientalism* (London: Routledge & Kegan Paul, 1978).

18. Edward W. Said, "Travelling Theory," in *The World, the Text and the Critic* (London: Vintage, 1991), 226.

19. See Richard Sorrenson, "The Ship as a Scientific Instrument in the Eighteenth Century," *Osiris* 2nd ser., 11 (1996): 221–36.

20. Ann Secord, "Science in the Pub: Artisan Botanists in Early Nineteenth-Century Lancashire," *History of Science* 32 (1994): 269–315.

21. Adrian Johns, "The Birth of Scientific Reading," *Nature* 409 (January 18, 2001): 287; see also his "Science and the Book in Early Modern Cultural Historiography," *Stud-*

ies in History and Philosophy of Science 29A (1998): 167–94; and *The Nature of the Book: Print and Knowledge in the Making* (Chicago: University of Chicago Press, 1998).

22. William Clark, "On the Bureaucratic Plots of the Research Library," in *Books and the Sciences in History*, ed. Marina Frasca-Spada and Nick Jardine (Cambridge: Cambridge University Press, 2000), 196, 200.

23. Some of these sites are treated in such publications as J. L. Heilbron, *The Sun in the Church: Cathedrals as Solar Observatories* (Cambridge, MA: Harvard University Press, 1999); Larry Stewart, "Public Lectures and Private Patronage in Newtonian England," *Isis* 75 (1986): 47–58; Steve Pincus, "'Coffee Politicians Does Create': Coffeehouses and Restoration Political Culture," *The Journal of Modern History* 67 (1995): 807–83; Lynette Schumaker, "A Tent with a View: Colonial Officers, Anthropologists, and the Making of the Field in Northern Rhodesia, 1937–1960," *Osiris* 2nd ser., 11 (1996): 237–58; James A. Secord, "Darwin and the Breeders: A Social History," in *The Darwinian Heritage*, ed. David Kohn (Princeton, NJ: Princeton University Press, 1985), 519–42; Mario Biagioli and Galileo Courtier, *The Practice of Science in the Culture of Absolutism* (Chicago: University of Chicago Press, 1993); and Iwan Rhys Morus, *Frankenstein's Children: Electricity, Exhibition, and Experiment in Early-Nineteenth Century London* (Princeton, NJ: Princeton University Press, 1998).

24. See Steven Shapin, "'The Mind Is Its Own Place': Science and Solitude in Seventeenth-Century England," *Science in Context* 4 (1990): 191–218.

25. See Deborah E. Harkness, "Managing an Experimental Household: The Dees of Mortlake and the Practice of Natural Philosophy," *Isis* 88 (1997): 247–62.

26. Steven Shapin, "The House of Experiment in Seventeenth-Century England," *Isis* 79 (1988): 373–404.

27. See Steven Shapin, *A Social History of Truth: Civility and Science in Seventeenth Century England* (Chicago: University of Chicago Press, 1994).

28. Alexander von Humboldt's experiments using his own body are recorded in Douglas Botting, *Humboldt and the Cosmos* (London: Sphere Books, 1973), 34, 101, 153–54.

29. Jean Antoine Nollet, "An Examination of Certain Phaenomena in Electricity, Published in Italy," *Philosophical Transactions* 46 (Jan.–Apr. 1750): 377.

30. Simon Schaffer, "Self Evidence," *Critical Inquiry* 18 (1992): 327–62.

31. See, e.g., Robert N. Proctor, *Racial Hygiene: Medicine Under the Nazis* (Cambridge, MA: Harvard University Press, 1988).

32. Dorinda Outram, "New Spaces in Natural History," in *Cultures of Natural History*, ed. N. Jardine, J. A. Secord, and E. C. Spary (Cambridge: Cambridge University Press, 1996), 249–65.

33. *Mountains of the Moon*, Bob Rafelson (1990; Carolco Pictures).

34. See Michael J. Heffernan, "'A Dream as Frail as Those of Ancient Time': the In-Credible Geographies of Timbuctoo," *Environment and Planning D: Society and Space* 19 (2001): 203–25.

35. This phrase comes from the introduction to the collection of essays drawn together by Kuklick and Kohler, *Science in the Field, Osiris* 2nd ser., 11 (1996).

36. Henrika Kuklick, *The Savage Within: The Social History of British Anthropology, 1885–1945* (Cambridge: Cambridge University Press, 1991); and "After Ishmael: The Fieldwork Tradition and Its Future," in *Anthropological Location: Boundaries and Grounds*

of a Field Science, ed. A. Gupta and J. Ferguson (Berkeley: University of California Press, 1998); George W. Stocking Jr, *After Tylor: British Social Anthropology 1888–1951* (Madison: University of Wisconsin Press, 1995).

37. George W. Stocking Jr., "The Ethnographer's Magic: Fieldwork in British Anthropology from Tylor to Malinowski," in *Observers Observed: Essays on Ethnographic Fieldwork*, ed. George W. Stocking Jr. (Madison: University of Wisconsin Press, 1983), 1:70–120.

38. See Oliver Impey and Arthur MacGregor, eds., *The Origins of Museums: The Cabinet of Curiosities in Sixteenth- and Seventeenth-Century Europe* (Oxford: Clarendon Press, 1985); Lorraine J. Daston, "Marvellous Facts and Miraculous Evidence in Early Modern Europe," *Critical Inquiry* 18 (1991): 93–124; Paula Findlen, *Possessing Nature: Museums, Collecting, and Scientific Culture in Early Modern Italy* (Berkeley: University of California Press, 1994).

39. See especially Francis Bacon, *Novum Organum*, Bk 2, §§21–52 in *The Works of Francis Bacon*, ed. James Spedding, Robert Ellis, and Douglas Heath (London: Longman, 1857–74), 4: 141–248.

40. See the discussion in Ronald Rainger, *An Agenda for Antiquity: Henry Fairfield Osborn and Vertebrate Paleontology at the American Museum of Natural History, 1890–1935* (Tuscaloosa: University of Alabama Press, 1991).

41. Phillips Verner Bradford and Harvey Blume, *Ota: The Pygmy in the Zoo* (New York: St. Martin's Press, 1992). For a comparable case, see Felix Driver, "Making Representations: From an African Exhibition to the High Court of Justice," in *Geography Militant: Cultures of Exploration and Empire* (Oxford: Blackwell, 2001), 146–69.

42. Herman Reichenbach, "A Tale of Two Zoos: The Hamburg Zoological Garden and Carl Hagenbeck's Tierpark," in *New World, New Animals: From Menagerie to Zoological Park in the Nineteenth Century*, ed. R.J. Hoage and William A. Deiss (Baltimore, MD: The Johns Hopkins University Press, 1996).

43. Michael A. Osborne, *Nature, the Exotic, and the Science of French Colonialism* (Bloomington: Indiana University Press, 1994).

44. See John Prest, *The Garden of Eden: The Botanic Garden and the Re-Creation of Paradise* (New Haven, CT: Yale University Press, 1981); Andrew Cunningham, "The Culture of Gardens" in Jardine, Secord, and Spary, *Cultures of Natural History*, 38–56.

45. Ray Desmond, *Kew: The History of the Royal Botanic Gardens* (London: Harvill Press, 1995). For the role of Kew Gardens in the "Banksian empire," see the essays in David Philip Miller and Peter Hanns Reill, eds., *Visions of Empire: Voyages, Botany, and Representations of Nature* (Cambridge: Cambridge University Press, 1996).

46. Owen Gingerich, "The Censorship of Copernicus's *De Revolutionibus*," in *The Eye of Heaven: Ptolemy, Copernicus, Kepler* (New York: American Institute of Physics, 1993), 269–85.

47. See, e.g., Ronald L. Numbers and John Stenhouse, eds., *Disseminating Darwinism: The Role of Place, Race, Religion, and Gender* (New York: Cambridge University Press, 1999).

48. See David N. Livingstone, "Situating Evangelical Reponses to Darwin," in *Evangelicals and Science in Historical Perspective*, ed. David N. Livingstone, D.G. Hart and Mark A. Noll (New York: Oxford University Press, 1999), 193–219.

49. See the discussion in Lester D. Stephens, *Science, Race, and Religion in the Ameri-*

can South: John Bachman and the Charleston Circle of Naturalists, 1815–1895 (Chapel Hill: University of North Carolina Press, 2000).

50. John Stenhouse, "Darwinism in New Zealand, 1859–1900," in Numbers and Stenhouse, *Disseminating Darwinism*, 61–89.

51. Daniel P. Todes, *Darwin Without Malthus: The Struggle for Existence in Russian Evolutionary Thought* (Oxford: Oxford University Press, 1989).

52. Nick Jardine, "Books, Texts, and the Making of Knowledge," in *Books and the Sciences in History*, ed. Marina Frasca-Spada and Nick Jardine (Cambridge: Cambridge University Press, 2000), 401.

53. Among many relevant works see Adrian Johns, *The Nature of the Book: Print and Knowledge in the Making* (Chicago: University of Chicago Press, 1998); and Jonathan R. Topham, "Scientific Publishing and the Reading of Science in Nineteenth-Century Britain: An Historiographical Survey and Guide to Sources," *Studies in History and Philosophy of Science* (2000): 31A.

54. James A. Secord, *Victorian Sensation: The Extraordinary Publication, Reception, and Secret Authorship of Vestiges of the Natural History of Creation* (Chicago: University of Chicago Press, 2001), 14, 23.

55. Nicolaas Rupke, "Translation Studies in the History of Science: the Example of Vestiges," *British Journal for the History of Science* 33 (2000): 209–22.

56. What follows is drawn, in part, from David N. Livingstone, "Science, Religion and the Geography of Reading: Sir William Whitla and the Editorial Staging of Isaac Newton's Writings on Biblical Prophecy," *British Journal for the History of Science* 36 (2003): 27–42

57. Earlier uses of Newton as a cultural resource are discussed in Margaret C. Jacob, *The Cultural Meaning of the Scientific Revolution* (New York: Knopf, 1988).

58. Sir William Whitla, *Sir Isaac Newton's Daniel and the Apocalypse, With an Introductory Study of the Nature and Cause of Unbelief, of Miracles and Prophecy* (London: John Murray, 1922), xi.

59. See the discussion in David N. Livingstone and Ronald A. Wells, *Ulster-American Religion: Episodes in the History of a Cultural Connection* (Notre Dame, IN: University of Notre Dame Press, 1999).

60. Whitla, *Sir Isaac Newton's Daniel*, 41–42.

61. "Science and Religion: Address by Sir Wm Whitla, M.P.," *Christian Advocate*, October 13, 1922.

62. Sir William Whitla, *The Methodists of Ireland and Home Rule. Message to English Nonconformists* (Belfast: Committee of the Methodist Demonstration against Home Rule, 1910), 3.

63. *Christian Advocate*, November 25, 1921.

64. *Witness*, November 25, 1921.

65. S. G. Kennedy, "Dangers Confronting Us Viewed in the Light of Church History," *Witness*, November 18, 1921.

66. Whitla, *Methodists of Ireland and Home Rule*, 8, 12.

67. Whitla, *Sir Isaac Newton's Daniel*, 103, 104.

68. See Alvin Jackson, *Ireland: 1798–1998* (Oxford: Blackwell, 1999).

CONTRIBUTORS

Peter Dear is professor of history and of science and technology studies at Cornell University. He has published widely on aspects of early modern science and is the author of *Mersenne and the Learning of the Schools* (Cornell, 1988), *The Intelligibility of Nature: How Science Makes Sense of the World* (Chicago, 2006), *Discipline and Experience: The Mathematical Way in the Scientific Revolution* (Chicago, 1995), and *Revolutionizing the Sciences: European Knowledge and its Ambitions, 1500–1700* (Princeton, 2009).

Peter Harrison is the Andreas Idreos Professor of Science and Religion at the University of Oxford. He is also director of the Ian Ramsey Centre and a Fellow of Harris Manchester College. He has published extensively in the area of cultural and intellectual history with a focus on the philosophical, scientific, and religious thought of the early-modern period. His publications include *'Religion' and the Religions in the English Enlightenment* (Cambridge, 1990); *The Bible, Protestantism, and the Rise of Natural Science* (Cambridge, 1990); *The Fall of Man and the Foundations of Science* (Cambridge, 2007); and *The Cambridge Companion to Science and Religion* (Cambridge, 2010). He is a founding member of the International Society for Science and Religion and a Fellow of the Australian Academy of the Humanities.

J. L. Heilbron is professor of history and vice-chancellor emeritus at the University of California, Berkeley, and honorary Fellow of Worcester College, Oxford. His work relevant to early-modern natural philosophy includes *Electricity in the 17th and 18th Centuries* (Berkeley, 1979), *Weighing Imponderables* (Berkeley, 1993), *The Sun in the church* (Harvard, 1999), *Galileo* (Oxford, 2010), and, as editor, *The Quantifying Spirit in the 18th Century* (with

Tore Frängsmyr and Robin Rider, Berkeley, 1990) and *The Oxford Companion to the History of Modern Science* (Oxford, 2003).

Ronald R. Kline is the Bovay Professor in History and Ethics of Engineering, with a joint appointment between the Science and Technology Studies Department and the School of Electrical and Computer Engineering at Cornell University. In 2008, he became vice president/president-elect of the Society for the History of Technology. He is the author of *Steinmetz: Engineer and Socialist* (Johns Hopkins, 1992) and *Consumers in the Country: Technology and Social Change in Rural America* (Johns Hopkins, 2000) and is currently completing a book on the history of cybernetics and information discourses in the United States and Britain during the cold war.

Daryn Lehoux is professor of classics at Queen's University, Kingston, Ontario. He is the author of *Astronomy, Weather, and Calendars in the Ancient World* (Cambridge, 2007); the forthcoming titles *What Did the Romans Know? An Inquiry into Worldmaking* and *Ancient Science*; and numerous articles on ancient science. He has been a member at the Institute for Advanced Study, Princeton, and the Max Planck Institute for the History of Science in Berlin.

Bernard Lightman is professor of humanities at York University and editor of the journal *Isis*. His research centers on the cultural history of Victorian science. He is author of *The Origins of Agnosticism* (Johns Hopkins, 1987) and *Victorian Popularizers of Science* (Chicago, 2007), editor of *Victorian Science in Context* (Chicago, 1997), general editor of the monograph series *Science and Culture in the Nineteenth Century* (Pickering and Chatto), and coeditor of *Figuring it Out* (with Ann Shteir, Dartmouth, 2006) and *Science in the Marketplace* (with Aileen Fyfe, Chicago, 2007). Currently he is working on a biography of the physicist John Tyndall.

David N. Livingstone is professor of geography and intellectual history at Queen's University Belfast. A Fellow of the British Academy, he works on the history of geographical knowledge, the spatiality of scientific culture, and the historical geography of science and religion. His books include *Nathaniel Southgate Shaler and the Culture of American Science* (Alabama, 1987), *Darwin's Forgotten Defenders* (Eerdmans, 1987), *The Geographical Tradition* (Blackwell, 1992), *Putting Science in its Place* (Chicago, 2003), and *Adam's Ancestors: Race, Religion, and the Politics of Human Origins* (Johns Hopkins, 2008). He is currently working two projects, one entitled *Locating Darwinism: Chapters in the Historical Geography of Darwinian Encounters*, the other on the history of environmental determinism under the title *The Empire of Climate*.

Jon McGinnis is associate professor of classical and medieval philosophy at the University of Missouri, St. Louis. His general research interest, on which he has published extensively, is the history and philosophy of physics as that science is represented within Aristotle and the Aristotelian commentary tradition extending from its early Greek roots up to the Latin scholastics. His particular focus is the influence of that tradition within the medieval

Arabic-speaking world and especially on the natural philosophy of Ibn Sīnā, or Avicenna. McGinnis has received two National Endowment for the Humanities fellowships, an Andrew Mellon grant, and has been a member of the Institute for Advanced Study at Princeton.

Ronald L. Numbers is the Hilldale Professor of the History of Science and Medicine and a member of the Department of Medical History and Bioethics at the University of Wisconsin–Madison. He has written or edited more than two dozen books, including, most recently, *When Science and Christianity Meet* (coedited with David Lindberg, Chicago, 2003) *The Creationists: From Scientific Creationism to Intelligent Design* (Harvard, 2006), *Science and Christianity in Pulpit and Pew* (Oxford, 2007), *Galileo Goes to Jail and Other Myths about Science and Religion* (Harvard, 2009), and *Biology and Ideology from Descartes to Dawkins* (coedited with Denis Alexander, Chicago, 2010). He is a previous editor of *Isis* and a past president of the History of Science Society, the American Society of Church History, and the International Union of History and Philosophy of Science.

Jon H. Roberts is the Tomorrow Foundation Professor of American Intellectual History at Boston University. He has written a number of articles dealing primarily with the history of the relationship between science and religion, as well as the book *Darwinism and the Divine in America: Protestant Intellectuals and Organic Evolution, 1859–1900* (Wisconsin, 1988), which received the Frank S. and Elizabeth D. Brewer Prize from the American Society of Church History. He also coauthored *The Sacred and the Secular University* (with James Turner, Princeton, 2001). He is currently working on a book dealing with American Protestant thinkers' treatment of the mind during the nineteenth and early twentieth centuries.

Francesca Rochberg is the Catherine and William L. Magistretti Distinguished Professor of Near Eastern Studies in the Department of Near Eastern Studies and the Office for the History of Science and Technology at the University of California, Berkeley. She is the author of several monographs on Babylonian science, including *The Heavenly Writing: Divination, Horoscopy and Astronomy in Mesopotamian Culture* (Cambridge, 2004) and *In the Path of the Moon: Babylonian Celestial Divination and Its Legacy* (Brill, 2010).

Michael H. Shank is professor of the history of science at the University of Wisconsin–Madison. His primary research interests focus on late medieval natural philosophy and astronomy, with special attention to the Viennese tradition and most specifically, of late, the work of Johannes Regiomontanus. He is author of *"Unless You Believe, You Shall Not Understand": Logic, University, and Society in Late Medieval Vienna* (Princeton, 1988), editor of *The Scientific Enterprise in Antiquity and the Middle Ages* (Chicago, 2000), and coeditor of the forthcoming *The Cambridge History of Science, Vol. 2: The Middle Ages* (with David Lindberg).

Daniel P. Thurs is a faculty fellow in science studies at New York University's John W. Draper Interdisciplinary Master's Program. He has researched public discussion of nanotechnology at Cornell University and has taught history of science and American history courses at Western Oregon University, Oregon State University, and the University of Portland. His first book, *Science Talk: Changing Notions of Science in American Culture* (Rutgers, 2007), examines debate about the nature of science and its implications in American popular literature during the nineteenth and twentieth centuries. His next project looks at the history of mass panic and its relationship to science and technology.

INDEX